Shigeji Fujita and Salvador V. Godoy
Mathematical Physics

Related Titles

Vaughn, M. T.

Introduction to Mathematical Physics

2007
ISBN 978-3-527-40627-2

Kelly, J.J.

Graduate Mathematical Physics
With MATHEMATICA Supplements

2006
ISBN 978-3-527-40637-1

Wang, F.Y.

Physics with MAPLE
The Computer Algebra Resource for Mathematical Methods in Physics

2006
ISBN 978-3-527-40640-1

Kusse, B. Westwig, E.A.

Mathematical Physics
Applied Mathematics for Scientists and Engineers

2006
ISBN 978-3-527-40672-2

Boas, M.L.

Mathematical Methods in the Physical Sciences
Internation Edition

2005
ISBN 978-0-471-36580-8

Masujima, M.

Applied Mathematical Methods in Theoretical Physics
Second Edition

2009
ISBN 978-3-527-40936-5

Courant, R., Hilbert, D.

Methods of Mathematical Physics
Volume 1

1989
ISBN 978-0-471-50447-4

Courant, R., Hilbert, D.

Methods of Mathematical Physics
Volume 2

1989
ISBN 978-0-471-50439-9

Shigeji Fujita and Salvador V. Godoy

Mathematical Physics

WILEY-VCH Verlag GmbH & Co. KGaA

The Authors

Prof. Shigeji Fujita
University of Buffalo
SUNY
Buffalo NY 14260
USA

fujita@buffalo.edu

Prof. Salvador V. Godoy
Universidad Nacional Autonoma México
Facultad de Ciencias
México

sgs@hp.fciencias.unam.mx

Cover
Spiesz-Design,
Neu-Ulm, Germany
Copyright for image-background
by Yang Ming Qi (Fotolia.com)

1st Reprint 2011

All books published by **Wiley-VCH** are carefully produced. Nevertheless, authors, editors, and publisher do not warrant the information contained in these books, including this book, to be free of errors. Readers are advised to keep in mind that statements, data, illustrations, procedural details or other items may inadvertently be inaccurate.

Library of Congress Card No.: applied for
British Library Cataloguing-in-Publication Data: A catalogue record for this book is available from the British Library.
Bibliographic information published by the Deutsche Nationalbibliothek
The Deutsche Nationalbibliothek lists this publication in the Deutsche Nationalbibliografie; detailed bibliographic data are available on the Internet at
<http://dnb.d-nb.de>.

© 2010 WILEY-VCH Verlag GmbH & Co. KGaA, Weinheim

All rights reserved (including those of translation into other languages). No part of this book may be reproduced in any form – by photoprinting, microfilm, or any other means – nor transmitted or translated into a machine language without written permission from the publishers. Registered names, trademarks, etc. used in this book, even when not specifically marked as such, are not to be considered unprotected by law.

Printed in the Federal Republic of Germany
Printed on acid-free paper

Typesetting le-tex publishing services GmbH, Leipzig
Printing and Binding Strauss GmbH, Mörlenbach

ISBN 978-3-527-40808-5

Contents

Preface *XIII*

Table of Contents and Categories *XV*

Constants, Signs, Symbols, and General Remarks *XVII*

1 Vectors *1*
1.1 Definition and Important Properties *1*
1.1.1 Definitions *1*
1.2 Product of a Scalar and a Vector *2*
1.3 Position Vector *2*
1.4 Scalar Product *3*
1.5 Vector Product *4*
1.6 Differentiation *6*
1.7 Spherical Coordinates *6*
1.8 Cylindrical Coordinates *7*

2 Tensors and Matrices *11*
2.1 Dyadic or Tensor Product *11*
2.2 Cartesian Representation *12*
2.3 Dot Product *13*
2.3.1 Unit Tensor *14*
2.4 Symmetric Tensor *15*
2.5 Eigenvalue Problem *15*

3 Hamiltonian Mechanics *17*
3.1 Newtonian, Lagrangian and Hamiltonian Descriptions *17*
3.1.1 Newtonian Description *17*
3.1.2 Lagrangian Description *19*
3.1.3 Hamiltonian Description *20*
3.2 State of Motion in Phase Space. Reversible Motion *22*
3.3 Hamiltonian for a System of many Particles *28*
3.4 Canonical Transformation *32*
3.5 Poisson Brackets *36*

Mathematical Physics. Shigeji Fujita and Salvador V. Godoy
Copyright © 2010 WILEY-VCH Verlag GmbH & Co. KGaA, Weinheim
ISBN: 978-3-527-40808-5

4 Coupled Oscillators and Normal Modes 43
4.1 Oscillations of Particles on a String and Normal Modes 43
4.2 Normal Coordinates 48

5 Stretched String 53
5.1 Transverse Oscillations of a Stretched String 53
5.2 Normal Coordinates for a String 58

6 Vector Calculus and the del Operator 65
6.1 Differentiation in Time 65
6.2 Space Derivatives 66
6.2.1 The Gradient 66
6.2.2 The Divergence 67
6.2.3 The Curl 67
6.2.4 Space Derivatives of Products 67
6.3 Space Derivatives in Curvilinear Coordinates 69
6.3.1 Spherical Coordinates (r, θ, φ) 69
6.3.2 Cylindrical Coordinates (ρ, φ, z) 69
6.4 Integral Theorems 70
6.4.1 The Line Integral of $\mathbf{V}\phi$ 70
6.4.2 Stokes's Theorem 70
6.5 Gauss's Theorem 72
6.6 Derivation of the Gradient, Divergence and Curl 73

7 Electromagnetic Waves 77
7.1 Electric and Magnetic Fields in a Vacuum 77
7.2 The Electromagnetic Field Theory 82

8 Fluid Dynamics 87
8.1 Continuity Equation 87
8.2 Fluid Equation of Motion 89
8.3 Fluid Dynamics and Statistical Mechanics 92

9 Irreversible Processes 97
9.1 Irreversible Phenomena, Viscous Flow, Diffusion 97
9.2 Collision Rate and Mean Free Path 100
9.3 Ohm's Law, Conductivity, and Matthiessen's Rule 103

10 The Entropy 107
10.1 Foundations of Thermodynamics 107
10.2 The Carnot Cycle 109
10.3 Carnot's Theorem 111
10.4 Heat Engines and Refrigerating Machines 113
10.5 Clausius's Theorem 114
10.6 The Entropy 119
10.7 The Exact Differential 122

11 Thermodynamic Inequalities 125
11.1 Irreversible Processes and the Entropy 125
11.2 The Helmholtz Free Energy 128
11.3 The Gibbs Free Energy 130
11.4 Maxwell Relations 132
11.5 Heat Capacities 136
11.6 Nonnegative Heat Capacity and Compressibility 140

12 Probability, Statistics and Density 147
12.1 Probabilities 147
12.2 Binomial Distribution 150
12.3 Average and Root-Mean-Square Deviation. Random Walks 153
12.4 Microscopic Number Density 156
12.5 Dirac's Delta Function 160
12.6 The Three-Dimensional Delta Function 161

13 Liouville Equation 165
13.1 Liouville's Theorem 165
13.2 Probability Distribution Function. The Liouville Equation 169
13.3 The Gibbs Ensemble 173
13.4 Many Particles Moving in Three Dimensions 175
13.5 More about the Liouville Equation 177
13.6 Symmetries of Hamiltonians and Stationary States 179

14 Generalized Vectors and Linear Operators 183
14.1 Generalized Vectors. Matrices 183
14.2 Linear Operators 188
14.3 The Eigenvalue Problem 192
14.4 Orthogonal Representation 196

15 Quantum Mechanics for a Particle 201
15.1 Quantum Description of a Linear Motion 201
15.2 The Momentum Eigenvalue Problem 207
15.3 The Energy Eigenvalue Problem 211

16 Fourier Series and Transforms 213
16.1 Fourier Series 213
16.2 Fourier Transforms 214
16.3 Bra and Ket Notations 215
16.4 Heisenberg's Uncertainty Principle 217

17 Quantum Angular Momentum 221
17.1 Quantum Angular Momentum 221
17.2 Properties of Angular Momentum 224

18 Spin Angular Momentum 229
18.1 The Spin Angular Momentum 229
18.2 The Spin of the Electron 231

18.3 The Magnetogyric Ratio *234*
18.3.1 A. Free Electron *235*
18.3.2 B. Free Proton *236*
18.3.3 C. Free Neutron *237*
18.3.4 D. Atomic Nuclei *237*
18.3.5 E. Atoms and Ions *237*

19 Time-Dependent Perturbation Theory *239*
19.1 Perturbation Theory 1; The Dirac Picture *239*
19.2 Scattering Problem; Fermi's Golden Rule *242*
19.3 Perturbation Theory 2. Second Intermediate Picture *245*

20 Laplace Transformation *249*
20.1 Laplace Transformation *249*
20.2 The Electric Circuit Equation *251*
20.3 Convolution Theorem *252*
20.4 Linear Operator Algebras *253*

21 Quantum Harmonic Oscillator *255*
21.1 Energy Eigenvalues *255*
21.2 Quantum Harmonic Oscillator *259*

22 Permutation Group *263*
22.1 Permutation Group *263*
22.2 Odd and Even Permutations *267*

23 Quantum Statistics *273*
23.1 Classical Indistinguishable Particles *273*
23.2 Quantum-Statistical Postulate. Symmetric States for Bosons *276*
23.3 Antisymmetric States for Fermions. Pauli's Exclusion Principle *278*
23.4 Occupation-Number Representation *280*

24 The Free-Electron Model *283*
24.1 Free Electrons and the Fermi Energy *283*
24.2 Density of States *287*
24.3 Qualitative Discussion *291*
24.4 Sommerfeld's Calculations *293*

25 The Bose–Einstein Condensation *297*
25.1 Liquid Helium *297*
25.2 The Bose–Einstein Condensation of Free Bosons *298*
25.3 Bosons in Condensed Phase *301*

26 Magnetic Susceptibility *307*
26.1 Introduction *307*
26.2 Pauli Paramagnetism *308*
26.3 Motion of a Charged Particle in Electromagnetic Fields *310*
26.4 Electromagnetic Potentials *313*
26.5 The Landau States and Energies *316*

26.6	The Degeneracy of the Landau Levels *317*
26.7	Landau Diamagnetism *321*

27 Theory of Variations *329*
- 27.1 The Euler–Lagrange Equation *329*
- 27.2 Fermat's Principle *331*
- 27.3 Hamilton's Principle *333*
- 27.4 Lagrange's Field Equation *333*

28 Second Quantization *335*
- 28.1 Boson Creation and Annihilation Operators *335*
- 28.2 Observables *338*
- 28.3 Fermions Creation and Annihilation Operators *340*
- 28.4 Heisenberg Equation of Motion *341*

29 Quantum Statistics of Composites *345*
- 29.1 Ehrenfest–Oppenheimer–Bethe's Rule *345*
- 29.2 Two-Particle Composites *346*
- 29.3 Discussion *351*

30 Superconductivity *357*
- 30.1 Basic Properties of a Superconductor *357*
- 30.1.1 Zero Resistance *357*
- 30.1.2 Meissner Effect *357*
- 30.1.3 Ring Supercurrent and Flux Quantization *359*
- 30.1.4 Josephson Effects *360*
- 30.1.5 Energy Gap *362*
- 30.1.6 Sharp Phase Change *363*
- 30.2 Occurrence of a Superconductor *363*
- 30.2.1 Elemental Superconductors *363*
- 30.2.2 Compound Superconductors *363*
- 30.2.3 High-T_c Superconductors *365*
- 30.3 Theoretical Survey *365*
- 30.3.1 The Cause of Superconductivity *365*
- 30.3.2 The Bardeen–Cooper–Schrieffer Theory *366*
- 30.4 Quantum-Statistical Theory *367*
- 30.4.1 The Full Hamiltonian *367*
- 30.4.2 Summary of the Results *369*

31 Complex Numbers and Taylor Series *375*
- 31.1 Complex Numbers *375*
- 31.2 Exponential and Logarithmic Functions *377*
- 31.2.1 Laws of Exponents *377*
- 31.2.2 Natural Logarithm *377*
- 31.2.3 Relationship between Exponential and Trigonometric Functions *377*
- 31.3 Hyperbolic Functions *378*
- 31.3.1 Definition of Hyperbolic Functions *378*

31.3.2 Addition Formulas *378*
31.3.3 Double-Angle Formulas *379*
31.3.4 Sum, Difference and Product of Hyperbolic Functions *379*
31.3.5 Relationship between Hyperbolic and Trigonometric Functions *379*
31.4 Taylor Series *380*
31.4.1 Derivatives *380*
31.4.2 Taylor Series *381*
31.4.3 Binomial Series *381*
31.4.4 Series for Exponential and Logarithmic Functions *382*
31.5 Convergence of a Series *382*

32 Analyticity and Cauchy–Riemann Equations *385*
32.1 The Analytic Function *385*
32.2 Poles *387*
32.3 Exponential Functions *387*
32.4 Branch Points *388*
32.5 Function with Continuous Singularities *389*
32.6 Cauchy–Riemann Relations *390*
32.7 Cauchy–Riemann Relations Applications *391*

33 Cauchy's Fundamental Theorem *395*
33.1 Cauchy's Fundamental Theorem *395*
33.2 Line Integrals *398*
33.3 Circular Integrals *400*
33.4 Cauchy's Integral Formula *402*

34 Laurent Series *405*
34.1 Taylor Series and Convergence Radius *405*
34.2 Uniform Convergence *406*
34.3 Laurent Series *407*

35 Multivalued Functions *411*
35.1 Square-Root Functions. Riemann Sheets and Cut *411*
35.2 Multivalued Functions *413*

36 Residue Theorem and Its Applications *415*
36.1 Residue Theorem *415*
36.2 Integrals of the Form $\int_{-\infty}^{\infty} dx\, f(x)$ *417*
36.3 Integrals of the Type $\int_{-\infty}^{\infty} dx\, e^{ix} f(x)$ *419*
36.4 Integrals of the Type $\int_{0}^{2\pi} d\theta\, f(\cos\theta, \sin\theta)$ *420*
36.5 Miscellaneous Integrals *421*

Appendix A Representation-Independence of Poisson Brackets *423*

Appendix B Proof of the Convolution Theorem *427*

Appendix C Statistical Weight for the Landau States *431*

Appendix D Useful Formulas *433*

    References *435*

    Index *439*

Preface

We see stars in the dark night sky. This is a relativistic quantum effect. The light travels in vacuum with the speed of light, $c = 3 \times 10^8$ m s^{-1}. A photon (light quantum) travels as a lump (quantum) and hits the eye ball and disappears, causing a reaction with a transfer of energy $h\nu$, where ν is the light frequency, and h is Planck's constant: $h = 6.63 \times 10^{-34}$ J s. The daylight comes from the Sun in 8 min. The Sun shines indefinitely, using nuclear fusion reactions based on Einstein's famous mass–energy relation: $E = mc^2$, where m is the mass. Electrons, protons and neutrons are massive elementary particles, which compose atoms and molecules. These particles (electrons, protons, neutrons, atoms, molecules) move, following quantum laws of motion.

Today's physics graduate students are required to take core courses on quantum mechanics, quantum statistical mechanics, dynamics, thermodynamics, electrodynamics before completing advanced degrees programs (MA, MS, Ph.D.). These core subjects require a fair amount of mathematics specially dealing with quantum theories. The present authors believe that the students should learn the mathematics necessary to deal with quantum physics but also study basic physics at the same time. Learning mathematics and physics together will save time. Besides, the memory of learning in this way may last longer than otherwise. We have selected topics (chapters) in three categories: (1) primarily mathematics, uncircled and unstarred, (2) primarily physics, circled (◦), and (3) more advanced materials, starred (∗). In a typical semester course all material in (1) and a selection of (2) may be covered.

Recently, a superconductor with the critical temperature as high as 1275 K was discovered in multiwalled carbon nanotubes. Nanotubes are very strong and technologically important materials. Physicists and engineers working with nanomaterials must know the basics of superconductivity. The Bardeen–Cooper–Schrieffer theory that solves the mystery of superconductivity starts with a Hamiltonian expressed in second quantization, which may or may not be covered in the first-year graduate courses. We have included second quantization, basic elements of superconductivity, Dirac's time-dependent perturbation theory in the starred chapters. These are more advanced materials but they are indispensable for quantum field theory and many-body physics. The readers may omit the starred chapters for the first time around. Photons and phonons are bosons. Bosons can be created or annihilated, and hence the conservation of the number of particles does not hold. Then, no Schrödinger-like wavefunctions exist for bosons. The second-quantization formalism can be used to treat them fully. This is another reason why we shoul learn second quantization.

If all materials including the starred chapters are read or given as a part of a two-semester course, the reader should be ready to take on any advanced quantum topics, such as superconductivity, quantum Hall effect and quantum field theory.

The book is written in a self-contained manner. Thus, nonphysics majors who want to learn mathematical physics and elementary quantum theory step by step with no particular hurry may find it useful as a self-study reference. Experimental and theoretical physics and engineering researchers are also invited to examine the text. Problems at the end of a section are usually of the straightforward exercise type, directly connected with the material presented in that section. By doing these problems one by one, the reader may be able to grasp the meanings of the newly introduced subjects more firmly.

The authors thank: Sachiko, Amelia, Michio, Isao, Yoshiko, Eriko, George Redden and Kurt Borchardt for their encouragement and reading the drafts.

Shigeji Fujita and *Salvador Godoy*
Buffalo, NY, March 2009

Table of Contents and Categories

Chapter	Category	Partial contents
1		Vectors
2		Tensors and Matrices
3	○	Hamiltonian Mechanics
4	○	Coupled Harmonic Oscillator, Normal Modes
5	○	Stretched String
6		Vector Calculus. The del (nabla) Operator
7	○	Electromagnetic Wave Equations
8	○	Fluid Dynamics
9	○	Irreversible Processes and Kinetic Theory
10	○	The Entropy
11	○	Thermodynamic Inequalities
12		Probability, Statistics and Density
13	○	Liouville Theorem and Equilibrium
14		Generalized Vectors and Linear Operators
15	○	Quantum Mechanics for a Particle
16		Fourier Series and Transforms
17	○	Quantum Angular Momentum
18	○	Spin Angular Momentum
19	∗	Time-Dependent Perturbation Theory
20		Laplace Transformation
21	○	Quantum Harmonic Oscillator
22		Permutation Group
23	○	Quantum Statistics
24	○	The Free-Electron Model
25	○	Superfluids, The Bose–Einstein Condensation
26	○	Magnetism
27		Theory of Variations
28	∗	Second Quantization
29	∗	Composite Particles

Mathematical Physics. Shigeji Fujita and Salvador V. Godoy
Copyright © 2010 WILEY-VCH Verlag GmbH & Co. KGaA, Weinheim
ISBN: 978-3-527-40808-5

Chapter	Category	Partial contents
30	*	The Cooper Pairs and Superconductivity
31		Complex Numbers and Taylor Expansion
32		Analytic Functions, Cauchy–Riemann Relations
33		Cauchy's Fundamental Theorem
34		Laurent Expansion
35		Multivalued Functions
36		Residue Theorem: Calculation of Integrals

Constants, Signs, Symbols, and General Remarks

Useful Physical Constants

Quantity	Symbol	Value
Absolute zero on Celsius scale		$-273.16\,°\text{C}$
Avogadro's number	N_0	$6.02 \times 10^{23}\,\text{mol}^{-1}$
Boltzmann constant	k_B	$1.38 \times 10^{-16}\,\text{erg}\,\text{K}^{-1}$
Bohr magneton	μ_B	$9.22 \times 10^{-21}\,\text{erg}\,\text{gauss}^{-1}$
Bohr radius	a_0	$5.29 \times 10^{-9}\,\text{cm}$
Electron mass	m	$9.11 \times 10^{-28}\,\text{g}$
Electron charge (magnitude)	e	$4.80 \times 10^{-10}\,\text{esu}$
Gas constant	R	$8.314\,\text{J}\,\text{mol}^{-1}\,\text{K}^{-1}$
Molar volume (gas at STP)		$2.24 \times 10^4\,\text{cm}^3 = 22.4\,\text{l}$
Mechanical equivalent of heat		$4.186\,\text{J}\,\text{cal}^{-1}$
Permeability constant	μ_0	$1.26 \times 10^{-6}\,\text{H}\,\text{m}^{-1}$
Permittivity constant	ε_0	$8.854 \times 10^{-12}\,\text{F}\,\text{m}^{-1}$
Planck's constant	h	$6.63 \times 10^{-27}\,\text{erg}\,\text{s}$
Planck's constant/2π	$\hbar$	$1.05 \times 10^{-27}\,\text{erg}\,\text{s}$
Proton mass	m_p	$1.67 \times 10^{-24}\,\text{g}$
Speed of light	c	$3.00 \times 10^{10}\,\text{cm}\,\text{s}^{-1}$

Mathematical Signs

$=$	equal to
$\simeq$	approximately equal to
$\neq$	not equal to
$\equiv$	identical to, defined as
$>$	greater than
$\gg$	much greater than
$<$	less than
$\ll$	much less than
$\geq$	greater than or equal to
$\leq$	less than or equal to
$\propto$	proportional to
$\sim$	represented by, of the order
$\langle x \rangle, \bar{x}$	average value of x
$\ln$	natural logarithm
Δx	increment in x
dx	infinitesimal increment in x
z^*	complex conjugate of a number z
$\alpha^\dagger$	Hermitean conjugate of operator (matrix) α
α^T	transpose of matrix α
P^{-1}	inverse of P
$\delta_{ab} = \begin{cases} 1 & \text{if } a = b \\ 0 & \text{if } a \neq b \end{cases}$	Kronecker's delta
$\delta(x)$	Dirac's delta function
∇	nabla (or del) operator
$\dot{x} \equiv dx/dt$	time derivative
$\text{grad}\, \phi \equiv \nabla \phi$	gradient of ϕ
$\text{div}\, \mathbf{A} \equiv \nabla \cdot \mathbf{A}$	divergence of $\mathbf{A}$
$\text{curl}\, \mathbf{A} \equiv \nabla \times \mathbf{A}$	curl of $\mathbf{A}$
∇^2	Laplacian operator

List of Symbols

The following list is not intended to be exhaustive. It includes symbols of special importance.

Å	Ångstrom (= 10^{-8} cm = 10^{-10} m)
A	vector potential
B	magnetic field (magnetic flux density)
C	heat capacity
c	velocity of light
c	specific heat
$\mathcal{D}(p)$	density of states in momentum space
$\mathcal{D}(\omega)$	density of states in angular frequency
E	total energy
E	internal energy
E	electric field
e	base of natural logarithm
e	electron charge (absolute value)
F	Helmholtz free energy
f	one-body distribution function
f_B	Bose distribution function
f_F	Fermi distribution function
f_0	Planck distribution function
G	Gibbs free energy
H	Hamiltonian
H_c	critical magnetic field
$\mathbf{H}_a$	applied magnetic field
$\mathcal{H}$	Hamiltonian density
h	Planck's constant
h	single-particle Hamiltonian
$\hbar$	Dirac's h
$i \equiv \sqrt{-1}$	imaginary unit
i, j, k	Cartesian unit vectors
J	Jacobian of transformation
J	total current
j	single-particle current
j	current density
k	angular wave vector $\equiv$ k-vector
k_B	Boltzmann constant
L	Lagrangian function
L	normalization length
ln	natural logarithm
$\mathcal{L}$	Lagrangian density
l	mean free path

M	molecular mass
m	electron mass
m^*	effective mass
N	number of particles
$\hat{N}$	number operator
$\mathcal{N}(\varepsilon)$	density of states in energy
n	particle-number density
P	pressure
$\mathbf{P}$	total momentum
$\mathbf{p}$	momentum vector
p	momentum (magnitude)
Q	quantity of heat
R	resistance
$\mathbf{R}$	position of center of mass
r	radial coordinate
$\mathbf{r}$	position vector
S	entropy
T	kinetic energy
T	absolute temperature
T_c	critical (condensation) temperature
T_F	Fermi temperature
t	time
TR	sum of N particle trace $\equiv$ grand ensemble trace
Tr	many-particle trace
tr	one-particle trace
V	potential energy
V	volume
$\mathbf{v}$	velocity (field)
W	work
Z	partition function
$e^\alpha \equiv z$	fugacity
$\beta \equiv (k_B T)^{-1}$	reciprocal temperature
Δx	small variation in x
$\delta(x)$	Dirac delta function
$\delta_P = \begin{cases} +1 & \text{if } P \text{ is even} \\ -1 & \text{if } P \text{ is odd} \end{cases}$	parity sign of permutation P
ε	energy
ε_F	Fermi energy
η	viscosity coefficient
Θ_D	Debye temperature
Θ_E	Einstein temperature
θ	polar angle
λ	wavelength
λ	penetration depth

κ	curvature
μ	linear mass density of a string
μ	chemical potential
μ_B	Bohr magneton
ν	frequency
Ξ	grand partition function
ξ	dynamical variable
ξ	coherence length
ρ	mass density
ρ	(system) density operator
ρ	many-particle distribution function
σ	total cross section
σ	electrical conductivity
$\sigma_x, \sigma_y, \sigma_z$	Pauli spin matrices
τ	tension
τ_d	duration of collision
τ_c	average time between collisions
ϕ	azimuthal angle
ϕ	scalar potential
Ψ	quasi-wavefunction for condensed bosons
ψ	wavefunction for a quantum particle
$d\Omega = \sin\theta\, d\theta\, d\phi$	element of solid angle
$\omega \equiv 2\pi\nu$	angular frequency
ω_c	rate of collision
ω_D	Debye frequency
$[,] \equiv [,]_-$	commutator brackets
$\{,\} \equiv [,]_+$	anticommutator brackets
$\{,\}$	Poisson brackets
$[A]$	dimension of A

Compacted Expression

If A is B (non-B), C is D (non-D) means that if A is B, C is D *and* if A is non-B, C is non-D.

Crystallographic Notation

This is mainly used to denote a direction, or the orientation of a plane, in a cubic metal. A plane (hkl) intersects the orthogonal Cartesian axes, coinciding with the cube edges, at a/h, a/k, and a/l from the origin, a being a constant, usually the length of a side of the unit cell. The direction of a line is denoted by $[hkl]$, the direction cosines with respect to the Cartesian axes being h/N, k/N, and l/N, where

$N^2 = h^2 + k^2 + l^2$. The indices may be separated by commas to avoid ambiguity. Only occasionally will the notation be used precisely; thus [100] or [001] usually mean any cube axis and [111] any diagonal.

B and H

When an electron is described in quantum mechanics, its intersection with a magnetic field is determined by **B** rather than **H**; that is, if the permeabilty μ is not unity the elctron motion is determined by μ**H**. It is preferable to forget **H** altogether and use **B** to define all field strengths. The vector potential **A** is correspondingly defined such that $\nabla \times \mathbf{A} = \mathbf{B}$. **B** is effectively the same inside and outside the metal sample.

Units

In much of the literature quoted, the unit of magnetic field **B** is the gauss. Electric fields are frequently expressed in V/cm and reistivities in Ω cm.

$$1 \text{ Tesla (T)} = 10 \text{ kilogauss}, \quad 1 \, \Omega \, \text{m} = 10^2 \, \Omega \, \text{cm}$$

The Planck's constant over 2π, $\hbar = h/2\pi$ is used in dealing with an electron. The original Planck's constant h is used in dealing with a photon.

1
Vectors

We enumerate definitions and important properties of vectors in this chapter. A vector **A** has magnitude $A = |\mathbf{A}|$ and a direction $\mathbf{A}/A$. Two vectors **A** and **B** can be summed: $\mathbf{A} + \mathbf{B}$. A scalar product of **A** and **B** is denoted by $\mathbf{A} \cdot \mathbf{B}$. The magnitude of **A** is equal to $\sqrt{\mathbf{A} \cdot \mathbf{A}} = |\mathbf{A}| \equiv A$. A vector product of **A** and **B** is denoted by $\mathbf{A} \times \mathbf{B}$, which is noncommutative: $\mathbf{B} \times \mathbf{A} = -\mathbf{A} \times \mathbf{B}$.

1.1
Definition and Important Properties

1.1.1
Definitions

A *vector* **A** is a quantity specified by a *magnitude*, denoted by $|\mathbf{A}| \equiv A$ and a *direction* in space $\mathbf{A}/A$. A vector will be denoted by a letter in bold face in the text. The vector **A** may be represented geometrically by an arrow of length A pointing in the prescribed direction.

Addition. The sum $\mathbf{A} + \mathbf{B}$ of two vectors **A** and **B** is defined geometrically by drawing vector **A** originating from the tip of vector **B** as shown in Figure 1.1a. The same result is obtained if we draw the vector **B** from the tip of the vector **A** as shown in Figure 1.1b. This is expressed mathematically by

$$\mathbf{A} + \mathbf{B} = \mathbf{B} + \mathbf{A} \tag{1.1}$$

which expresses the *commutative* rule for addition.

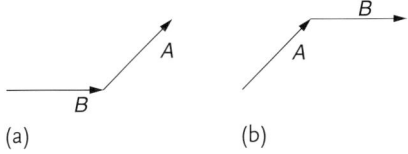

(a)   (b)

Fig. 1.1 The sum $\mathbf{A} + \mathbf{B}$, represented by (a) is equal to the sum $\mathbf{B} + \mathbf{A}$, represented by (b).

Mathematical Physics. Shigeji Fujita and Salvador V. Godoy
Copyright © 2010 WILEY-VCH Verlag GmbH & Co. KGaA, Weinheim
ISBN: 978-3-527-40808-5

Vectors also satisfy the *associative* rule:

$$(A + B) + C = A + (B + C) \tag{1.2}$$

The quantity represented by an ordinary (positive or negative) number is called a *scalar*, to distinguish it from a vector.

1.2
Product of a Scalar and a Vector

The product of a vector **A** and a positive scalar c is a vector, denoted by $c\mathbf{A}$, whose magnitude is equal to $c|\mathbf{A}|$ and whose direction is the same as that of **A**. If c is negative, then $c\mathbf{A}$, by definition, is a vector of magnitude $|c||\mathbf{A}|$ pointing in the direction opposite to **A**. The following rules of computation hold:

$$|c\mathbf{A}| = |c||\mathbf{A}| \tag{1.3}$$

$$(cd)\mathbf{A} = c(d\mathbf{A}) \tag{1.4}$$

$$\mathbf{A}c = c\mathbf{A} \tag{1.5}$$

$$c(\mathbf{A} + \mathbf{B}) = c\mathbf{A} + c\mathbf{B} \tag{1.6}$$

$$(c + d)\mathbf{A} = c\mathbf{A} + d\mathbf{A} \tag{1.7}$$

Equation (1.5) means that the same product is obtained irrespective of the order of c and **A**. We say that the product $c\mathbf{A}$ is *commutative*. The properties represented by (1.6)–(1.7) are called *distributive*.

1.3
Position Vector

The position of an arbitrary point P in space with respect to a given origin 0 may be specified by the position vector **r** drawn from 0 to P. If x, y, z are the Cartesian coordinates of the point P, then we can express the vector **r** by

$$\mathbf{r} = x\mathbf{i} + y\mathbf{j} + z\mathbf{k} \tag{1.8}$$

where **i, j**, and **k** are vectors of unit length pointing along the positive x-, y-, and z-axes. See Figure 1.2. For the fixed Cartesian unit vectors, the position vector **r** is specified by a set of three real numbers, (x, y, z). We represent this by

$$\mathbf{r} = (x, y, z) \tag{1.9}$$

The distance r of point P from the origin, is given by

$$r \equiv |\mathbf{r}| = (x^2 + y^2 + z^2)^{1/2} \tag{1.10}$$

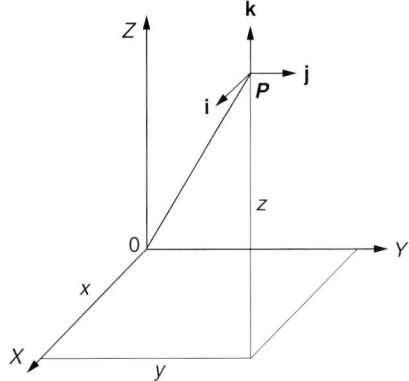

Fig. 1.2 The Cartesian coordinates (x, y, z). The orthonormal vectors **i**, **j** and **k** point in the directions of increasing x, y and z, respectively.

When point P coincides with the origin 0, we have, by definition, the *zero vector* or *null vector*, which is denoted by **0**, and can be represented by $(0, 0, 0)$. The null vector has zero magnitude and no definite direction.

1.4 Scalar Product

The *dot product*, also called the *scalar product*, $\mathbf{A} \cdot \mathbf{B}$, of two vectors **A** and **B**, is by definition a number equal to the product of their magnitudes times the cosine of the angle θ between them.

$$\mathbf{A} \cdot \mathbf{B} \equiv AB \cos \theta, \quad 0 \leq \theta \leq \pi \tag{1.11}$$

From this definition, the following properties can be derived:

$$\mathbf{A} \cdot \mathbf{B} = \mathbf{B} \cdot \mathbf{A}$$
$$\mathbf{A} \cdot (c\mathbf{B}) = (c\mathbf{A}) \cdot \mathbf{B} = c(\mathbf{A} \cdot \mathbf{B})$$
$$\mathbf{A} \cdot (\mathbf{B} + \mathbf{C}) = (\mathbf{A} \cdot \mathbf{B}) + (\mathbf{A} \cdot \mathbf{C}) \tag{1.12}$$

The last two equations show that the dot product is a *linear operation*. That is, given a vector **B**, the dot product with a vector **A** generates a scalar $\mathbf{A} \cdot \mathbf{B}$, which is a linear function of **B**. For example, if **B** is multiplied by 2, the scalar product $\mathbf{A} \cdot \mathbf{B}$ is also doubled.

The set of *Cartesian unit vectors* (**i**, **j**, **k**) satisfy the *orthonormality* relations:

$$\mathbf{i} \cdot \mathbf{j} = \mathbf{j} \cdot \mathbf{k} = \mathbf{k} \cdot \mathbf{i} = 0 \tag{1.13}$$

$$\mathbf{i} \cdot \mathbf{i} = \mathbf{j} \cdot \mathbf{j} = \mathbf{k} \cdot \mathbf{k} = 1 \tag{1.14}$$

The property (1.13) follows from the fact that the angles between any pair of $(\mathbf{i}, \mathbf{j}, \mathbf{k})$ are $90°$, and that $\cos 90° = 0$. We will say that the vectors $(\mathbf{i}, \mathbf{j}, \mathbf{k})$ are *orthogonal* to each other. The normalization property (1.14) holds because each of $(\mathbf{i}, \mathbf{j}, \mathbf{k})$ has unit length.

An arbitrary vector $\mathbf{A}$ can be decomposed as follows:

$$\mathbf{A} = A_x \mathbf{i} + A_y \mathbf{j} + A_z \mathbf{k} \tag{1.15}$$

where A_x, A_y and A_z are the projections of the vector $\mathbf{A}$ along the positive x-, y-, and z-axes, respectively, and are given numerically by

$$A_x = \mathbf{i} \cdot \mathbf{A}, \quad A_y = \mathbf{j} \cdot \mathbf{A}, \quad A_z = \mathbf{k} \cdot \mathbf{A} \tag{1.16}$$

Given the Cartesian unit vectors, the vector $\mathbf{A}$ can be represented by the set of the projections (A_x, A_y, A_z) called Cartesian components:

$$\mathbf{A} = (A_x, A_y, A_z) \tag{1.17}$$

Using the Cartesian decomposition (1.15), we obtain

$$\begin{aligned}
\mathbf{A} \cdot \mathbf{B} &= (A_x \mathbf{i} + A_y \mathbf{j} + A_z \mathbf{k}) \cdot (B_x \mathbf{i} + B_y \mathbf{j} + B_z \mathbf{k}) \\
&= A_x B_x \mathbf{i} \cdot \mathbf{i} + A_x B_y \mathbf{i} \cdot \mathbf{j} + A_x B_z \mathbf{i} \cdot \mathbf{k} \\
&\quad + A_y B_x \mathbf{j} \cdot \mathbf{i} + A_y B_y \mathbf{j} \cdot \mathbf{j} + A_y B_z \mathbf{j} \cdot \mathbf{k} \\
&\quad + A_z B_x \mathbf{k} \cdot \mathbf{i} + A_z B_y \mathbf{k} \cdot \mathbf{j} + A_z B_z \mathbf{k} \cdot \mathbf{k} \\
&= A_x B_x + A_y B_y + A_z B_z
\end{aligned}$$

or

$$\mathbf{A} \cdot \mathbf{B} = A_x B_x + A_y B_y + A_z B_z \tag{1.18}$$

By setting $\mathbf{A} = \mathbf{B}$ here, we obtain

$$\mathbf{A} \cdot \mathbf{A} = A_x^2 + A_y^2 + A_z^2 \geq 0 \tag{1.19}$$

The magnitude of the vector, $|\mathbf{A}|$, can be expressed by the square root of this quantity:

$$|\mathbf{A}| = (\mathbf{A} \cdot \mathbf{A})^{1/2} = (A_x^2 + A_y^2 + A_z^2)^{1/2} \tag{1.20}$$

We note that the properties of any vector $\mathbf{A}$ can be visualized analogously to the position vector $\mathbf{r}$ except for the difference in the physical dimension.

1.5
Vector Product

The *vector product*, $\mathbf{A} \times \mathbf{B}$, of two vectors $\mathbf{A}$ and $\mathbf{B}$ is by definition a vector having a magnitude equal to the area of the parallelogram with $\mathbf{A}$ and $\mathbf{B}$ as sides, and pointing in a direction perpendicular to the plane comprising $\mathbf{A}$ and $\mathbf{B}$. The direction of

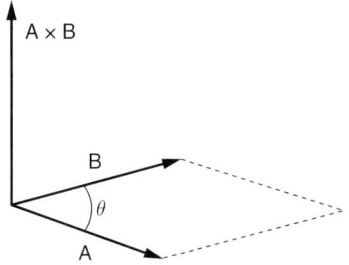

Fig. 1.3

$\mathbf{A} \times \mathbf{B}$ is, by convention, that direction in which a *right hand screw* would advance when turned from $\mathbf{A}$ to $\mathbf{B}$, as indicated in Figure 1.3.

$$|\mathbf{A} \times \mathbf{B}| = AB \sin \theta, \quad 0 \leqq \theta \leqq \pi \tag{1.21}$$

The vector product is a linear operation:

$$\mathbf{A} \times (c\mathbf{B}) = c\mathbf{A} \times \mathbf{B}$$

$$\mathbf{A} \times (\mathbf{B} + \mathbf{C}) = \mathbf{A} \times \mathbf{B} + \mathbf{A} \times \mathbf{C} \tag{1.22}$$

The following properties are observed:

$$\mathbf{B} \times \mathbf{A} = -\mathbf{A} \times \mathbf{B} \tag{1.23}$$

$$\mathbf{A} \times \mathbf{B} = 0 \quad \text{if } \mathbf{A} \parallel \mathbf{B} \tag{1.24}$$

$$\mathbf{A} \times \mathbf{A} = 0 \tag{1.25}$$

$$\mathbf{A} \times \mathbf{B} = \begin{vmatrix} \mathbf{i} & \mathbf{j} & \mathbf{k} \\ A_x & A_y & A_z \\ B_x & B_y & B_z \end{vmatrix} \tag{1.26}$$

$$\mathbf{i} \times \mathbf{j} = \mathbf{k}, \quad \mathbf{j} \times \mathbf{k} = \mathbf{i}, \quad \mathbf{k} \times \mathbf{i} = \mathbf{j} \tag{1.27}$$

$$(\mathbf{A} \times \mathbf{B}) \cdot \mathbf{C} = \mathbf{A} \cdot (\mathbf{B} \times \mathbf{C}) \tag{1.28}$$

$$\mathbf{A} \times (\mathbf{B} \times \mathbf{C}) = \mathbf{B}(\mathbf{A} \cdot \mathbf{C}) - \mathbf{C}(\mathbf{A} \cdot \mathbf{B}) \tag{1.29}$$

The last relation (1.29) may be verified by writing out the Cartesian components of both sides explicitly.

Problem 1.5.1

By writing out the Cartesian components of both sides, show that
1. $\mathbf{A} \times (\mathbf{B} \times \mathbf{C}) = \mathbf{B}(\mathbf{A} \cdot \mathbf{C}) - \mathbf{C}(\mathbf{A} \cdot \mathbf{B})$,
2. $(\mathbf{A} \times \mathbf{B}) \times \mathbf{C} = \mathbf{B}(\mathbf{A} \cdot \mathbf{C}) - \mathbf{A}(\mathbf{B} \cdot \mathbf{C})$

Problem 1.5.2

Show that

$$(\mathbf{A} \times \mathbf{B}) \cdot (\mathbf{C} \times \mathbf{D}) = (\mathbf{A} \cdot \mathbf{C})(\mathbf{B} \cdot \mathbf{D}) - (\mathbf{A} \cdot \mathbf{D})(\mathbf{B} \cdot \mathbf{C})$$

1.6
Differentiation

When a vector $\mathbf{A}$ depends on the time t, the derivative of $\mathbf{A}$ with respect to t, $d\mathbf{A}/dt$, is defined by

$$\frac{d\mathbf{A}}{dt} = \lim_{\Delta t \to 0} \frac{\mathbf{A}(t + \Delta t) - \mathbf{A}(t)}{\Delta t} \tag{1.30}$$

The following rules are observed for scalar and vector products:

$$\frac{d}{dt}(\mathbf{A} \cdot \mathbf{B}) = \frac{d\mathbf{A}}{dt} \cdot \mathbf{B} + \mathbf{A} \cdot \frac{d\mathbf{B}}{dt} \tag{1.31}$$

$$\frac{d}{dt}(\mathbf{A} \times \mathbf{B}) = \frac{d\mathbf{A}}{dt} \times \mathbf{B} + \mathbf{A} \times \frac{d\mathbf{B}}{dt} \tag{1.32}$$

Note that the operational rules as well as the definition are similar to the those of a scalar function.

A function $F(\mathbf{r})$ of the position $\mathbf{r} = (x, y, z)$ is called a *point function* or similarly a *field*. The space derivatives are discussed in Chapter 6.

1.7
Spherical Coordinates

For problems with special symmetries, it is convenient to use non-Cartesian coordinates. In particular, if the system under consideration has spherical symmetry, we may then use spherical coordinates (r, θ, φ), shown in Figure 1.4. These coordinates are related to the Cartesian coordinates by

$$x = r \sin\theta \cos\varphi$$
$$y = r \sin\theta \sin\varphi$$
$$z = r \cos\varphi \tag{1.33}$$

A system of orthogonal unit (orthonormal) vectors (**l**, **m**, **n**) in the directions of increasing θ, φ, and r, respectively, is also shown in Figure 1.4. These unit vectors

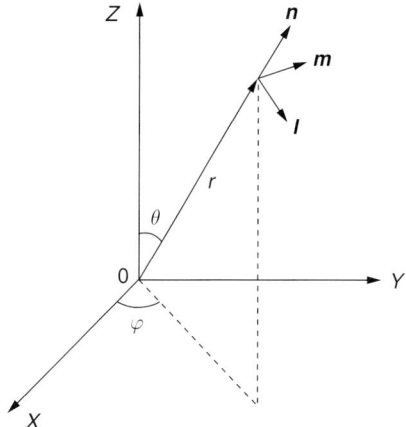

Fig. 1.4 The spherical polar coordinates (r, θ, φ). The orthonormal vectors **n**, **l**, and **m** point in the direction of increasing r, θ, and φ, respectively.

are related to the Cartesian unit vectors (**i**, **j**, **k**) as follows:

$$\mathbf{l} = -\mathbf{k}\sin\theta + \mathbf{i}\cos\theta\cos\varphi + \mathbf{j}\cos\theta\sin\varphi$$
$$\mathbf{m} = -\mathbf{i}\sin\varphi + \mathbf{j}\cos\varphi$$
$$\mathbf{n} = \mathbf{k}\cos\theta + \mathbf{i}\sin\theta\cos\varphi + \mathbf{j}\sin\theta\sin\varphi \tag{1.34}$$

An arbitrary vector **A** can be decomposed as follows:

$$\mathbf{A} = A_r \mathbf{n} + A_\theta \mathbf{l} + A_\varphi \mathbf{m} \tag{1.35}$$

where A_r, A_θ, A_φ are the components of **A** along **n**, **l**, and **m**, respectively. Further, they are given by

$$A_r = \mathbf{n} \cdot \mathbf{A}, \quad A_\theta = \mathbf{l} \cdot \mathbf{A}, \quad A_\varphi = \mathbf{m} \cdot \mathbf{A} \tag{1.36}$$

Problem 1.7.1

Two vectors point in directions (θ_1, φ_1) and (θ_2, φ_2). The angle between the two vectors is denoted by ψ. Show that

$$\cos\psi = \sin\theta_1 \sin\theta_2 \cos(\varphi_1 - \varphi_2) + \cos\theta_1 \cos\theta_2$$

1.8
Cylindrical Coordinates

For problems with axial symmetry, *cylindrical coordinates* (ρ, ϕ, z), as shown in Figure 1.5 are used. These cylindrical polar coordinates are related to the Cartesian

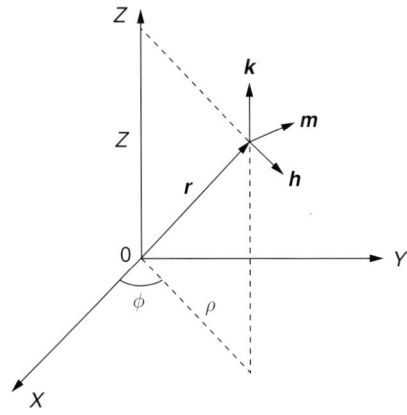

Fig. 1.5 The cylindrical polar coordinates (ρ, ϕ, z). The orthonormal vectors **h**, **m**, and **k** point in the direction of increasing ρ, ϕ, and z, respectively.

coordinates by

$$x = \rho \cos \phi$$
$$y = \rho \sin \phi$$
$$z = z \qquad (1.37)$$

The set of orthonormal vectors **h**, **m**, **k** in the direction of increasing ρ, ϕ, and z are shown in Figure 1.5. They are related to the Cartesian unit vectors (**i**, **j**, **k**) as follows:

$$\mathbf{h} = \mathbf{i} \cos \phi + \mathbf{j} \sin \phi$$
$$\mathbf{m} = -\mathbf{i} \sin \phi + \mathbf{j} \cos \phi$$
$$\mathbf{k} = \mathbf{k} \qquad (1.38)$$

An arbitrary vector **A** can be decomposed in the following form:

$$\mathbf{A} = A_\rho \mathbf{h} + A_\phi \mathbf{m} + A_z \mathbf{k} \qquad (1.39)$$

where A_ρ, A_ϕ, A_z are the components of **A** along **h**, **m**, **k**, given by

$$A_\rho = \mathbf{h} \cdot \mathbf{A}, \quad A_\phi = \mathbf{m} \cdot \mathbf{A}, \quad A_z = \mathbf{k} \cdot \mathbf{A} \qquad (1.40)$$

We note that an arbitrary vector **A** can be decomposed as follows:

$$\mathbf{A} = \mathbf{i}(\mathbf{i} \cdot \mathbf{A}) + \mathbf{j}(\mathbf{j} \cdot \mathbf{A}) + \mathbf{k}(\mathbf{k} \cdot \mathbf{A})$$
$$= \mathbf{n}(\mathbf{n} \cdot \mathbf{A}) + \mathbf{l}(\mathbf{l} \cdot \mathbf{A}) + \mathbf{m}(\mathbf{m} \cdot \mathbf{A})$$
$$= \mathbf{h}(\mathbf{h} \cdot \mathbf{A}) + \mathbf{m}(\mathbf{m} \cdot \mathbf{A}) + \mathbf{k}(\mathbf{k} \cdot \mathbf{A}) \qquad (1.41)$$

where $(\mathbf{i}, \mathbf{j}, \mathbf{k})$, $(\mathbf{n}, \mathbf{l}, \mathbf{m})$ and $(\mathbf{h}, \mathbf{m}, \mathbf{k})$ are orthonormal vectors in Cartesian, spherical and cylindrical coordinates, respectively. We note that the three equations (1.41) can be written as

$$\mathbf{A} = \mathbf{e}_1(\mathbf{e}_1 \cdot \mathbf{A}) + \mathbf{e}_2(\mathbf{e}_2 \cdot \mathbf{A}) + \mathbf{e}_3(\mathbf{e}_3 \cdot \mathbf{A})$$

$$= \sum_{j=1}^{3} \mathbf{e}_j (\mathbf{e}_j \cdot \mathbf{A}) \tag{1.42}$$

where $(\mathbf{e}_1, \mathbf{e}_2, \mathbf{e}_3)$ is a set of orthonormal vectors satisfying

$$\mathbf{e}_i \cdot \mathbf{e}_j = \delta_{ij} = \begin{cases} 1 & \text{if } i = j \\ 0 & \text{if } i \neq j \end{cases} \tag{1.43}$$

The symbol δ_{ij} is called Kronecker's delta.

2
Tensors and Matrices

Tensors are introduced in terms of a direct or dyadic product of vectors. A tensor can be represented by a 3×3 matrix.

2.1
Dyadic or Tensor Product

Definition Given two vectors **A** and **B**, the *direct* or *tensor product* **AB** of the two vectors **A** and **B** is defined in terms of the dot-product equation:

$$(\mathbf{AB}) \cdot \mathbf{C} = \mathbf{A}(\mathbf{B} \cdot \mathbf{C}) \tag{2.1}$$

where **C** is any third vector.

Note that the right-hand term is a vector proportional to the first vector **A**. From definition (2.1), we can show that

$$(\mathbf{AB}) \cdot (\mathbf{C} + \mathbf{D}) = \mathbf{A}[\mathbf{B} \cdot (\mathbf{C} + \mathbf{D})]$$
$$= \mathbf{A}(\mathbf{B} \cdot \mathbf{C} + \mathbf{B} \cdot \mathbf{D}) = (\mathbf{AB}) \cdot \mathbf{C} + (\mathbf{AB}) \cdot \mathbf{D}$$

or

$$(\mathbf{AB}) \cdot (\mathbf{C} + \mathbf{D}) = (\mathbf{AB}) \cdot \mathbf{C} + (\mathbf{AB}) \cdot \mathbf{D} \tag{2.2}$$

In a similar manner, we can show that

$$(\mathbf{AB}) \cdot (c\mathbf{C}) = c[(\mathbf{AB}) \cdot \mathbf{C}] \tag{2.3}$$

where c is a real number.

Equations (2.2) and (2.3) mean that the dot product underlying the definition of a tensor is a *linear* operation. In other words, the tensor product **AB** acting on a vector **C**, generates a new vector $(\mathbf{AB}) \cdot \mathbf{C}$, which is a linear function of **C**. For example, if **C** is multiplied by two, the resulting vector is doubled.

Mathematical Physics. Shigeji Fujita and Salvador V. Godoy
Copyright © 2010 WILEY-VCH Verlag GmbH & Co. KGaA, Weinheim
ISBN: 978-3-527-40808-5

Dyadic or tensor products will simply be called *tensors*, and be indicated by $\mathbb{T}$, $\mathbb{S}$, and so on. If we denote the tensor product **AB** by $\mathbb{T}$: **AB** $\equiv \mathbb{T}$, we can re-express (2.2) and (2.3) as follows:

$$\mathbb{T} \cdot (\mathbf{A} + \mathbf{B}) = \mathbb{T} \cdot \mathbf{A} + \mathbb{T} \cdot \mathbf{B} \tag{2.4}$$

$$\mathbb{T} \cdot (c\mathbf{C}) = c\mathbb{T} \cdot \mathbf{C} \tag{2.5}$$

These two equations indicate the *linearity* in a more transparent manner.

The sum of two tensors $\mathbb{T}$ and $\mathbb{S}$ generates a tensor that is defined in terms of the dot-product equation:

$$(\mathbb{S} + \mathbb{T}) \cdot \mathbf{C} \equiv \mathbb{S} \cdot \mathbf{C} + \mathbb{T} \cdot \mathbf{C} \quad \text{for any } \mathbf{C} \tag{2.6}$$

The algebraic rules for the sum are similar to those for the sum of vectors:

$$\mathbb{S} + \mathbb{T} = \mathbb{T} + \mathbb{S} \quad \text{(commutative)} \tag{2.7}$$

$$(\mathbb{S} + \mathbb{T}) + \mathbb{U} = \mathbb{S} + (\mathbb{T} + \mathbb{U}) \quad \text{(associative)} \tag{2.8}$$

Definition It is possible to define a dot product of the tensor **AB** with a vector on the left such that

$$\mathbf{C} \cdot (\mathbf{AB}) \equiv (\mathbf{C} \cdot \mathbf{A})\mathbf{B} \tag{2.9}$$

The resulting product is a vector but this vector in general differs from the vector obtained by the dot product of the reversed order:

$$\mathbf{C} \cdot (\mathbf{AB}) \neq (\mathbf{AB}) \cdot \mathbf{C} \tag{2.10}$$

or

$$\mathbf{C} \cdot \mathbb{T} \neq \mathbb{T} \cdot \mathbf{C} \tag{2.11}$$

2.2
Cartesian Representation

A vector can be decomposed in terms of Cartesian components, for example,

$$\mathbf{A} = A_x \mathbf{i} + A_y \mathbf{j} + A_z \mathbf{k} \tag{2.12}$$

Applying such decomposition to (2.1) and collecting terms, we can show that the tensor **AB** can be decomposed as follow:

$$\begin{aligned} \mathbf{AB} = &A_x B_x \mathbf{ii} + A_x B_y \mathbf{ij} + A_x B_z \mathbf{ik} \\ &+ A_y B_x \mathbf{ji} + A_y B_y \mathbf{jj} + A_y B_z \mathbf{jk} \\ &+ x\mathbf{ki} + A_z B_y \mathbf{kj} + A_z B_z \mathbf{kk} \end{aligned} \tag{2.13}$$

The quantities **ii**, **ij**, **ik**, and so on, are called *unit dyads*. Note that, for example, **ij** is not the same as **ji**. Any tensor $\mathbb{T}$ can be expressed in the form:

$$\mathbb{T} = T_{xx}\mathbf{ii} + T_{xy}\mathbf{ij} + T_{xz}\mathbf{ik}$$
$$+ T_{yx}\mathbf{ji} + T_{yy}\mathbf{jj} + T_{yz}\mathbf{jk}$$
$$+ T_{zx}\mathbf{ki} + T_{zy}\mathbf{kj} + T_{zz}\mathbf{kk} \qquad (2.14)$$

Alternatively, we may represent the tensor $\mathbb{T}$ by the *matrix* with nine components as follows:

$$\mathbb{T} = \begin{pmatrix} T_{xx} & T_{xy} & T_{xz} \\ T_{yx} & T_{yy} & T_{yz} \\ T_{zx} & T_{zy} & T_{zz} \end{pmatrix} \qquad (2.15)$$

In order to simplify writing the tensor components, it is convenient to number the coordinate axes (x_1, x_2, x_3) instead of (x, y, z):

$$x_1 \equiv x, \quad x_2 \equiv y, \quad x_3 \equiv z \qquad (2.16)$$

We also write the corresponding unit vectors as

$$\mathbf{i} \equiv \mathbf{e}_1, \quad \mathbf{j} \equiv \mathbf{e}_2, \quad \mathbf{k} \equiv \mathbf{e}_3 \qquad (2.17)$$

In the new notations, (2.12), (2.14) and (2.15) can be written as

$$\mathbf{A} = \sum_{j=1}^{3} A_j \mathbf{e}_j \qquad (2.18)$$

$$\mathbb{T} = \sum_{j=1}^{3} \sum_{k=1}^{3} T_{jk} \mathbf{e}_j \mathbf{e}_k \qquad (2.19)$$

$$\mathbb{T} = \begin{pmatrix} T_{11} & T_{12} & T_{13} \\ T_{21} & T_{22} & T_{23} \\ T_{31} & T_{32} & T_{33} \end{pmatrix} \equiv (T_{jk}) \qquad (2.20)$$

2.3 Dot Product

Definition The dot product of two tensors $\mathbb{T}$ and $\mathbb{S}$ generates a tensor, and is defined by the equation

$$(\mathbb{T} \cdot \mathbb{S}) \cdot \mathbf{C} \equiv \mathbb{T} \cdot (\mathbb{S} \cdot \mathbf{C}) \qquad \text{for any } \mathbf{C} \qquad (2.21)$$

Writing this equation in terms of the components, we find that

$$\mathbb{T} \cdot \mathbb{S} = \sum_j \sum_k \sum_l T_{jk} S_{kl} \mathbf{e}_j \mathbf{e}_l$$
$$= \sum_j \sum_l (TS)_{jl} \mathbf{e}_j \mathbf{e}_l \qquad (2.22)$$

This equation shows that if $\mathbb{T}$ and $\mathbb{S}$ are represented by matrices (T_{jk}) and (S_{jk}) the dot product $\mathbb{T} \cdot \mathbb{S}$ can be represented by the matrix whose elements are obtained by the usual rule of the *matrix multiplication*:

$$(\mathbb{T} \cdot \mathbb{S})_{jl} = \sum_{k=1}^{3} T_{jk} S_{kl} \equiv (TS)_{jl} \tag{2.23}$$

The following properties can be verified from the definition (2.21)

$$\mathbb{T} \cdot (\mathbb{S} + \mathbb{U}) = \mathbb{T} \cdot \mathbb{S} + \mathbb{T} \cdot \mathbb{U} \quad \text{(distributive)} \tag{2.24}$$

$$\mathbb{T} \cdot (\mathbb{S} \cdot \mathbb{U}) = (\mathbb{T} \cdot \mathbb{S}) \cdot \mathbb{U} \quad \text{(associative)} \tag{2.25}$$

However, the dot product is not in general commutative

$$\mathbb{T} \cdot \mathbb{S} \neq \mathbb{S} \cdot \mathbb{T} \tag{2.26}$$

2.3.1
Unit Tensor

The special tensor defined by

$$\mathbb{E} \equiv \mathbf{ii} + \mathbf{jj} + \mathbf{kk} \equiv \mathbf{e}_1\mathbf{e}_1 + \mathbf{e}_2\mathbf{e}_2 + \mathbf{e}_3\mathbf{e}_3 \tag{2.27}$$

satisfies the relation

$$\mathbb{E} \cdot \mathbb{T} = \mathbb{T} \cdot \mathbb{E} = \mathbb{T} \tag{2.28}$$

This tensor is called the *unit tensor*. The unit tensor can be represented by the unit matrix:

$$\mathbb{E} = \begin{pmatrix} 1 & 0 & 0 \\ 0 & 1 & 0 \\ 0 & 0 & 1 \end{pmatrix} \tag{2.29}$$

By choosing $\mathbb{T} = \mathbb{E}$, we obtain from (2.28)

$$\mathbb{E} \cdot \mathbb{E} = \mathbb{E} \tag{2.30}$$

If c is a number, the product $c\mathbb{E}$ is called a *constant tensor*, and has the following property

$$(c\mathbb{E}) \cdot \mathbb{T} = \mathbb{T} \cdot (c\mathbb{E}) = c\mathbb{T} \tag{2.31}$$

By putting $\mathbb{T} = \mathbb{E}$ here, we have

$$(c\mathbb{E}) \cdot \mathbb{E} = \mathbb{E} \cdot (c\mathbb{E}) = c\mathbb{E} \tag{2.32}$$

From definition (2.27), we can easily show that, for any vector **A**,

$$\mathbb{E} \cdot \mathbf{A} = \mathbf{A} \cdot \mathbb{E} = \mathbf{A} \tag{2.33}$$

The last several equations involving the unit tensor $\mathbb{E}$ can be summarized as follows. The dot product of the unit tensor $\mathbb{E}$ and any quantity, which may be a tensor or vector, generates the very same quantity. That is, the dot product with the unit tensor $\mathbb{E}$ is the *identity operation*, and can be looked upon as the multiplication by unity:

$$\mathbb{E} \cdot \alpha = (1)(\alpha) = \alpha \tag{2.34}$$

where α is a vector or tensor. Conversely, the multiplication by unity can be regarded as the dot product with the unit tensor $\mathbb{E}$. Combining this result with (2.31), we may also say that multiplication by a constant c can be viewed as the dot product with the constant tensor $c\mathbb{E}$.

2.4
Symmetric Tensor

From two vectors **A** and **B**, we may construct the two tensor products **AB** and **BA**. These products are different from each other. One product is the transpose of the other product. Using mathematical symbols, we have

$$(\mathbf{AB})^\mathsf{T} \equiv \mathbf{BA} \tag{2.35}$$

Equivalently for any tensor $\mathbb{S}$, we can define the transposed tensor $\mathbb{S}^\mathsf{T}$ by the matrix-element relations:

$$(\mathbb{S}^\mathsf{T})_{jk} = S_{kj} \tag{2.36}$$

This corresponds to reflecting the matrix elements about the leading diagonal.

The tensor $\mathbb{T}$ is called *symmetric* if

$$(\mathbb{T})^\mathsf{T} = \mathbb{T} \tag{2.37}$$

or equivalently

$$T_{jk} = T_{kj} \tag{2.38}$$

In this case, the matrix (2.20) is unchanged by reflection about the leading diagonal.

2.5
Eigenvalue Problem

Given a symmetric tensor $\mathbb{T}$, we can set up the *eigenvalue* (vector) *equation* by

$$\mathbb{T} \cdot \mathbf{A} = \lambda \mathbf{A} \tag{2.39}$$

where λ is a number called an *eigenvalue* of $\mathbb{T}$, and **A** is a nonzero vector called an *eigenvector*. The vector equation (2.39) may be written as

$$(\mathbb{T} - \lambda \mathbb{E}) \cdot \mathbf{A} = 0 \qquad (2.40)$$

or written out in components,

$$(T_{11} - \lambda)A_1 + T_{12}A_2 + T_{13}A_3 = 0$$
$$T_{21}A_1 + (T_{22} - \lambda)A_2 + T_{23}A_3 = 0$$
$$T_{31}A_1 + T_{32}A_2 + (T_{33} - \lambda)A_3 = 0 \qquad (2.41)$$

The solution of the eigenvalue problem and its importance are discussed later in Section 14.3.

Problem 2.5.1

Using the Cartesian representation, show that

1. $\nabla \cdot (\mathbf{AB}) = (\nabla \cdot \mathbf{A})\mathbf{B} + (\mathbf{A} \cdot \nabla)\mathbf{B}$
2. $\nabla \cdot (\mathbf{A}\mathbf{r}) = (\nabla \cdot \mathbf{A})\mathbf{r} + \mathbf{A}$
3. $\nabla \cdot (\mathbf{A}\mathbf{rr}) = (\nabla \cdot \mathbf{A})\mathbf{rr} + \mathbf{Ar} + \mathbf{rA}$

3
Hamiltonian Mechanics

We describe the Hamiltonian formulation of mechanics in this chapter. This serves as the preliminaries for the discussion of quantum and statistical mechanics.

3.1
Newtonian, Lagrangian and Hamiltonian Descriptions

In the present section, we discuss the linear motion of a particle by three equivalent methods.

3.1.1
Newtonian Description

Let us consider a particle of mass m moving along a straight line and acted upon by a force F. See Figure 3.1. Newton's equation of motion is

$$\boxed{m\ddot{x} \equiv \frac{d^2x}{dt^2} = F} \qquad (3.1)$$

If force F is given as a function of the position x, the velocity $v \equiv \dot{x} \equiv dx/dt$, and time t:

$$F = F(x, v, t) \qquad (3.2)$$

then (3.1) reads

$$m\frac{dv}{dt} = F(x, v, t) \qquad (3.3)$$

Since

$$v(t + dt) \simeq v(t) + \frac{dv(t)}{dt} dt$$
$$= v(t) + m^{-1} F(x, v, t) dt$$

Mathematical Physics. Shigeji Fujita and Salvador V. Godoy
Copyright © 2010 WILEY-VCH Verlag GmbH & Co. KGaA, Weinheim
ISBN: 978-3-527-40808-5

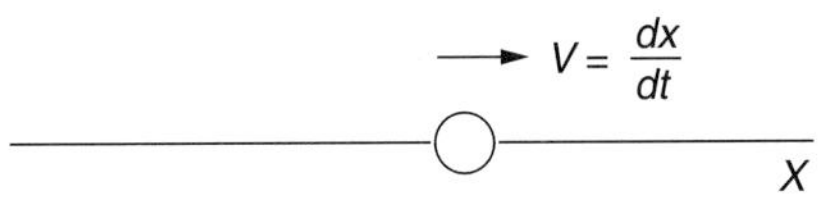

Fig. 3.1 A particle moving in one dimension.

and

$$x(t+dt) \simeq x(t) + \frac{dx(t)}{dt}dt \equiv x(t) + v(t)dt \qquad (3.4)$$

the knowledge of the pair (x, v) at time t uniquely determines the same pair at time $t + dt$, where dt is an infinitesimal increment in time. Repeating such infinitesimal processes, we can conclude that if the position and velocity (x, v) are given at an initial time and the force is known, then the position and velocity at any later time can be uniquely determined. This is an important general consequence, which follows from the mathematical structure of Newton's equation of motion: an ordinary differential equation of second order requires two independent constants for its unique solution. In the above case the two constants were given by the initial position and velocity.

In many cases of interest force F is a function of position x only. The equation of motion is, then

$$m\frac{dv}{dt} = F(x) \qquad (3.5)$$

Multiplying this equation by $v \equiv \dot{x}$, we have

$$m\frac{dv}{dt}v = F(x)\frac{dx}{dt} \qquad (3.6)$$

The left-hand side (lhs) can be written as

$$m\frac{dv}{dt}v = \frac{d}{dt}\left(\frac{1}{2}mv^2\right) \equiv \frac{dT}{dt} \qquad (3.7)$$

where

$$T \equiv \frac{1}{2}mv^2 \qquad (3.8)$$

is the *kinetic energy* of the particle. If we introduce a *potential energy* (function) $V(x)$, which satisfies the relation

$$F(x) = -\frac{dV}{dx} \qquad (3.9)$$

then we can rewrite the right-hand side (rhs) of (3.6) as

$$F(x)\frac{dx}{dt} = -\frac{dV}{dx}\frac{dx}{dt} = -\frac{dV(x)}{dt} \qquad (3.10)$$

Using (3.6), (3.7) and (3.9), we obtain

$$\frac{d}{dt}(T+V) = 0 \tag{3.11}$$

Integrating this, we have

$$T + V = \text{constant} = E \tag{3.12}$$

This expresses the *law of conservation of energy*; the constant E, which is the sum of kinetic and potential energies, is called the *total energy*.

For a force F that depends on position x alone, we can integrate (3.9) and obtain the potential energy V by

$$V(x) - V(x_0) = -\int_{x_0}^{x} F(x)dx \tag{3.13}$$

where x_0 is some fixed position.

As an example let us consider a harmonic force, which is directed toward the origin, given by

$$F = -kx \tag{3.14}$$

where k is a positive constant. Introducing this into (3.13), we obtain

$$V(x) - V(x_0) = -\int_{x_0}^{x} dx(-kx) = \frac{1}{2}k\left(x^2 - x_0^2\right)$$

By choosing $x_0 = 0$, we have

$$V(x) = \frac{1}{2}kx^2 \tag{3.15}$$

The conservation of energy is then expressed by

$$E = \frac{1}{2}mv^2 + \frac{1}{2}kx^2 = \text{constant} \tag{3.16}$$

3.1.2
Lagrangian Description

In Lagrange's formulation, we define the *Lagrangian function* L of position x and velocity $\dot{x}$ by

$$L(x, \dot{x}) \equiv T - V(x) \tag{3.17}$$

We then set up the equation of motion through *Lagrange's equation*:

$$\boxed{\frac{d}{dt}\left(\frac{\partial L}{\partial \dot{x}}\right) - \frac{\partial L}{\partial x} = 0} \tag{3.18}$$

The *partial derivatives* are defined as follows. Given a function $f(x, y)$, the partial derivative of f with respect to x with y fixed, is defined by

$$\frac{\partial f(x, y)}{\partial x} \equiv \lim_{\Delta x \to 0} \frac{f(x + \Delta x, y) - f(x, y)}{\Delta x} \tag{3.19}$$

In the case of the harmonic oscillator, the Lagrangian L is obtained as follows:

$$L = T - V$$
$$= \frac{1}{2}m\dot{x}^2 - \frac{1}{2}kx^2, \quad \text{[use of (3.8) and (3.15)]} \tag{3.20}$$

From this, we obtain

$$\frac{\partial L(x, \dot{x})}{\partial x} = \frac{\partial}{\partial x}\left(\frac{1}{2}m\dot{x}^2 - \frac{1}{2}kx^2\right) = -kx$$

$$\frac{\partial L(x, \dot{x})}{\partial \dot{x}} = \frac{\partial}{\partial \dot{x}}\left(\frac{1}{2}m\dot{x}^2 - \frac{1}{2}kx^2\right) = m\dot{x}$$

Notice the importance of recognizing the fixed variables while computing partial derivatives.

Introducing these results in (3.18), we obtain

$$\frac{d}{dt}(m\dot{x}) - (-kx) = m\ddot{x} + kx = 0$$

which is equivalent to Newton's equation of motion:

$$m\ddot{x} = \text{force} = -kx \tag{3.21}$$

The equivalence between Lagrange's and Newton's equations of motion can be established for a wide class of forces, and again for two- and three-dimensional motion. Readers who are interested in the detailed theory and applications of Lagrangian formulation of mechanics, should refer to textbooks of intermediate mechanics such as Symon [1] and Kibble [2].

3.1.3
Hamiltonian Description

We define the *canonical momentump* by

$$p \equiv \frac{\partial L}{\partial \dot{x}} \tag{3.22}$$

If the potential energy V is a function of x only, we obtain

$$p \equiv \frac{\partial}{\partial \dot{x}}(T - V) = \frac{\partial T}{\partial \dot{x}} = \frac{\partial}{\partial \dot{x}}\left(\frac{1}{2}m\dot{x}^2\right) = m\dot{x} \tag{3.23}$$

In this case, the canonical momentum equals the linear momentum defined as mass times velocity.

We define a *Hamiltonian* (function) by

$$H \equiv p\dot{x} - L \tag{3.24}$$

and express it in terms of the coordinate x and the momentum p. For the harmonic oscillator, we start with

$$H = p\dot{x} - \left(\frac{1}{2}m\dot{x}^2 - \frac{1}{2}kx^2\right)$$

After eliminating $\dot{x}$ with the aid of (3.23), we then obtain

$$H = \frac{1}{2m}p^2 + \frac{1}{2}kx^2 = H(x, p) \tag{3.25}$$

This quantity, now regarded as a *function of coordinate x and momentum p*, is called the Hamiltonian of the harmonic oscillator.

For a general system with a potential energy $V(x)$, the Hamiltonian H is given by

$$H = \dot{x}p - (T - V) = \left(\frac{p}{m}\right)p - \frac{1}{2}m\left(\frac{p}{m}\right)^2 + V$$

$$= \frac{1}{2m}p^2 + V$$

or

$$H = \frac{p^2}{2m} + V(x) \tag{3.26}$$

It is interesting to note that the Hamiltonian H numerically equals the total energy E given by (3.12). However, the Hamiltonian H must be regarded as a function of position x and momentum p.

The variables (x, p) change according to Hamilton's equations of motion given as follows:

$$\boxed{\dot{x} = \frac{\partial H(x, p)}{\partial p}, \quad \dot{p} = -\frac{\partial H(x, p)}{\partial x}} \tag{3.27}$$

In fact, using (3.26),

$$\frac{\partial H}{\partial p} = \frac{\partial}{\partial p}\left(\frac{p^2}{2m} + V(x)\right) = \frac{p}{m}$$

$$\frac{\partial H}{\partial x} = \frac{\partial}{\partial x}\left(\frac{p^2}{2m} + V(x)\right) = \frac{dV(x)}{dx}$$

We therefore obtain

$$\dot{x} = \frac{p}{m} \tag{3.28}$$

$$\dot{p} = -\frac{dV}{dx} = F \tag{3.29}$$

Note that the first equation generates (3.23) and the second corresponds to Newton's equation of motion, (3.5).

In the present section, we introduced and discussed three equivalent descriptions of linear motion. No clear advantages of one formulation over the others are seen for one-dimensional systems. The concepts and most formulas can, however, be generalized in a straightforward manner for three-dimensional motion and also for a system composed of a number of particles. We will see later that the dynamics formulated in terms of Hamilton's equations of motion is the most suitable one for the general formulation of quantum mechanics and statistical mechanics.

Problem 3.1.1

Evaluate the partial derivatives $(\partial f/\partial x, \partial f/\partial y)$ of the following functions:
1. $f = ax + by$
2. $f = a(x + y)^2$
3. $f = (x + y)^{-1/2}$
4. $f = ax^3 + by^2$

Problem 3.1.2

Evaluate $\partial^2 f/\partial x^2 + \partial^2 f/\partial y^2$ for each function f given in Problem 3.1.1.

3.2
State of Motion in Phase Space. Reversible Motion

We have seen in the preceding section that if the force is known, then the position and velocity pair, (x, v), can be uniquely determined at a later time. In a similar manner, the position and momentum pair (x, p) changes with time uniquely as prescribed by Hamilton's equations of motion. We can then look at the motion of the particle as the *deterministic* change in the *state of motion* or *dynamical state* specified by either (x, v) or (x, p). It should be noted that knowledge of the position of the particle itself is not enough by itself to predict the position at a later time; therefore, position alone does not characterize the state of motion. For reasons that will become clear later, specification of the dynamical state by position and momentum (x, p) is more convenient than that by position and velocity (x, v). We will, therefore, specify the state of motion by position and momentum from now on.

The state of motion can be conveniently represented by a point in the *position–momentum space*, also called the *phase space*. Since (x, p) change with time according to Hamilton's equations of motion, the point representing the state will move on a continuous curve in the phase space. This curve will not branch out into two or more lines if the Hamiltonian is a single-valued function of x and p.

For illustration, let us consider some simple examples.

1. Free particle.
   The Hamiltonian for the system is

   $$H = \frac{p^2}{2m} \tag{3.30}$$

   Equations of motion can be obtained from (3.27). They are

   $$\dot{x} = \frac{\partial H}{\partial p} = \frac{p}{m} \tag{3.31}$$

   $$\dot{p} = -\frac{\partial H}{\partial x} = 0 \tag{3.32}$$

   The point representing the state of motion moves along a straight line parallel to the x-axis with a constant speed p/m in the phase space. If the momentum p is positive, then the point moves toward the right, that is, in the positive x-direction. If $p < 0$, then the point moves toward the left. This behavior is indicated in Figure 3.2.

2. Simple harmonic oscillator.
   The Hamiltonian for this system is given by

   $$H = \frac{p^2}{2m} + \frac{1}{2}kx^2 \tag{3.33}$$

   which is numerically equal to the total energy E [see (3.16)]:

   $$H(x, p) = \frac{p^2}{2m} + \frac{1}{2}kx^2 = E \tag{3.34}$$

   This numerical coincidence between Hamiltonian and total energy often happens in practice. It is stressed that while the total energy may be expressed in terms of (x, v) or (x, p), the Hamiltonian must, by definition, be regarded as a function of x and p.

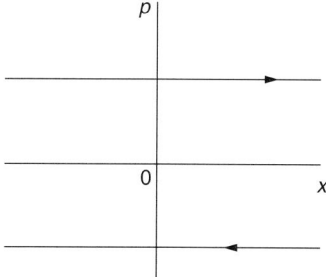

Fig. 3.2 Free particle in motion in phase space.

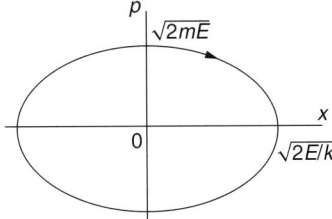

Fig. 3.3 The point representing a simple harmonic oscillator describes an ellipse in phase space.

Solving (3.34) with respect to x, we obtain

$$x = \pm\sqrt{\frac{2}{k}\left(E - \frac{p^2}{2m}\right)} \tag{3.35}$$

The maximum and minimum of p, both of which occur at $x = 0$, are given by $\pm\sqrt{2mE}$. In summary, the point representing the state of motion lies on the ellipse with semi-axes ($\sqrt{2E/k}$, $\sqrt{2mE}$) and center at origin as shown in Figure 3.3. This point moves with the "velocities" given by Hamilton's equations of motion:

$$\dot{x} = \frac{p}{m}, \quad \dot{p} = -kx \tag{3.36}$$

The point completes a cycle clockwise. The period T_0, that is, the time required to complete a cycle, can be obtained as follows. Eliminating p from (3.36), we obtain

$$\ddot{x} = -\frac{k}{m}x \tag{3.37}$$

whose solutions are

$$x = A\sin(\omega_0 t + \phi) \tag{3.38}$$

$$\omega_0 \equiv \sqrt{\frac{k}{m}} \tag{3.39}$$

where A and ϕ are constants. From the figure we see that the *amplitude A* equals $\sqrt{2E/K}$. The quantity ω_0 is called the *angular frequency*. The period T_0 is given by

$$T_0 = 2\pi/\omega_0 = 2\pi\left(\frac{m}{k}\right)^{1/2} \tag{3.40}$$

We now consider the general case of a particle subjected to an arbitrary force.

Suppose that the system (particle) moves from a state $A = (x_a, p_a)$ to another state $B = (x_b, p_b)$ in a time t. Let A' and B' be the states obtained from A and B, respectively, by changing the sign of the momenta. Thus, $A' = (x_a, -p_a)$, and $B' = (x_b, -p_b)$. If the motion from the state B' to the state A' in the same time t is possible, then the system is said to have a *reversible* motion.

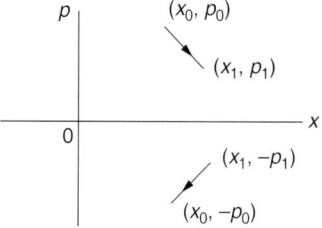

Fig. 3.4 The pair of paths corresponding to the reversible motion.

The two systems discussed above, that is the free particle and the simple harmonic oscillator, clearly allow such reversible motions. In fact, many systems treated in mechanics allow reversible motions. Let us look into this in more detail.

Let us assume that the system under consideration is characterized by a Hamiltonian $H(x, p)$, which does not depend on time explicitly. Let us further assume that the Hamiltonian H is even in p, that is,

$$H(x,-p) = H(x, p) \tag{3.41}$$

which is normally satisfied because the p-dependence of the Hamiltonian H arises from the kinetic energy $p^2/2m$. Assume that the particle is in the state (x_0, p_0) at a certain time. See Figure 3.4. At an infinitesimal time dt the particle will move to the neighboring state (x_1, p_1), which, from Hamilton's equations (3.27), is given by

$$x_1 = x_0 + \dot{x}\,dt = x_0 + \left(\frac{\partial H}{\partial p}\right)_0 dt$$

$$p_1 = p_0 + \dot{p}\,dt = p_0 - \left(\frac{\partial H}{\partial x}\right)_0 dt \tag{3.42}$$

Here, the suffix 0 means that the partial derivatives are to be calculated at (x_0, p_0). We now wish to see if the system will move (in phase space) from $(x_1, -p_1)$ to $(x_0, -p_0)$ in the same time interval dt. Starting from $-p_1$, the momentum will change to

$$-p_1 + dt(\dot{p})_{x=x_1, p=-p_1}$$

$$= -\left[p_0 - dt\left(\frac{\partial H}{\partial x}\right)_0\right] - dt\left(\frac{\partial H}{\partial x}\right)_{x=x_1, p=-p_1} \quad \text{[by (3.22) and (3.42)]}$$

$$= -p_0 + dt\left\{\left(\frac{\partial H}{\partial x}\right)_0 - \left(\frac{\partial H}{\partial x}\right)_{x=x_1, p=-p_1}\right\} \tag{3.43}$$

If H is an even function of p, then so is $\partial H/\partial x$. Therefore, we have

$$\left(\frac{\partial H}{\partial x}\right)_{x=x_1, p=-p_1} = \left(\frac{\partial H}{\partial x}\right)_{x=x_1, p=p_1} \tag{3.44}$$

3 Hamiltonian Mechanics

The quantity in the curly brackets in (3.43),

$$\left(\frac{\partial H}{\partial x}\right)_0 - \left(\frac{\partial H}{\partial x}\right)_{x=x_1, p=-p_1}$$

is equal to the difference in the derivative $\partial H/\partial x$ calculated at neighboring points, (x_0, p_0) and (x_1, p_1), and will be proportional to the small distance between the two points:

$$\left(\frac{\partial H}{\partial x}\right)_0 - \left(\frac{\partial H}{\partial x}\right)_{x=x_1, p=-p_1}$$

$$= \left(\frac{\partial H}{\partial x}\right)_{x=x_0, p=p_0} - \left(\frac{\partial H}{\partial x}\right)_{x=x_1, p=p_1}$$

$$= \text{linear in } |x_1 - x_0| \text{ or in } |p_1 - p_0| \tag{3.45}$$

Since, from (3.42),

$$|x_1 - x_0| \propto dt, \quad |p_1 - p_0| \propto dt \tag{3.46}$$

we then obtain

$$\left(\frac{\partial H}{\partial x}\right)_0 - \left(\frac{\partial H}{\partial x}\right)_{x=x_1, p=-p_1} \propto dt \tag{3.47}$$

Substituting this result in (3.43), we obtain

$$-p_1 + dt\, (\dot{p})_{x=x_1, p=p_1}$$

$$= -p_0 + \text{terms of the order } dt^2$$

$$\equiv -p_0 + O(dt^2) \tag{3.48}$$

That is, the momentum changes from $-p_1$ to $-p_0$ in the time interval dt with the neglect of terms of higher orders in dt. We can further show that the position will change from x_1 to x_0 in the same interval. In fact, the position reached in dt can be calculated as follows:

$$x_1 + dt\, (\dot{x})_{x=x_1, p=-p_1}$$

$$= x_1 + dt \left(\frac{\partial H}{\partial p}\right)_{x=x_1, p=-p_1}$$

$$= \left[x_0 + dt \left(\frac{\partial H}{\partial p}\right)_0\right] + dt \left(\frac{\partial H}{\partial p}\right)_{x=x_1, p=-p_1} \quad \text{[use (3.42)]} \tag{3.49}$$

3.2 State of Motion in Phase Space. Reversible Motion

Since the Hamiltonian H is even in p, the p-derivative $\partial H/\partial p$ is odd. Therefore, we get

$$\left(\frac{\partial H}{\partial p}\right)_{x=x_1, p=-p_1} = -\left(\frac{\partial H}{\partial p}\right)_{x=x_1, p=p_1} \quad (3.50)$$

Substituting this result in (3.49), we obtain

$$x_0 + dt\left[\left(\frac{\partial H}{\partial p}\right)_0 - \left(\frac{\partial H}{\partial p}\right)_{x=x_1, p=p_1}\right] \quad (3.51)$$

$$= x_0 + O(dt^2) \quad (3.52)$$

Therefore, the system allows a reversible motion for infinitesimal time. Repeating the infinitesimal processes, we can then deduce the same conclusion for a finite time. Note that no explicit form of the Hamiltonian $H(x, p)$ has been assumed in the above arguments. We may, therefore restate our findings as follows:

Theorem
If a system is characterized by a Hamiltonian that is independent of time and even in p, then the system allows a reversible motion.

The vertical motion of a falling body in the air can be regarded as a typical linear motion. In the real environment, however, the body is subjected to the air resistance in addition to the gravitational force. If we assume that the resistance force is proportional to the velocity $v = p/m$, then we may write the resistance as

$$F(\text{resistance}) = -bv = -(b/m)p \quad (3.53)$$

where b is a constant. Such a force cannot be derived from any energy-conserving potential. The system under the action of such a force cannot be characterized by a usual Hamiltonian and does not allow a reversible motion.

Problem 3.2.1

Consider a particle moving in one dimension. When the potential V has the form show in Figure 3.5, discuss the motion of the point in phase space, representing the dynamical state of the particle with different total energies say $E_1, E_2, \ldots$.

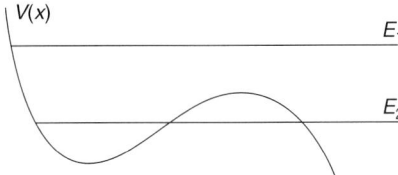

Fig. 3.5 Problem 3.2.1

3.3
Hamiltonian for a System of many Particles

In the last few sections we have looked at the linear motion of a particle. Many theoretical developments can be generalized straightforwardly to more complicated cases. We will discuss such generalizations in the present section.

Let us take a particle moving in three-dimensional space. The position of the particle will be denoted by the position vector **r**. Many important properties of vectors were collected and discussed in Chapter 1. Ordinary three-dimensional *vectors* will will be indicated by bold face letters. The vector **r** can be represented by the set of coordinates (x, y, z) after Cartesian axes are chosen, or it can be represented by other sets of coordinates such as spherical coordinates.

Newton's equation of motion is given by the vector equation:

$$m\ddot{\mathbf{r}} \equiv m\frac{d^2\mathbf{r}}{dt^2} = \mathbf{F} \tag{3.54}$$

where **F** represents a vector force acting on the particle. In many cases of interest, the force **F** can be expressed in terms of the gradient of a *scalar potential-energy* function V

$$\mathbf{F} = -\text{grad}\, V \equiv -\nabla V$$

$$\equiv -\left(\mathbf{i}\frac{\partial}{\partial x} + \mathbf{j}\frac{\partial}{\partial y} + \mathbf{k}\frac{\partial}{\partial z}\right) V(x, y, z) \tag{3.55}$$

where $(\mathbf{i}, \mathbf{j}, \mathbf{k})$ are *orthonormal* (or *orthogonal unit*) vectors pointing in the positive x-, y- and z-axes. The symbol ∇, called *nabla* or *del*, represents a partial-differential vector operator as indicated.

The kinetic energy T of the particle is defined by

$$T \equiv \frac{1}{2} m\, (\dot{\mathbf{r}} \cdot \dot{\mathbf{r}}) \equiv \frac{1}{2} m\dot{r}^2$$
$$= \frac{1}{2} m\left(\dot{x}^2 + \dot{y}^2 + \dot{z}^2\right) \tag{3.56}$$

where the the center dot between two vectors indicates the *scalar product* of the two

$$\mathbf{A} \cdot \mathbf{B} \equiv A_x B_x + A_y B_y + A_z B_z$$

When the particle is acted on by a force derivable from a potential V as given by (3.55), the total energy, that is, the sum of the kinetic and potential energies is conserved [Problem 3.3.1]:

$$T + V = E \tag{3.57}$$

Because of this property, such a force is called a *conservative force*.

In the Lagrangian description of the motion, (a) we choose a set of *generalized coordinates* (q_1, q_2, q_3), (b) construct the Lagrangian

$$L \equiv T - V = L(q_1, q_2, q_3, \dot{q}_1, \dot{q}_2, \dot{q}_3) \tag{3.58}$$

as a function of coordinates $\{q_k\}$ and velocities $\{\dot{q}_k\}$, and (c) derive the equations of motion from *Lagrange's equations*:

$$\frac{d}{dt}\left(\frac{\partial L}{\partial \dot{q}_k}\right) - \frac{\partial L}{\partial q_k} = 0 \quad k = 1, 2, 3 \tag{3.59}$$

This method works especially well if we can recognize a symmetry in the dynamics of the system in advance, and choose a suitable set of coordinates.

For example, if the applied force is a *central force* whose derelictions toward or away from a fixed center, and can be derived from the potential $V(r)$, which depends on the distance r from the center of force, then we may choose the spherical polar coordinates (r, θ, ϕ) for the coordinates of description (q_1, q_2, q_3). The kinetic energy T can then be expressed by

$$T = \frac{1}{2} m(\dot{r}^2 + r^2 \dot{\theta}^2 + r^2 \sin^2\theta \, \dot{\phi}^2)$$

and the Lagrangian L is given by

$$L(r, \theta, \phi, \dot{r}, \dot{\theta}, \dot{\phi}) = \frac{1}{2} m(\dot{r}^2 + r^2 \dot{\theta}^2 + r^2 \sin^2\theta \, \dot{\phi}^2) - V(r) \tag{3.60}$$

Since this function L does not depend on ϕ,

$$\partial L / \partial \phi = 0$$

From Lagrange's equation (3.59) we obtain

$$\frac{d}{dt}\left(\frac{\partial L}{\partial \dot{\phi}}\right) - 0 = 0$$

which implies the existence of a constant of motion:

$$\frac{\partial L}{\partial \dot{\phi}} = \text{constant}$$

Let us now turn to the Hamiltonian description of the motion. In this case (a) we select coordinates (q_1, q_2, q_3) and "momenta" (p_1, p_2, p_3), which are defined by

$$\boxed{p_k \equiv \frac{\partial L}{\partial \dot{q}_k}} \tag{3.61}$$

(b) Construct the Hamiltonian

$$H = \sum_k p_k \dot{q}_k - L = H(q_1, q_2, \ldots, p_3) \tag{3.62}$$

as a function of $(q_1, q_2, \ldots, p_3)$ and (c) derive the equation of motion from Hamilton's equations of motion:

$$\boxed{\dot{q}_k = \frac{\partial H}{\partial p_k}, \quad \dot{p}_k = -\frac{\partial H}{\partial q_k}} \tag{3.63}$$

The "momenta" defined by (3.61) are called *canonical momenta*. If we choose the Cartesian coordinates (x, y, z) to represent the position vector $\mathbf{r}$, the canonical momenta are equal to the components of the linear momentum, (p_x, p_y, p_z), [see Problem 3.3.3]. In general, the canonical momenta may have a dimension different from that of a linear momentum. For example, the canonical momenta for the spherical polar coordinates (r, θ, ϕ) are given by

$$p_r \equiv \frac{\partial L}{\partial \dot{r}} = m\dot{r}$$

$$p_\theta \equiv \frac{\partial L}{\partial \dot{\theta}} = mr^2 \dot{\theta}$$

$$p_\phi \equiv \frac{\partial L}{\partial \dot{\phi}} = mr^2 \sin^2 \theta \, \dot{\phi} \tag{3.64}$$

where (3.60) was used. We note that the canonical momenta p_θ and p_ϕ have dimensions of angular momentum (moment of inertia × angular velocity) while p_r has dimensions of linear momentum (mass × linear velocity).

The set of generalized coordinates (q_1, q_2, q_3) and momenta (p_1, p_2, p_3) is said to form *canonical variables*. Each of these variables change following Hamilton's equations of motion (3.63), which are also called.

Again, the degree of difficulty associated with solving the equations of motion depends on the choice of coordinates. The proper choice therefore is very important in practice.

Let us now consider a system of N particles moving in three dimensions. To describe a dynamical state of this system we need $3N$ coordinates $(q_1, q_2, \ldots, q_{3N})$ and $3N$ momenta $(p_1, p_2, \ldots, p_{3N})$. The Hamiltonian H of the system depends on $6N$ canonical variables. These variables change, following Hamilton's equation of motion of the form (3.63), where the index k now applies to $1, 2, \ldots, 3N$.

At this point we emphasize the generality of the Hamiltonian formulation of mechanics. From the linear motion of a single particle to the motion of many particles moving in three dimensions, we can formulate the dynamics of the system in question in a unified manner.

Problem 3.3.1

Derive the conservation of the total energy $T + V$ = constant from (3.54), (3.55) and (3.56).

Problem 3.3.2

The necessary and sufficient condition for the existence of a conservative force is that the force $\mathbf{F}(\mathbf{r})$ satisfies

$$\nabla \times \mathbf{F} \equiv \mathbf{i}\left(\frac{\partial F_z}{\partial y} - \frac{\partial F_y}{\partial z}\right) + \mathbf{j}\left(\frac{\partial F_x}{\partial z} - \frac{\partial F_z}{\partial x}\right) + \mathbf{k}\left(\frac{\partial F_y}{\partial x} - \frac{\partial F_x}{\partial y}\right) = 0$$

Show that a central force whose magnitude depends only on the distance from the center is a conservative force.

Problem 3.3.3

Consider a particle of mass m moving under a conservative force $\mathbf{F} = -\nabla V(\mathbf{r})$ in three-dimensional space.

1. Write down the kinetic energy T in terms of the velocity $\mathbf{v} \equiv \dot{\mathbf{r}}$.
2. Express the Lagrangian function $L = T - V$ in terms of Cartesian coordinates and velocities $(x, y, z, \dot{x}, \dot{y}, \dot{z}) \equiv (q_1, q_2, q_3, \dot{q}_1, \dot{q}_2, \dot{q}_3)$.
3. Derive the equations of motion by means of Lagrange's equations:

$$\frac{d}{dt}\left(\frac{\partial L}{\partial \dot{q}_k}\right) - \frac{\partial L}{\partial q_k} = 0$$

Carry out the partial differentiation explicitly.

4. Compare the resulting equations with Newton's equation of motion

$$m\frac{d\mathbf{v}}{dt} = \mathbf{F} = -\nabla V$$

5. Define the canonical momenta through the formula

$$p_k \equiv \frac{\partial L}{\partial \dot{q}_k}$$

Notice that the canonical momenta (p_1, p_2, p_3) agree with the Cartesian components (p_x, p_y, p_z) of the vector momentum $\mathbf{p} \equiv m\mathbf{r}$.

6. Construct the Hamiltonian H through

$$H = \sum_{k=1}^{3} p_k \dot{q}_k - L$$

and express it as a function of the coordinates and momenta $(q_1, q_2, q_3, p_1, p_2, p_3)$. Observe that this Hamiltonian H has the same numerical value as the total energy $E = T + V$.

7. Derive the equations of motion through Hamilton's equations:

$$\dot{q}_k = \frac{\partial H}{\partial p_k}, \quad \dot{p}_k = -\frac{\partial H}{\partial q_k}$$

Carry out the partial differentiation explicitly.

Problem 3.3.4

Consider a system composed of an electron and a proton (a hydrogen atom).
1. Write down the Hamiltonian H in Cartesian coordinates and momenta.
2. Write down Hamilton's equations of motion explicitly.
3. Show that they are equivalent to Newton's equations of motion.

3.4
Canonical Transformation

Hamilton's equations of motion, (3.63) hold for *any* set of canonical variables (q, p). In the present section we will study the relations between different sets of canonical variables in depth.

Let us consider a particle whose motion is constrained to a plane. The motion can be described in terms of Cartesian coordinates and momenta (x, y, p_x, p_y) or in terms of polar coordinates and momenta $(r, \theta, p_r, p_\theta)$. These coordinates are related by

$$x = r\cos\theta, \quad y = r\sin\theta$$
$$p_x \equiv m\dot{x} = m\left(\dot{r}\cos\theta - r\dot{\theta}\sin\theta\right) = p_r\cos\theta - \frac{p_\theta}{r}\sin\theta$$
$$p_y \equiv m\dot{y} = m\left(\dot{r}\sin\theta - r\dot{\theta}\cos\theta\right) = p_r\sin\theta - \frac{p_\theta}{r}\cos\theta \quad (3.65)$$

and conversely,

$$r = (x^2 + y^2)^{1/2}, \quad \theta = \tan^{-1}(y/x)$$
$$p_r \equiv m\dot{r} = m\frac{x\dot{x} + y\dot{y}}{(x^2 + y^2)^{1/2}} = \frac{xp_x + yp_y}{(x^2 + y^2)^{1/2}}$$
$$p_\theta = mr^2\dot{\theta} = mr^2\frac{\dot{y}x - \dot{x}y}{x^2 + y^2} = xp_y - yp_x \quad (3.66)$$

In general, the differential volume elements $dx\,dy\,dp_x\,dp_y$ and $dr\,d\theta\,dp_r\,dp_\theta$ are related by

$$dx\,dy\,dp_x\,dp_y = J\,dr\,d\theta\,dp_r\,dp_\theta \quad (3.67)$$

where J denotes the *Jacobian of transformation* given by

$$J \equiv \frac{\partial(x, y, p_x, p_y)}{\partial(r, \theta, p_r, p_\theta)}$$

$$\equiv \begin{vmatrix} \frac{\partial x}{\partial r} & \frac{\partial x}{\partial \theta} & \frac{\partial x}{\partial p_r} & \frac{\partial x}{\partial p_\theta} \\ \frac{\partial y}{\partial r} & \frac{\partial y}{\partial \theta} & \frac{\partial y}{\partial p_r} & \frac{\partial y}{\partial p_\theta} \\ \frac{\partial p_x}{\partial r} & \frac{\partial p_x}{\partial \theta} & \frac{\partial p_x}{\partial p_r} & \frac{\partial p_x}{\partial p_\theta} \\ \frac{\partial p_y}{\partial r} & \frac{\partial p_y}{\partial \theta} & \frac{\partial p_y}{\partial p_r} & \frac{\partial p_y}{\partial p_\theta} \end{vmatrix} \qquad (3.68)$$

After explicit calculations using (3.67) [Problem 3.4.1], we can show that

$$J = \frac{\partial(x, y, p_x, p_y)}{\partial(r, \theta, p_r, p_\theta)} = 1 \qquad (3.69)$$

We can also show that the Jacobian of the inverse transformation from (x, y, p_x, p_y) to $(r, \theta, p_r, p_\theta)$ is unity:

$$\frac{\partial(r, \theta, p_r, p_\theta)}{\partial(x, y, p_x, p_y)} = 1 \qquad (3.70)$$

These two properties are related to each other. From the general property of Jacobians of transformation from $(z_1, z_2, \ldots, z_n)$ to $(Z_1, Z_2, \ldots, Z_n)$ and vice versa, we must have

$$\frac{\partial(z_1, z_2, \ldots, z_n)}{\partial(Z_1, Z_2, \ldots, Z_n)} \times \frac{\partial(Z_1, Z_2, \ldots, Z_n)}{\partial(z_1, z_2, \ldots, z_n)} = 1 \qquad (3.71)$$

Selecting $(z_1, z_2, z_3, z_4) = (r, \theta, p_r, p_\theta)$ and $(Z_1, Z_2, Z_3, Z_4) = (x, y, p_x, p_y)$, we have

$$\frac{\partial(r, \theta, p_r, p_\theta)}{\partial(x, y, p_x, p_y)} \times \frac{\partial(x, y, p_x, p_y)}{\partial(r, \theta, p_r, p_\theta)} = 1$$

Combining this with (3.69) we then obtain (3.70).

It is possible to generalize these results, (3.69) and (3.70), as follows. A transformation of dynamical variables in which canonical (Hamilton's) equations of motion are kept in the form (3.63) is called a canonical transformation.

☐ **Theorem**

For any canonical transformation, the Jacobian is unity.

$$\boxed{J = 1 \quad \text{for canonical transformations}} \qquad (3.72)$$

We wish to prove this theorem. Assume that the set of canonical variables (q_1, q_2, p_1, p_2) are related to the Cartesian coordinates and momenta (x, y, p_x, p_y) by

$$x = x(q_1, q_2) \equiv x(q), \quad y = y(q)$$

$$p_x = p_x(q_1, q_2, p_1, p_2) \equiv p_x(q, p), \quad p_y = p_y(q, p) \tag{3.73}$$

Note that the coordinates (x, y) are regarded as functions of the coordinates (q_1, q_2) alone. Inverting these equations, we have the relations:

$$q_1 = q_1(x, y), \quad q_2 = q_2(x, y)$$

$$p_1 = p_1(x, y, p_x, p_y), \quad p_2 = p_2(x, y, p_x, p_y) \tag{3.74}$$

Using (3.73), we obtain

$$\frac{\partial x}{\partial p_1} = \frac{\partial x}{\partial p_2} = \frac{\partial y}{\partial p_1} = \frac{\partial y}{\partial p_2} = 0 \tag{3.75}$$

Using these results, the Jacobian J of the transformation from (q_1, q_2, p_1, p_2) to (x, y, p_x, p_y) can be calculated as follows:

$$J \equiv \frac{\partial(x, y, p_x, p_y)}{\partial(q_1, q_2, p_1, p_2)}$$

$$= \begin{vmatrix} \frac{\partial x}{\partial q_1} & \frac{\partial x}{\partial q_2} & 0 & 0 \\ \frac{\partial y}{\partial q_1} & \frac{\partial y}{\partial q_2} & 0 & 0 \\ \frac{\partial p_x}{\partial q_1} & \frac{\partial p_x}{\partial q_2} & \frac{\partial p_x}{\partial p_1} & \frac{\partial p_x}{\partial p_2} \\ \frac{\partial p_y}{\partial q_1} & \frac{\partial p_y}{\partial q_2} & \frac{\partial p_y}{\partial p_1} & \frac{\partial p_y}{\partial p_2} \end{vmatrix}$$

$$= \begin{vmatrix} \frac{\partial x}{\partial q_1} & \frac{\partial x}{\partial q_2} \\ \frac{\partial y}{\partial q_1} & \frac{\partial y}{\partial q_2} \end{vmatrix} \times \begin{vmatrix} \frac{\partial p_x}{\partial p_1} & \frac{\partial p_x}{\partial p_2} \\ \frac{\partial p_y}{\partial p_1} & \frac{\partial p_y}{\partial p_2} \end{vmatrix} \tag{3.76}$$

The Hamiltonian $H(x, y, p_x, p_y)$ depends on (x, y, p_x, p_y), which, according to (3.73), can themselves be regarded as functions of (q_1, q_2, p_1, p_2). A change in p_1 generates the changes in $p_x(q, p)$ and $p_y(q, p)$:

$$dp_x = \frac{\partial p_x}{\partial p_1} dp_1, \quad dp_y = \frac{\partial p_y}{\partial p_1} dp_1 \tag{3.77}$$

These changes induce the change in the Hamiltonian H. We may express this by the following differential equation:

$$dH[x(q), y(q), p_x(q, p), p_y(q, p)]$$
$$= \frac{\partial H}{\partial p_x} \frac{\partial p_x}{\partial p_1} dp_1 + \frac{\partial H}{\partial p_y} \frac{\partial p_y}{\partial p_1} dp_1 \tag{3.78}$$

3.4 Canonical Transformation

Here, we tacitly assumed that the variables q_1, q_2 and p_2 are fixed so that

$$dq_1 = dq_2 = dp_2 = 0$$

If we divide (3.78) by dp_1 the lhs is, by definition, equal to the partial derivative $\partial H(q_1, q_2, p_1, p_2)/\partial p_1$:

$$\left(\frac{dH}{dp_1}\right)_{q_1, q_2, p_2 \text{ fixed}} \equiv \frac{\partial}{\partial p_1} H(q_1, q_2, p_1, p_2) \tag{3.79}$$

We then obtain

$$\frac{\partial H}{\partial p_1} = \frac{\partial H}{\partial p_x}\frac{\partial p_x}{\partial p_1} + \frac{\partial H}{\partial p_y}\frac{\partial p_y}{\partial p_1} \tag{3.80}$$

By using the canonical equations of motion (3.63) on both sides, we obtain

$$\frac{dq_1}{dt} = \frac{dx}{dt}\frac{\partial p_x}{\partial p_1} + \frac{dy}{dt}\frac{\partial p_y}{\partial p_1} \tag{3.81}$$

Multiplying by dt, we get

$$dq_1 = \frac{\partial p_x}{\partial p_1}dx + \frac{\partial p_y}{\partial p_1}dy \tag{3.82}$$

On the other hand, we obtain directly from (3.74),

$$dq_1(x, y) = \frac{\partial q_1}{\partial x}dx + \frac{\partial q_1}{\partial y}dy \tag{3.83}$$

Comparison of the last two equations yields

$$\frac{\partial p_x}{\partial p_1} = \frac{\partial q_1}{\partial x}, \quad \frac{\partial p_y}{\partial p_1} = \frac{\partial q_1}{\partial y} \tag{3.84}$$

In a similar manner, we obtain

$$\frac{\partial p_x}{\partial p_2} = \frac{\partial q_2}{\partial x}, \quad \frac{\partial p_y}{\partial p_2} = \frac{\partial q_2}{\partial y} \tag{3.85}$$

Using the last four equations, we obtain

$$\begin{vmatrix} \frac{\partial p_x}{\partial p_1} & \frac{\partial p_x}{\partial p_2} \\ \frac{\partial p_y}{\partial p_1} & \frac{\partial p_y}{\partial p_2} \end{vmatrix} = \begin{vmatrix} \frac{\partial q_1}{\partial x} & \frac{\partial q_2}{\partial x} \\ \frac{\partial q_1}{\partial y} & \frac{\partial q_2}{\partial y} \end{vmatrix} \tag{3.86}$$

Substituting this result into (3.76) and noting (3.70), we then obtain the desired result, namely that the Jacobian J equals unity.

The above proof was executed without using an explicit form for the transformation. The critical use of canonical equations in the proof should be noted.

In (3.73) we assumed that coordinates alone are related to each other. The theorem holds for *any canonical* transformation in which two sets of variables, (q, p) and (Q, P) are related in a general manner:

$$q_k = q_k(Q, P), \quad p_k = p_k(Q, P) \tag{3.87}$$

The proof of this case, however, is more involved. Interested readers should refer to a textbook on advanced classical mechanics such as Goldstein [3].

The theorem (3.72) holds for higher dimensions and for cases where many particles are present in the system. The proof given above may simply be extended for a subclass of canonical transformations called point transformations, in which coordinates $\{q_k\}$ and $\{Q_k\}$ are related to each other.

Let us now consider a system of N particles moving in three dimensions. The dynamical state of this system can be specified by a set of $6N$ canonical variables $(Q_1, \ldots, Q_{3N}, P_1, \ldots, P_{3N}) \equiv (q, p)$ or again by another set of canonical variables $(Q_1, \ldots, Q_{3N}, P_1, \ldots, P_{3N}) \equiv (Q, P)$. Because of theorem (3.72), the phase-space volume elements satisfy

$$dQ_1 \ldots dQ_{3N} dP_1 \ldots dP_{3N} = dq_1 \ldots dq_{3N} dp_1 \ldots dp_{3N}$$

or, in shorthand notation,

$$d^{3N}Q d^{3N}P = d^{3N}q d^{3N}p \tag{3.88}$$

The significance of this result will be discussed later in Chapter 13.

Problem 3.4.1

Prove (3.69).

3.5
Poisson Brackets

In the Hamiltonian mechanics, the basic variables of description, called the *dynamical variables*, are the canonical coordinates and momenta, (q, p). Quantities such as the Hamiltonian, total momentum, and total angular momentum are regarded as functions of the dynamical variables (q, p) and are called *dynamical functions*. Although the dynamical variables (q, p) can be chosen in many equivalent ways, the equations of motion for these dynamical variables and functions can be written in a unified manner. This topic will be discussed in the present section. The material presented here is of utmost importance to the development of quantum mechanics.

Let us take two dynamical functions $u(q, p)$ and $v(q, p)$. We introduce the quantity $\{u, v\}$ as

$$\{u, v\} \equiv \sum_{k=1}^{f} \left[\frac{\partial u}{\partial q_k} \frac{\partial v}{\partial p_k} - \frac{\partial v}{\partial q_k} \frac{\partial u}{\partial p_k} \right] \tag{3.89}$$

where f is the *number of degrees of freedom*. For example, $f = 3N$ for a system of N particles moving in three dimensions. The quantity $\{u, v\}$ is called the *Poisson bracket* of u, v.

From the definition in (3.89), we see that

$$\{u, v\} = -\{v, u\} \tag{3.90}$$
$$\{u, c_1 v + c_2 w\} = c_1 \{u, v\} + c_2 \{u, w\} \tag{3.91}$$

where c_1 and c_2 are numbers and w is another dynamical function. We can also show that

$$\{u v, w\} = \{u, w\} v + u \{v, w\}$$
$$\{u, v w\} = \{u, v\} w + v \{u, w\} \tag{3.92}$$

In fact, the first of (3.92) may by proved as follows:

$$\{u v, w\} = \sum_k \left[\frac{\partial(uv)}{\partial q_k} \frac{\partial w}{\partial p_k} - \frac{\partial w}{\partial q_k} \frac{\partial(uv)}{\partial p_k} \right]$$

$$= \sum_k \left[\left(\frac{\partial u}{\partial q_k} v + u \frac{\partial v}{\partial q_k} \right) \frac{\partial w}{\partial p_k} - \frac{\partial w}{\partial q_k} \left(\frac{\partial u}{\partial p_k} v + u \frac{\partial v}{\partial p_k} \right) \right]$$

$$= \sum_k \left[\left(\frac{\partial u}{\partial q_k} \frac{\partial w}{\partial p_k} - \frac{\partial w}{\partial q_k} \frac{\partial u}{\partial p_k} \right) v + u \left(\frac{\partial v}{\partial q_k} \frac{\partial w}{\partial p_k} - \frac{\partial w}{\partial q_k} \frac{\partial v}{\partial p_k} \right) \right]$$

$$= \{u, w\} v + u \{v, w\}$$

Putting $u = q_r$ and $v = p_s$ in (3.89), we have

$$\{q_r, p_s\} \equiv \sum_k \left[\frac{\partial q_r}{\partial q_k} \frac{\partial p_s}{\partial p_k} - \frac{\partial p_s}{\partial q_k} \frac{\partial q_r}{\partial p_k} \right] \tag{3.93}$$

Clearly,

$$\frac{\partial q_r}{\partial q_k} = \delta_{rk}, \quad \frac{\partial p_s}{\partial p_k} = \delta_{sk}, \quad \frac{\partial p_s}{\partial q_k} = \frac{\partial q_r}{\partial p_k} = 0 \tag{3.94}$$

where δ_{rs} is the *Kronecker delta* defined by

$$\delta_{rs} = \begin{cases} 1 & \text{if } r = s \\ 0 & \text{otherwise} \end{cases} \tag{3.95}$$

Using (3.94), we obtain from (3.93)

$$\{q_r, p_s\} = \delta_{rs} \tag{3.96}$$

In a similar manner, we can show that

$$\{q_r, q_s\} = \{p_r, p_s\} = 0 \tag{3.97}$$

In summary, we have

$$\{q_r, p_s\} = \delta_{rs}, \quad \{q_r, q_s\} = \{p_r, p_s\} = 0 \tag{3.98}$$

These relations are called the *fundamental Poisson bracket relations*.

Although we obtained these relations by calculating the Poisson brackets in terms of the special canonical variables (q, p), we can obtain the same results by using any set of canonical variables.

To see this, let us introduce new canonical variables (Q, P) and assume the relations

$$Q_r = Q_r(q, p), \quad P_r = P_r(q, p) \tag{3.99}$$

Conversely,

$$q_s = q_s(Q, P), \quad p_s = p_s(Q, P) \tag{3.100}$$

By a method similar to that used for deriving (3.84) and (3.85), we can show (Problem 3.5.2) that

$$\frac{\partial Q_r(q, p)}{\partial p_s} = -\frac{\partial q_s(Q, P)}{\partial P_r}, \quad \frac{\partial Q_r}{\partial q_s} = \frac{\partial p_s}{\partial P_r}$$

$$\frac{\partial P_r}{\partial p_s} = \frac{\partial q_s}{\partial Q_r}, \quad \frac{\partial P_r}{\partial q_s} = -\frac{\partial p_s}{\partial Q_r} \tag{3.101}$$

It is noted that these relations are the extensions of (3.84) and (3.85).

Let us now take q_r and differentiate it with respect to q_s, with all other q's and p's fixed. Naturally, we obtain

$$\frac{\partial q_r}{\partial q_s} = \delta_{rs} \tag{3.102}$$

On the other hand, we can regard q_r as a function of Q's and P's, which are functions of q's and p's as implied by (3.99). Looking at q_r in this way and differentiating it with respect to q_s, we obtain

$$\frac{\partial}{\partial q_s} q_r[Q(q, p), P(q, p)]$$
$$= \sum_k \left[\frac{\partial q_r}{\partial Q_k} \frac{\partial Q_k(q, p)}{\partial q_s} + \frac{\partial q_r}{\partial P_k} \frac{\partial P_k(q, p)}{\partial q_s} \right] \tag{3.103}$$

Comparing the last two equations, we find that

$$\sum_k \left[\frac{\partial q_r}{\partial Q_k} \frac{\partial Q_k}{\partial q_s} + \frac{\partial q_r}{\partial P_k} \frac{\partial P_k}{\partial q_s} \right] = \delta_{rs} \tag{3.104}$$

We now transform the lhs using (3.101), and obtain

$$\sum_k \left[\frac{\partial q_r}{\partial Q_k} \frac{\partial Q_k}{\partial q_s} + \frac{\partial q_r}{\partial P_k} \frac{\partial P_k}{\partial q_s} \right]$$

$$= \sum_k \left[\frac{\partial q_r}{\partial Q_k} \left(\frac{\partial p_s}{\partial P_k} \right) + \frac{\partial q_r}{\partial P_k} \left(-\frac{\partial p_s}{\partial Q_k} \right) \right]$$

$$= \sum_k \left[\frac{\partial q_r}{\partial Q_k} \frac{\partial p_s}{\partial P_k} - \frac{\partial p_s}{\partial Q_k} \frac{\partial q_r}{\partial P_k} \right] \equiv \{q_r, p_s\}_{(Q,P)} \qquad (3.105)$$

which is just the Poisson bracket $\{q_r, p_s\}$, calculated in terms of (Q, P). We therefore obtain

$$\{q_r, p_s\}_{(Q,P)} = \delta_{rs} \qquad (3.106)$$

In a similar manner, we can show that

$$\{q_r, q_s\}_{(Q,P)} = \{p_r, p_s\}_{(Q,P)} = 0 \qquad (3.107)$$

In fact,

$$\frac{\partial q_r}{\partial p_s} = 0, \quad \text{[by definition of the partial differentiation]}$$

$$\frac{\partial q_r}{\partial p_s} = \sum_k \left[\frac{\partial q_r}{\partial Q_k} \frac{\partial Q_k}{\partial p_s} + \frac{\partial q_r}{\partial P_k} \frac{\partial P_k}{\partial p_s} \right]$$

$$= \sum_k \left[-\frac{\partial q_r}{\partial Q_k} \frac{\partial q_s}{\partial P_k} + \frac{\partial q_r}{\partial P_k} \frac{\partial q_s}{\partial Q_k} \right]$$

$$\equiv -\{q_r, q_s\}_{(Q,P)} = 0$$

$$\frac{\partial p_r}{\partial q_s} = \sum_k \left[\frac{\partial p_r}{\partial Q_k} \frac{\partial Q_k}{\partial q_s} + \frac{\partial p_r}{\partial P_k} \frac{\partial P_k}{\partial q_s} \right]$$

$$= \sum_k \left[\frac{\partial p_r}{\partial Q_k} \frac{\partial p_s}{\partial P_k} - \frac{\partial p_r}{\partial P_k} \frac{\partial p_s}{\partial Q_k} \right]$$

$$\equiv \{p_r, p_s\}_{(Q,P)} = 0$$

In the derivations of (3.106) and (3.107), no explicit functional relationships between (q, p) and (Q, P) are required. Therefore, the Poisson brackets $\{q_r, p_s\}$, $\{q_r, q_s\}$, and $\{p_r, p_s\}$ have the values δ_{rs}, 0, and 0, respectively, no matter what canonical variables are used to calculate them. As it turns out, we can generalize this property in a significant manner:

☐ **Theorem**

The Poisson brackets $\{u, v\}$ of any dynamical quantities u and v have the same value for any and every canonical representation.

Symbolically we may represent this by

$$\{u, v\}_{(q,p)} = \{u, v\}_{(Q,P)} \equiv \{u, v\} \tag{3.108}$$

A proof of this theorem can be worked out with the aid of the fundamental bracket relations in (3.106) and (3.107). But since it is somewhat lengthy, it is given in Appendix A. The above theorem means that we can calculate Poisson brackets using any set of canonical variables. Because of this property, it is not necessary to specify the variables of computations. We will hereafter drop the subindices (q, p) and (Q, P).

The canonical equations of motion can be expressed in terms of Poisson brackets as follows:

$$\frac{dq_k}{dt} = \frac{\partial H}{\partial p_k} = \{q_k, H\}$$

$$\frac{dp_k}{dt} = -\frac{\partial H}{\partial q_k} = \{p_k, H\} \tag{3.109}$$

which can be established by direct calculation. We note that both equations can be expressed in a unified form by

$$\boxed{\frac{d\xi}{dt} = \{\xi, H\}} \tag{3.110}$$

where $\xi = q_k$ or p_k. It is possible to show that (3.110) is valid for *any* dynamical function ξ. In fact, we obtain

$$\frac{d\xi(q, p)}{dt} = \sum_k \left(\frac{\partial \xi}{\partial q_k} \frac{dq_k}{dt} + \frac{\partial \xi}{\partial p_k} \frac{dp_k}{dt} \right)$$

$$= \sum_k \left[\frac{\partial \xi}{\partial q_k} \frac{\partial H}{\partial p_k} - \frac{\partial \xi}{\partial p_k} \frac{\partial H}{\partial q_k} \right]$$

$$= \{\xi, H\}$$

In practical calculations, Poisson brackets must be evaluated by using a special representation. We most often use Cartesian coordinates and momenta.

The central importance of Poisson brackets lies in the fact that their study helps find the fundamental commutation relations, the most substantial hypotheses in quantum mechanics. This method of constructing a quantum theory in analogy with classical mechanics was due to Dirac [4], and will be reviewed in Chapter 15.

Problem 3.5.1

Verify (3.90), (3.91), and (3.92).

Problem 3.5.2

Prove (3.101).

Problem 3.5.3

Show that the Poisson brackets of the components of the angular momentum satisfy: $\{J_x, J_y\} = J_z$
  1. for a single particle
  2. for a many-body system.

Problem 3.5.4

Prove the Jacobi identity:

$$\{u, \{v, w\}\} + \{v, \{w, u\}\} + \{w, \{u, v\}\} = 0 \tag{3.111}$$

References

1 Symon, K.R. (1953) *Mechanics*, Addison-Wesley, Reading, USA.
2 Kibble, T.W.B. (1966) *Classical Mechanics*, McGraw-Hill, London.
3 Goldstein, H., Poole, C., and Safko, J. (2002) *Classical Mechanics*, San Francisco, Addison Wesley, Reading MA.
4 Dirac, P.A.M. (1958) *Principles of Quantum Mechanics*, 4th edn, Oxford University Press, London, UK.

4
Coupled Oscillators and Normal Modes

A system of particles on a string is treated in this chapter. The normal modes of oscillations and the normal coordinates are introduced and discussed.

4.1
Oscillations of Particles on a String and Normal Modes

Let us consider a set of N particles of equal mass m attached to a light string of length $(N + 1)l$ (in equilibrium), stretched to a tension τ. The particles are equally spaced with a regular separation. We now look at transverse oscillations of the particles. We denote the transverse displacements by y_j as indicated in Figure 4.1. The kinetic energy of this system is

$$T = \frac{m}{2}\left(\dot{y}_1^2 + \dot{y}_2^2 + \cdots + \dot{y}_N^2\right) \tag{4.1}$$

The potential energy for small displacements can be calculated as follows. We look at the length of the string between the jth and $(j + 1)$th particles. In equilibrium its length is l, but when the particles are vertically displaced the increment of the length, Δl, is given by

$$\Delta l \equiv \left[l^2 + (y_{j+1} - y_j)^2\right]^{1/2} - l = l\left\{\left[1 + \frac{1}{l^2}(y_{j+1} - y_j)^2\right]^{1/2} - 1\right\}$$

$$= l\left\{1 + \frac{1}{2l^2}(y_{j+1} - y_j)^2 - \ldots - 1\right\} \cong \frac{1}{2l}(y_{j+1} - y_j)^2 \tag{4.2}$$

where we assumed that the displacements are small:

$$\left|y_{j+1} - y_j\right| \ll l \tag{4.3}$$

and we used

$$(1 + x)^{1/2} \simeq 1 + \frac{1}{2}x + \cdots$$

Mathematical Physics. Shigeji Fujita and Salvador V. Godoy
Copyright © 2010 WILEY-VCH Verlag GmbH & Co. KGaA, Weinheim
ISBN: 978-3-527-40808-5

4 Coupled Oscillators and Normal Modes

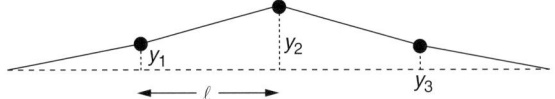

Fig. 4.1 Transverse displacements of three particles on a string are represented by y_1, y_2 and y_3.

The same expression applies to the sections of the string at both ends if we set the *fixed-end boundary conditions*:

$$y_0 = y_{N+1} = 0 \tag{4.4}$$

The work done against the tension τ in increasing the length by Δl is $\tau \Delta l$. (We assume that the tension does not change for small displacements.) Thus, adding the contributions from each section of the string, we find the potential energy

$$V = \frac{\tau}{2l}\left[y_1^2 + (y_2 - y_1)^2 + \cdots + (y_N - y_{N-1})^2 + y_N^2\right] \tag{4.5}$$

The Lagrangian of our system is then given by

$$L(y_1, y_2, \ldots, \dot{y}_1, \dot{y}_2, \ldots) \equiv T - V$$

$$= \frac{m}{2}\sum_{j=1}^{N}\dot{y}_j^2 - \frac{\tau}{2l}\left[\sum_{j=1}^{N-1}(y_{j+1} - y_j)^2 + y_1^2 + y_N^2\right] \tag{4.6}$$

From Lagrange's equations

$$\frac{d}{dt}\left(\frac{\partial L}{\partial \dot{y}_j}\right) - \frac{\partial L}{\partial y_j} = 0 \quad j = 1, 2, \ldots, N$$

we obtain

$$\ddot{y}_1 = \frac{\tau}{ml}(-2y_1 + y_2)$$

$$\ddot{y}_2 = \frac{\tau}{ml}(y_1 - 2y_2 + y_3)$$

$$\cdots\cdots\cdots\cdots$$

$$\ddot{y}_N = \frac{\tau}{ml}(y_{N-1} - 2y_N) \tag{4.7}$$

[Problem 4.1.1]. Here, we have a set of coupled linear homogeneous differential equations.

Clearly,

$$y_1 = y_2 = \cdots = y_N = 0 \tag{4.8}$$

is one possible set of solutions. This set corresponds to the trivial case in which all displacements remain zero at all time.

However, (4.7) allow other solutions, which will be discussed next. Let us assume that the solutions of (4.7) are of the form

$$y_j(t) = A_j \exp(i\omega t), \quad j = 1, 2, \ldots, N \tag{4.9}$$

where $i \equiv \sqrt{-1}$ denotes the *imaginary unit*. Substituting these into (4.7) and dividing the results by $\exp(i\omega t)$, we obtain

$$(2\omega_0^2 - \omega^2) A_1 - \omega_0^2 A_2 = 0$$
$$-\omega_0^2 A_1 + (2\omega_0^2 - \omega^2) A_2 - \omega_0^2 A_3 = 0$$
$$\ldots\ldots\ldots\ldots\ldots\ldots\ldots\ldots\ldots\ldots\ldots$$
$$-\omega_0^2 A_{N-1} + (2\omega_0^2 - \omega^2) A_N = 0 \tag{4.10}$$

where

$$\omega_0 \equiv \sqrt{\frac{\tau}{ml}} \tag{4.11}$$

We note that quantity ω_0 has the dimensions of a frequency since

$$[\omega_0] = \left[\sqrt{\frac{\tau}{ml}}\right] = \left[\sqrt{\frac{MLT^{-2}}{ML}}\right] = \left[\frac{1}{T}\right] \tag{4.12}$$

where the angular brackets indicate the *dimensions* of the quantity inside (M = mass, L = length, T = time.)

Let us look at the case of $N = 2$. We have from (4.10)

$$(2\omega_0^2 - \omega^2) A_1 - \omega_0^2 A_2 = 0$$

$$-\omega_0^2 A_1 + (2\omega_0^2 - \omega^2) A_2 = 0 \tag{4.13}$$

If we exclude the case in which $A_1 = A_2 = 0$, the solutions of (4.13) can exist only if the determinant constructed from the coefficients vanishes

$$\begin{vmatrix} 2\omega_0^2 - \omega^2 & -\omega_0^2 \\ -\omega_0^2 & 2\omega_0^2 - \omega^2 \end{vmatrix} = 0 \tag{4.14}$$

[Problem 4.1.3]. This equation is called the *characteristic frequency equation*, from which we obtain two solutions:

$$\omega = \omega_0 \quad \text{and} \quad \omega = \sqrt{3}\omega_0 \tag{4.15}$$

In the first case, after substitution of $\omega = \omega_0$ in (4.13), we find that the ratio of the coefficients, A_1/A_2, equals unity. In the second case of $\omega = \sqrt{3}\omega_0$, the ratio A_1/A_2 is found to be -1. In summary, we have

$$\omega = \omega_0, \quad A_1 : A_2 = 1 : 1$$

$$\omega = \sqrt{3}\omega_0, \quad A_1 : A_2 = 1 : -1 \tag{4.16}$$

(a) $\omega = \omega_0$

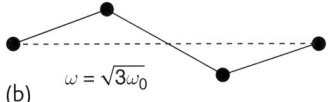

(b) $\omega = \sqrt{3}\omega_0$

Fig. 4.2 The amplitude relations of the two normal modes for a system of two particles.

In general, we can characterize these special modes of oscillation, called the *normal modes*, by their *characteristic frequencies* and. The amplitude relations for the present case are schematically shown in Figure 4.2, where we chose a positive A_1. The choice of a negative A_1 generates different figures. Both cases, however, represent the same normal modes because only the relative amplitude relation $A_1 : A_2$ is significant.

Let us now examine the case of $N = 3$. From (4.10), we have

$$(2\omega_0^2 - \omega^2)A_1 - \omega_0^2 A_2 = 0$$
$$-\omega_0^2 A_1 + (2\omega_0^2 - \omega^2)A_2 - \omega_0^2 A_3 = 0$$
$$-\omega_0^2 A_2 + (2\omega_0^2 - \omega^2)A_3 = 0 \tag{4.17}$$

The characteristic equation is

$$\begin{vmatrix} 2\omega_0^2 - \omega^2 & -\omega_0^2 & 0 \\ -\omega_0^2 & 2\omega_0^2 - \omega^2 & -\omega_0^2 \\ 0 & -\omega_0^2 & 2\omega_0^2 - \omega^2 \end{vmatrix} = 0 \tag{4.18}$$

Solving this equation, we obtain three characteristic frequencies:

$$\sqrt{2 - \sqrt{2}}\omega_0, \quad \sqrt{2}\omega_0, \quad \text{and} \quad \sqrt{2 + \sqrt{2}}\omega_0$$

In each case, after introducing the characteristic frequency in (4.17), we can find the amplitude relation. The result is as follows:

$$\omega = \sqrt{2 - \sqrt{2}}\omega_0; \quad A_1 : A_2 : A_3 = 1 : \sqrt{2} : 1$$
$$\omega = \sqrt{2}\omega_0; \quad A_1 : A_2 : A_3 = 1 : 0 : -1$$
$$\omega = \sqrt{2 + \sqrt{2}}\omega_0; \quad A_1 : A_2 : A_3 = 1 : -\sqrt{2} : 1 \tag{4.19}$$

[Problem 4.1.5]. The amplitude relations of these normal modes are illustrated in Figure 4.3. For large N, the characteristics equation, being an algebraic equation of Nth order, becomes very difficult to solve. We can, however, note the following general features:

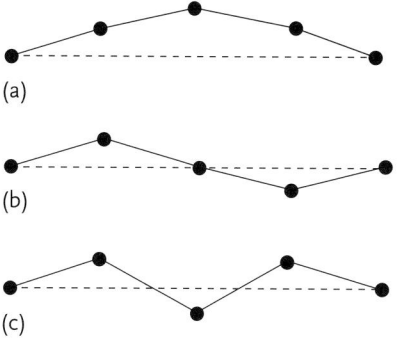

Fig. 4.3 The three normal modes with the characteristic amplitude relations as given by (4.19). The corresponding characteristic frecuencies from top to bottom are: $\sqrt{2-\sqrt{2}}\omega_0$, $\sqrt{2}\omega_0$, and $\sqrt{2+\sqrt{2}}\omega_0$.

For a general N, there will be N distinct normal frequencies. The mode with the lowest frequency is the one in which all of the particles oscillate in the same direction, while the mode with the highest frequency is the one in which successive masses oscillate in opposite directions.

Problem 4.1.1

1. Derive the equation of motion in (4.7) from the Lagrangian L given in (4.6).
2. Derive (4.10) by substituting (4.9) into (4.7).

Problem 4.1.2

Consider a set of two equations: $ax + by = 0$, and $cx + dy = 0$. Show that the necessary and sufficient condition for the existence of a solution other than $x = y = 0$ is

$$\begin{vmatrix} a & b \\ c & d \end{vmatrix} = 0 \tag{4.20}$$

Remark

To show that it is sufficient, assume (4.20) and find a solution other than $x = y = 0$. To show that it is necessary, assume a solution for which at least one of (x, y) is nonzero and verify (4.20).

Problem 4.1.3

1. Solve (4.14) and find the characteristic frequencies.
2. Find the amplitude relation corresponding to each characteristic frequency obtained in (a). The answers are given in (4.16).

Problem 4.1.4

Find the normal mode solutions of (4.17), and verify (4.19). Proceed in the same manner as for Problem 4.1.3.

Problem 4.1.5

We assume the periodic boundary condition such that $y_{j+N}(t) = y_j(t)$. Then, the potential energy V takes a symmetric form:

$$V = \frac{\tau}{2l}\left[(y_2 - y_1)^2 + (y_3 - y_2)^2 + \cdots + (y_1 - y_N)^2\right]$$

Find the normal modes of oscillation for the system.

HINT: Take the cases of $N = 2$ and $N = 3$. If succesful, then try the general case.

4.2
Normal Coordinates

We continue to discuss the oscillations of particles on a string. Let us take the case of two particles: $N = 2$. The Lagrangian L of the system is, from (4.6),

$$\begin{aligned} L &= \frac{m}{2}\left(\dot{y}_1^2 + \dot{y}_2^2\right) - \frac{\tau}{2l}\left[y_1^2 + (y_2 - y_1)^2 + y_2^2\right] \\ &= \frac{m}{2}\left(\dot{y}_1^2 + \dot{y}_2^2\right) - \frac{1}{2}m\omega_0^2\left[y_1^2 + (y_2 - y_1)^2 + y_2^2\right] \end{aligned} \quad (4.21)$$

The characteristic frequencies and amplitude relations are given in (4.16).

Let us now introduce a new set of *generalized coordinates*

$$q_1 \equiv \sqrt{\frac{m}{2}}(y_1 + y_2), \quad q_2 \equiv \sqrt{\frac{m}{2}}(y_1 - y_2) \quad (4.22)$$

which have the amplitude relations as prescribed in (4.16). Note that although these generalized coordinates do not have the dimensions of length, they are nonetheless convenient, as we will see later. Solving with respect to y_1 and y_2 we obtain

$$y_1 = \left(\frac{2}{m}\right)^{1/2}(q_1 + q_2)$$

$$y_2 = \left(\frac{2}{m}\right)^{1/2}(q_1 - q_2) \tag{4.23}$$

By differentiation, we obtain

$$\dot{y}_1 = \left(\frac{2}{m}\right)^{1/2}(\dot{q}_1 + \dot{q}_2)$$

$$\dot{y}_2 = \left(\frac{2}{m}\right)^{1/2}(\dot{q}_1 - \dot{q}_2) \tag{4.24}$$

Using (4.23) and (4.24), we can transform the Lagrangian L from (4.21) to:

$$\begin{aligned}L &= \frac{1}{2}\dot{q}_1^2 - \frac{1}{2}\omega_0^2 q_1^2 + \frac{1}{2}\dot{q}_2^2 - \frac{3}{2}\omega_0^2 q_2^2 \\ &= \left(\frac{1}{2}\dot{q}_1^2 - \frac{1}{2}\omega_1^2 q_1^2\right) + \left(\frac{1}{2}\dot{q}_2^2 - \frac{1}{2}\omega_2^2 q_2^2\right) \equiv L(q_1, q_2, \dot{q}_1, \dot{q}_2)\end{aligned} \tag{4.25}$$

where

$$\omega_1 \equiv \omega_0, \quad \omega_2 \equiv \sqrt{3}\omega_0 \tag{4.26}$$

are characteristic frequencies for the system (Problem 4.2.1).

The Lagrangian L in (4.25) can be regarded as a function of the new set of generalized coordinates and velocities $(q_1, q_2, \dot{q}_1, \dot{q}_2)$. From Lagrange's equations

$$\frac{d}{dt}\left(\frac{\partial L}{\partial \dot{q}_j}\right) - \frac{\partial L}{\partial q_j} = 0$$

we obtain

$$\frac{d}{dt}(\dot{q}_j) + \omega_j^2 q_j = \ddot{q}_j + \omega_j^2 q_j = 0 \tag{4.27}$$

which indicates that each coordinate q_j oscillates sinusoidally with the characteristic frequency ω_j. The coordinates (q_1, q_2) that describe the normal modes in this special manner are called the *normal coordinates*.

We note that the choice of a set normal coordinates is not unique; the factors $(m/2)^{1/2}$ in (4.22) can be chosen in an arbitrary manner. As long as the correct amplitude relations are selected in agreement with (4.16), the generalized coordinates oscillate independently and with the correct normal-mode frequencies. (Problem 4.2.2). Our particular choice, however, makes the Lagrangian (and Hamiltonian) simplest.

Let us now introduce a set a of canonical momenta in the standard form:

$$p_j \equiv \frac{\partial}{\partial \dot{q}_j} L(q_1, \dot{q}_1, q_2, \dot{q}_2), \quad j = 1, 2 \tag{4.28}$$

Using (4.25), we obtain

$$p_j = \dot{q}_j \tag{4.29}$$

We can derive the Hamiltonian H in the following manner:

$$H \equiv \sum_j p_j \dot{q}_j - L = \sum_j \left(p_j \dot{q}_j - \frac{1}{2}\dot{q}_j^2 + \frac{1}{2}\omega_j^2 q_j^2 \right)$$

$$= \left(\frac{1}{2}p_1^2 + \frac{1}{2}\omega_1^2 q_1^2 \right) + \left(\frac{1}{2}p_2^2 + \frac{1}{2}\omega_2^2 q_2^2 \right) \tag{4.30}$$

Here, we see that the Hamilton H, which is numerically equal to the total energy, is the sum of normal-mode Hamiltonians.

We now take up the case of three particles. We know the amplitude relations in (4.19), and may introduce the following set of normal coordinates:

$$q_1 \equiv \frac{\sqrt{m}}{2}(y_1 + \sqrt{2}y_2 + y_3), \quad q_2 \equiv \sqrt{m}(y_1 - y_3)$$

$$q_3 \equiv \frac{\sqrt{m}}{2}(y_1 - \sqrt{2}y_2 + y_3) \tag{4.31}$$

After simple algebra, we obtain

$$L \equiv \frac{m}{2}(\dot{y}_1^2 + \dot{y}_2^2 + \dot{y}_3^2) - \frac{m}{2}\omega_0^2 \left[y_1^2 + (y_2 - y_1)^2 + (y_3 - y_2)^2 + y_3^2 \right]$$

$$= \frac{1}{2}\sum_{j=1}^{3} \left(\dot{q}_j^2 - \omega_j^2 q_j^2 \right) \tag{4.32}$$

where

$$\omega_1 \equiv \sqrt{2 - \sqrt{2}}\,\omega_0, \quad \omega_2 \equiv \sqrt{2}\,\omega_0, \quad \omega_3 \equiv \sqrt{2 + \sqrt{2}}\,\omega_0 \tag{4.33}$$

are the characteristic frequencies, see (4.19), [Problem 4.2.3].

The *canonical momenta* are defined in the standard manner, and are given by

$$p_j \equiv \frac{\partial L}{\partial \dot{q}_j} = \dot{q}_j, \quad j = 1, 2, 3 \tag{4.34}$$

The Hamiltonian H, constructed from $\sum_j p_j \dot{q}_j - L$, is given by

$$H = \frac{1}{2}\sum_{j=1}^{3} \left(p_j^2 + \omega_j^2 q_j^2 \right) \tag{4.35}$$

Many results, all expressed in simple forms, can be extended to a general case of N particles. The Lagrangian L for the systems was given in (4.6):

$$L \equiv \frac{m}{2}(\dot{y}_1^2 + \dot{y}_2^2 + \dot{y}_3^2) - \frac{m}{2}\omega_0^2 \left[y_1^2 + (y_2 - y_1)^2 + \cdots + y_N^2 \right] \quad m\omega_0^2 \equiv \tau/l \tag{4.36}$$

Note that this function L is *quadratic* in $\{y_j, \dot{y}_j\}$. By a suitable choice of normal coordinates $\{q_j\}$, such a Lagrangian can in principle be transformed into

$$L = \sum_{j=1}^{N} \frac{1}{2}\left(\dot{q}_j^2 - \omega_j^2 q_j\right) \tag{4.37}$$

where $\{\omega_j\}$ represent the characteristic frequencies. As stated in the last section, it is not easy to find characteristic frequencies $\{\omega_j\}$ for the system when N is large. Also, finding the relation between $\{y_j\}$ and $\{q_j\}$, like (4.22) or (4.23), is tedious work. The transformation from (4.36) to (4.37) is known as the *principal-axis transformation*. Here, we simply assume (4.37) and continue our discussions.

Equation (4.37) indicates, first, that there are N characteristic frequencies. This number N is equal to the number of degrees of freedom of the system. Second, each normal coordinate q_j oscillates sinusoidally with its characteristic frequency ω_j.

Let us define the canonical momenta by

$$p_j \equiv \frac{\partial L}{\partial \dot{q}_j} = \dot{q}_j \tag{4.38}$$

Now, the Hamiltonian H for the system can be written in the simple form

$$H = \sum_{j=1}^{N} \frac{1}{2}\left(p_j + \omega_j^2 q_j^2\right) \tag{4.39}$$

The theory treated in this section is useful in the discussion of lattice vibrations in solids.

Problem 4.2.1

Using the coordinate (q_1, q_2) in (4.22), derive the expression for the Lagrangian L, (4.25).

Problem 4.2.2

Consider the case of two particles ($N = 2$);
  1. Introduce a set of coordinates

$$Q_1 \equiv y_1 + y_2, \quad Q_2 \equiv y_1 - y_2$$

Find the expression for the Lagrangian L as a function of coordinates and velocities $(Q_1, Q_2, \dot{Q}_1, \dot{Q}_2)$.
  2. Verify that Q_1 and Q_2 oscillates with frequencies ω_1 and ω_2 given in (4.26).

Problem 4.2.3

Verify (4.32).

5
Stretched String

The transverse vibration of a stretched string is treated by using the normal modes in this chapter. The normal coordinates are introduced and discussed.

5.1
Transverse Oscillations of a Stretched String

Let us consider a string of length L and mass per unit length, μ, stretched to a tension τ. We will discuss small transverse oscillations of this string.

In place of a finite set of coordinates $\{y_j(t)\}$ we have a continuous function $y(x, t)$ to describe the transverse displacement of the string. The equation of motion for the string can be obtained as follows. Consider a small section of the string between x and $x + dx$ as shown in Figure 5.1. The mass of this portion is μdx. The acceleration along the y-axis is $\partial^2 y/\partial t^2$. The forces acting on the section are the two tensions shown in the diagram. By assumption, these two tensions have the same magnitude, and are directed along the tangents of the curve $y = y(x)$ at x and $x + dx$. We now set up Newton's equation (mass × acceleration = net force) for the transverse motion. From the diagram, we obtain

$$(\mu dx)\frac{\partial^2 y}{\partial t^2} = \tau \sin\theta(x + dx, t) - \tau \sin\theta(x, t) \tag{5.1}$$

where $\theta(x, t)$ is the angle between the tangent and the positive x-axis; $\sin\theta$ appears because we take the y-components of the tensions.

We apply to $\sin\theta(x + dx, t)$ Taylor's expansions formula:

$$f(x + dx) = f(x) + dx\frac{df(x)}{dx} + \cdots \tag{5.2}$$

and obtain

$$\sin\theta(x + dx, t) = \sin\theta(x, t) + dx\cos\theta\frac{\partial}{\partial x}\theta(x, t) \tag{5.3}$$

where we retained terms of the first order in dx. (This is sufficient since the term on the lhs of (5.1) is of the first order.) Introducing (5.3) in (5.1) and dropping the

5 Stretched String

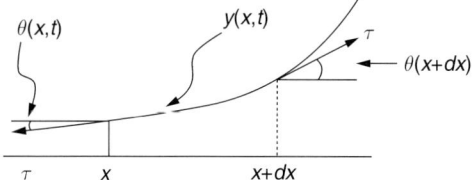

Fig. 5.1 Section of the string between x and x + dx is acted upon by the two tensions shown.

common factor dx, we obtain

$$\mu \frac{\partial^2 y}{\partial t^2} = \tau \cos\theta \frac{\partial \theta}{\partial x} \qquad (5.4)$$

From the diagram, we observe that

$$\frac{\partial y(x,t)}{\partial x} = \tan\theta(x,t) \qquad (5.5)$$

Differentiating with respect to x, we obtain

$$\frac{\partial y^2}{\partial x^2} = \frac{\partial}{\partial x}\tan\theta(x,t) = \frac{1}{\cos^2\theta}\frac{\partial \theta}{\partial x}$$

or

$$\frac{\partial \theta}{\partial x} = \cos^2\theta \frac{\partial^2 y}{\partial x^2} \qquad (5.6)$$

Substituting this into (5.4), we obtain

$$\mu \frac{\partial^2 y}{\partial t^2} = \tau \cos^3\theta \frac{\partial y^2}{\partial x^2} \qquad (5.7)$$

If we assume that the slope $\partial y(x,t)/\partial x = \tan\theta(x,t)$ is small everywhere and at all times, so that

$$\theta(x,t) \ll 1 \qquad (5.8)$$

then we can approximate $\cos\theta \simeq 1$, and obtain, from (5.6),

$$\mu \frac{\partial^2 y}{\partial t^2} = \tau \frac{\partial^2 y}{\partial x^2} \qquad (5.9)$$

It is convenient to introduce

$$c \equiv (\tau/\mu)^{1/2} \qquad (5.10)$$

We can then rewrite (5.9) in the form:

$$\boxed{\frac{\partial^2 y}{\partial x^2} = \frac{1}{c^2}\frac{\partial^2 y}{\partial t^2}} \qquad (5.11)$$

This is the *wave equation* in one dimension. Similar equations occur in many branches of physics, wherever wave phenomena are encountered.

Using dimensional analysis:

$$[\tau/\mu] = [\text{force}/\{\text{mass}/\text{length}\}] = [MT^{-2}L/(M/L)]$$
$$= (L/T)^2 \tag{5.12}$$

we see that the constant c has the dimensions of a velocity. In fact, as we will show later, the constant c is the velocity of the wave propagating along the string.

We now wish to find the normal-mode solutions of the wave equation (5.11) with the fixed-end boundary condition, which are represented by

$$y(0,t) = 0 \quad y(L,t) = 0 \tag{5.13}$$

(The partial differentiation equation (5.11) can be solved by various methods. In this section, we will use the method of separation of variables [3]. Another method will be described in the following section.)

Let us assume solutions of the form

$$y(x,t) = A(x)\exp(i\omega t) \tag{5.14}$$

Notice that the x and t dependencies are separated on the rhs. Introducing (5.14) in (5.11) and dropping the common factor $\exp(i\omega t)$, we obtain

$$\frac{d^2 A}{dx^2} + k^2 A = 0 \tag{5.15}$$

where

$$k \equiv \pm \omega/c \tag{5.16}$$

The general solution of (5.15) is

$$A(x) = a\cos(kx) + b\sin(kx) \tag{5.17}$$

where a and b are constants.

Substitution of (5.14) and (5.17) into (5.13) yields

$$[a\cos(0) + b\sin(0)]e^{i\omega t} = 0$$

From this we obtain

$$a = 0 \tag{5.18}$$

Thus, we find that

$$A(x) = b\sin(kx) \tag{5.19}$$

Substitution of this expression in the first part of (5.13) yields

$$b \sin(kL) = 0$$

whose solution is either

$$b = 0 \tag{5.20}$$

or

$$\sin(kL) = 0 \tag{5.21}$$

The first case ($b = 0$) corresponds to the trivial solutions, where the displacement A remains zero at all times along the line. In the second case, the value of k is limited to

$$\frac{\pi}{L} n \equiv k_n \quad n = 1, 2, \ldots \tag{5.22}$$

The angular frequency ω is related to k by $\omega = ck$. The corresponding frequencies are therefore also discrete. These frequencies are given by

$$\frac{\pi c}{L} n \equiv \omega_n \tag{5.23}$$

which are all multiples of the *fundamental* (or *base*) *frequency* $\pi c/L$.

In summary, we have obtained solutions of the form

$$b_n \sin(k_n x) \exp(\pm i \omega_n t), \quad n = \text{integer} \tag{5.24}$$

where b_n are constants. The solutions with k_n and k_{-n} ($n > 0$) are *linearly dependent*; they can be obtained from each other by multiplying by a constant factor. For definiteness, we will take

$$b_n \sin(k_n x) \exp(-i \omega_n t), \quad n = 1, 2, \ldots \tag{5.25}$$

as a set of linearly independent solutions. Each of these solutions is characterized by the angular frequency ω_n and the sinusoidal space dependence with wave number k_n. These solutions are called the *normal-mode solutions* for the string. As we will see below, these normal modes have many similarities with the normal modes for the particles on a stretched string found in Section 4.2.

The space dependence of the first few normal modes,

$$\sin(k_n x) \equiv \sin\left(\frac{n \pi x}{L}\right) \tag{5.26}$$

is shown in Figure 5.2. According to (5.25), these normal-mode amplitudes change in time with the characteristic periods T_n given by

$$T_n \equiv 2\pi \omega_n^{-1} \tag{5.27}$$

Thus, the normal modes represent *standing waves* with a number of fixed nodes, including those at both ends. In general, the wavelength λ and the wave number k

5.1 Transverse Oscillations of a Stretched String

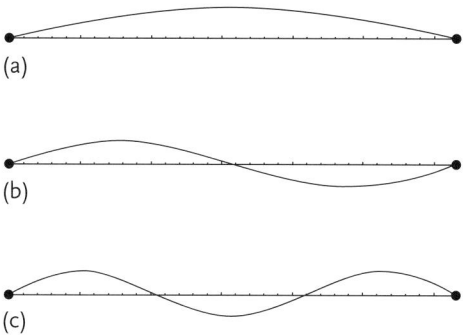

Fig. 5.2 The first three normal modes of the stretched string corresponding to (5.26). The corresponding characteristic frequencies are given by $\omega_n \equiv \pi c n/L$, $n = 1, 2$ and 3.

for a sinusoidal wave are related by $\lambda = 2\pi/k$. In the present case the wavelengths λ_n are therefore restricted to

$$\lambda_n \equiv \frac{2\pi}{k_n} = (2\pi)\left(\frac{L}{\pi n}\right) = \frac{2L}{n} \tag{5.28}$$

Let us now look at the case of $n = 1$. This case corresponds to the minimum k-number (π/L), the maximum wavelength ($2L$), and the minimum frequency ($\pi c/L$). As we see in Figure 5.2a, for all the range: $0 < x < L$, this normal-mode amplitude is similar to that of the minimum frequency for the N particles on the string. Compare Figure 5.2a with Figures 4.2a and 4.3a.

A more detailed correspondence may be established in the following manner. Let us take the case of a large number of particles on a stretched string, as indicated in Figure 5.3a. Let us imagine that the mass concentrated on each particle is uniformly distributed along the string. The resulting case of the mass–string system, show in Figure 5.3b, can be regarded as the stretched homogeneous string that we have been discussing in the present section. An important distinction between the two cases is that the number of normal modes for the stretched string is infinite since n can be any positive integer. This property arises from the fact that the string is characterized by a continuous function $y(x, t)$, and therefore, has an infinite number of degrees of freedom. Another difference related to this is that the stretched string can have a normal mode of extremely great frequency, ($\omega_n = c n\pi/L$), and extremely short wavelength ($\lambda = 2L/n$). Therefore, the analogy between the two systems must be discussed with care.

Fig. 5.3 If we redistribute the mass concentrated on particles in the mass–string system (a) along the string uniformly, we can obtain the homogeneous string shown in (b).

Problem 5.1.1

Assume the periodic boundary condition represented by $y(x, t) = y(x + L, t)$ (for any x), and find the normal modes of oscillation from the wave equation (5.11).

Problem 5.1.2

Show that the periodic boundary condition introduced in the preceding problem can also be re-expressed as

$$y(0, t) = y(L, t), \quad y'(0, t) = y'(L, t)$$
$$y''(0, t) = y''(L, t)$$

You can use any two of these relations in the present problem (why?).

5.2
Normal Coordinates for a String

In the last section, we used the method of separation of variables to find the normal-mode solutions of the wave equation

$$\frac{\partial^2 y}{\partial x^2} = \frac{1}{c^2} \frac{\partial^2 y}{\partial t^2} \tag{5.29}$$

subject to the boundary condition

$$y(0, t) = 0 \tag{5.30}$$

$$y(L, t) = 0 \tag{5.31}$$

We will look at the same problem from a different angle in the present section.

We first note that (5.29) is a linear homogeneous partial differential equation of the second order in both x and t. It is easy to verify by direct substitution that an arbitrary function f_1 of $x - ct$:

$$y = f_1(x - ct) \tag{5.32}$$

is a solution of (5.29) [Problem 5.2.1]. This represents a wave traveling along the positive x-axis with the velocity c. To see this, we choose an arbitrary real function f_1 and plot $y = f_1(x - ct)$ at two times, $t = 0$ and t, against x as indicated in Figure 5.4. We immediately observe that the shape of the function y at $t = t$ is identical to its shape at $t = 0$; the only change is that the function is shifted to the right by the distance ct.

We now take another arbitrary function f_2 of $x + ct$:

$$y = f_2(x + ct) \tag{5.33}$$

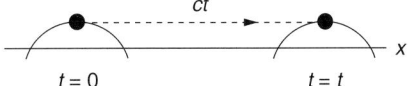

Fig. 5.4 A wave traveling along the positive x-axis with speed c can be represented by $y = f(x - ct)$. The curve $y = f(x - ct)$ is identical in shape to the curve $y(x)$, but is shifted to the right by the distance ct.

We can easily show that this is also a solution of the wave equation (5.29), and that it represents a wave traveling along a negative x-axis [Problem 5.2.2]. We can also show that if f and g are solutions of a second-order linear homogeneous differential equation, then any *linear combination* $c_1 f + c_2 g$, with c_1, c_2 constants, is also a solution. From these considerations, we may assume a general solution of (5.29) of the form

$$y(x, t) = f_1(x - ct) + f_2(x + ct) \tag{5.34}$$

Introducing (5.34) into (5.30), we obtain

$$f_1(-ct) + f_2(ct) = 0$$

By setting $ct = X$, we can express this relation by

$$f_2(X) = -f_1(-X) \equiv f(X) \tag{5.35}$$

Using this, we can rewrite the solution of y from (5.34) as

$$\begin{aligned} y &= f_1(x - ct) + f_2(x + ct) \\ &= -f(ct - x) + f(ct + x) \end{aligned} \tag{5.36}$$

Substitution of this y into (5.31) yields

$$-f(ct - L) + f(ct + L) = 0$$

or

$$f(ct + L) = f(ct - L)$$

Writing $ct - L = X$, we can express this relation as

$$f(X + 2L) = f(X) \tag{5.37}$$

This means that the function f is *periodic*, with the period $2L$.

It is known in the theory of Fourier transformation [5] that such a periodic function can be expanded in a Fourier series as follows:

$$f(X) = \sum_{n \text{(integers)}} f_n \exp(in\pi X/L) \tag{5.38}$$

where the summation is over the entire set of integers $\{n\}$ and the coefficients $\{f_n\}$ are in general complex numbers. Using this expansion and (5.36), we obtain

$$y(x,t) = \sum_n f_n \left[e^{in\pi(ct+x)/L} - e^{in\pi(ct-x)/L}\right]$$
$$= \sum_n 2i f_n \exp(in\pi ct/L) \sin(n\pi x/L) \qquad (5.39)$$

Here, we note that the term with $n = 0$ drops out [since $\sin(0) = 0$]. Since the displacement y is real, the Fourier coefficients must satisfy the relation:

$$f_{-n} = f_n^*, \quad n = 1, 2, \ldots \qquad (5.40)$$

By setting $A_n = 2i f_n$, we can rewrite (5.39) in the form:

$$y(x,t) = \text{Re}\left\{\sum_{n=1}^{\infty} A_n \exp(in\pi ct/L) \sin(n\pi x/L)\right\}$$
$$= \text{Re}\left\{\sum_{n=1}^{\infty} A_n \exp(i\omega_n t) \sin(k_n x)\right\} \qquad (5.41)$$

where the symbol Re{} means the *real part* of the bracketed expression. Comparing this expression with (4.25), we see that a general solution for the vibration of the stretched string is a linear combination of all normal-mode solutions.

If we assume

$$\text{Re}\{A_n \exp(i\omega_n t)\} \equiv q_n(t) \qquad (5.42)$$

then we can rewrite (5.41) as

$$y(x,t) = \sum_{n=1}^{\infty} q_n(t) \sin(n\pi x/L) \qquad (5.43)$$

This represents a Fourier-series expansion of the displacement. The coefficients $q_n(t)$, which are real and in general depend on t, can be obtained from $y(x,t)$ by the inverse Fourier transformation [5]:

$$q_n(t) = \frac{2}{L} \int_0^L dx\, y(x,t) \sin(n\pi x/L) \qquad (5.44)$$

Equation (5.43) means that by knowing $\{q_n(t)\}$, we can calculate the displacement of the string, $y(x,t)$; conversely, given $y(x,t)$, we can calculate all coefficients $\{q_n\}$ from (5.44). This suggests the possibility of choosing the set $\{q_n\}$ as generalized coordinates and describing the motion by Lagrange's method.

In order to carry out this program, we will need the kinetic energy T and the potential energy V for the system. The kinetic energy of a small element in the

5.2 Normal Coordinates for a String

range $(x, x + dx)$ is

$$\frac{1}{2}(\text{mass})(\text{velocity})^2 = \frac{1}{2}(\mu\, dx)\left[\frac{\partial y(x,t)}{\partial t}\right]^2$$

$$\equiv \frac{1}{2}\mu \dot{y}^2 dx \tag{5.45}$$

where we indicated the partial time derivative by the super dot, so that $\partial y(x,t)/\partial t \equiv \dot{y}$. Integrating (5.45) from 0 to L, we obtain

$$T = \int_0^L dx\, \frac{1}{2}\mu \dot{y}^2 \tag{5.46}$$

which represents the total kinetic energy T for the string.

When the string is straight and in equilibrium, its length is L. However, when the string is displaced from the straight line, its length becomes greater. If the shape of the string is represented by $y = y(x,t)$, its total length $L + \Delta L$ can be obtained by integrating $dx\sqrt{1 + (\partial y/\partial x)^2} \equiv dx\sqrt{1 + y'^2}$ from 0 to L:

$$\int_0^L dx(1 + y'^2)^{1/2} \equiv L + \Delta L \tag{5.47}$$

where we indicated the partial x-derivative by the prime: $\partial y(x,t)/\partial x \equiv y'$. For small, smooth displacement, we may assume that

$$|y'| \ll 1 \tag{5.48}$$

We can then approximate $(1 + y'^2)^{1/2} = 1 + (1/2)y'^2 + \cdots$ by $1 + (1/2)y'^2$. Using this approximation, we obtain from (5.47)

$$\Delta L = \frac{1}{2}\int_0^L y'^2 dx \tag{5.49}$$

The work done against the tension τ when the length of the string is increased by L is $\tau \Delta L$, which can be viewed as the potential energy of the string. We thus obtain, for the potential energy,

$$V = \int_0^L \frac{1}{2}\tau y'^2 dx \tag{5.50}$$

The Lagrangian $\mathcal{L}$ of the system is now given by

$$\mathcal{L} \equiv T - V$$

$$= \int_0^L dx\left(\frac{1}{2}\mu \dot{y}^2 - \frac{1}{2}\tau y'^2\right) \tag{5.51}$$

Here, the Lagrangian is designated by $\mathcal{L}$ to distinguish it from the length L of the string in equilibrium. We wish to express this Lagrangian $\mathcal{L}$ in terms of the set of generalized coordinates, $\{q_n\}$.

Differentiating (5.43) with respect to t and squaring the result, we obtain

$$\dot{y}^2 \equiv \left(\frac{\partial y}{\partial t}\right)^2 = \sum_{n=1}^{\infty}\sum_{j=1}^{\infty} \dot{q}_n \dot{q}_j \sin(n\pi x/L)\sin(j\pi x/L)$$

Substituting this into (5.46), we obtain

$$T = \int_0^L dx \frac{1}{2}\mu \left(\frac{\partial y}{\partial t}\right)^2$$

$$= \frac{1}{2}\mu \sum_{n=1}^{\infty}\sum_{j=1}^{\infty} \dot{q}_n \dot{q}_j \int_0^L dx \sin(n\pi x/L)\sin(j\pi x/L) \tag{5.52}$$

The x-integrals have the values:

$$\int_0^L dx \sin(n\pi x/L)\sin(j\pi x/L) = \frac{L}{2}\delta_{n,j} \tag{5.53}$$

which may be shown as follows: If $n = j$ (≥ 1), with the aid of $\sin^2 y = (1-\cos 2y)/2$, we then have

$$\int_0^L dx \sin^2\left(\frac{n\pi x}{L}\right) = \int_0^L dx \frac{1}{2}\left[1 - \cos\left(\frac{2n\pi x}{L}\right)\right]$$

$$\frac{L}{2} - \frac{1}{2}\int_0^L dx \cos\left(\frac{2n\pi x}{L}\right) = \frac{L}{2} - 0 = \frac{L}{2}$$

If $n \neq j$, the integrals can all be decomposed in terms of vanishing integrals of the form

$$\int_0^L dx \cos\left(\frac{2k\pi x}{L}\right) = 0, \quad k = 1, 2, \ldots$$

Introducing (5.53) in (5.52), we obtain

$$T = \frac{1}{4}L\mu \sum_{n=1}^{\infty} \dot{q}_n^2 \tag{5.54}$$

It is noted that this simple form is obtained with the use of the remarkable property in (5.53), which is known as the *orthogonality relation*. In a similar manner, we

5.2 Normal Coordinates for a String

obtain from (5.50)

$$V = \frac{1}{2}\tau \int_0^L dx \left(\frac{\partial y}{\partial x}\right)^2 = \frac{1}{4}L\tau \sum_{n=1}^{\infty} k_n^2 q_n^2 \quad \left(k_n \equiv \frac{n\pi}{L}\right) \tag{5.55}$$

Using (5.54) and (5.55), we obtain from (5.51):

$$\mathcal{L} = T - V$$
$$= \sum_{n=1}^{\infty}\left[\frac{1}{4}L\mu\dot{q}_n^2 - \frac{1}{4}L\tau k_n^2 q_n^2\right] \tag{5.56}$$

From Lagrange's equations of motion:

$$\frac{d}{dt}\left[\frac{\partial \mathcal{L}}{\partial \dot{q}_n}\right] - \frac{\partial \mathcal{L}}{\partial q_n} = 0$$

we obtain

$$\frac{1}{2}L\mu\ddot{q}_n + \frac{1}{2}L\tau k_n^2 q_n = 0$$

or

$$\ddot{q}_n + \omega_n^2 q_n = 0 \quad \omega_n^2 = \frac{\tau}{\mu}k_n^2 \tag{5.57}$$

These are simply the equations of motion for simple harmonic oscillators. For the nth mode, the real solution can be written as

$$q_n(t) = \text{Re}\{A_n \exp(i\omega_n t)\} = \text{Re}\{A_n \exp(in\pi ct/L)\}$$

which is in agreement with (5.42).

The coordinates $\{q_n\}$ defined by (5.42) (A_n is a pure imaginary number: $A_n = -2if_n$, $f_n = real$) are called the *normal coordinates* for the vibrating string. Each normal coordinate describes one normal mode of vibration. Note that the normal coordinates have the remarkable property that the Lagrangian $\mathcal{L}$ becomes a sum of single-mode Lagrangians. Thus, in terms of normal coordinates, the problem of the vibrating string is decomposed into separate problems, one for each degree of freedom.

For later convenience, we now introduce a new set of normal coordinates:

$$Q_n \equiv \left(\frac{\mu L}{2}\right)^{1/2} q_n, \quad n = 1, 2, \ldots \tag{5.58}$$

We can then re-express the Lagrangian $\mathcal{L}$ in the form:

$$\mathcal{L} = \sum_{n=1}^{\infty}[\dot{Q}_n^2 - \omega_n^2 Q_n^2] \tag{5.59}$$

Since new Q_n are different from old q_n only by a constant factor, they obey the same equations of motion in (5.57).

Let us now define the set of canonical momenta by

$$P_n \equiv \frac{\partial \mathcal{L}}{\partial \dot{Q}_n} \tag{5.60}$$

Using the explicit form (5.59) for $\mathcal{L}$, we obtain

$$P_n = \dot{Q}_n \tag{5.61}$$

The Hamiltonian H for the system is obtained by expressing

$$H = \sum_{n=1}^{\infty} P_n \dot{Q}_n - \mathcal{L} \tag{5.62}$$

in terms of Q's and P's. We obtain

$$H = \sum_n P_n^2 - \frac{1}{2} \sum_n (P_N^2 - \omega_n^2 Q_n^2)$$

$$H = \sum_{n=1}^{\infty} \frac{1}{2} (P_n^2 + \omega_n^2 Q_n^2) \tag{5.63}$$

Note that this Hamiltonian is the sum of simple-harmonic-oscillator Hamiltonians over all normal modes.

Let us check if we get correct equations of motion from this Hamiltonian. Using Hamilton's equations of motion, we obtain

$$\dot{Q}_n = \frac{\partial H}{\partial P_n} = P_n$$

$$\dot{P}_n = -\frac{\partial H}{\partial Q_n} = -\omega_n^2 Q_n = \ddot{Q}_n \tag{5.64}$$

The first set of equations reproduce (5.61). The second set of equations represent the equations of motion for the normal coordinates Q_n [see (5.56)]. Finally, we note that our Hamiltonian H is numerically equal to the total energy E of the system:

$$H = T + V = E \tag{5.65}$$

The materials presented here are useful in the quantum-statistical-mechanical theory of the electromagnetic radiation field and the lattice vibration.

Problem 5.2.1

By direct substitution show that both $f_1(x - ct)$ and $f_2(x + ct)$ are solutions of the wave equation, (5.29).

Problem 5.2.2

Verify (5.55).

6
Vector Calculus and the del Operator

Space derivatives are introduced and discussed using the vector differential del operator ∇, in this chapter. Stokes's and Gauss's theorems are proved. Expressions for the gradient, divergence and curl in general curvilinear coordinates systems are derived.

6.1
Differentiation in Time

When a vector $\mathbf{A}$ depends on the time t, the derivative of $\mathbf{A}$ with respect to t, $d\mathbf{A}/dt$, is defined by

$$\frac{d\mathbf{A}}{dt} \equiv \lim_{\Delta t \to 0} \frac{\mathbf{A}(t+\Delta t) - \mathbf{A}(t)}{\Delta t} \tag{6.1}$$

The following rules are observed:

$$\frac{d}{dt}(\mathbf{A} \cdot \mathbf{B}) = \frac{d\mathbf{A}}{dt} \cdot \mathbf{B} + \mathbf{A} \cdot \frac{d\mathbf{B}}{dt} \tag{6.2}$$

$$\frac{d}{dt}(\mathbf{A} \times \mathbf{B}) = \frac{d\mathbf{A}}{dt} \times \mathbf{B} + \mathbf{A} \times \frac{d\mathbf{B}}{dt} \tag{6.3}$$

Note that the operational rules as well as the definition are similar to the case of a scalar function.

6.2
Space Derivatives

6.2.1
The Gradient

A function of the position $\mathbf{r} \equiv (x, y, z)$ is called a *point function* or a *field*.

If a *scalar field*, that is, a scalar function of the position, is denoted by $\phi(x, y.z)$, then the *gradient* of ϕ, denoted by grad ϕ, is defined by

$$\operatorname{grad} \phi \equiv \mathbf{i}\frac{\partial \phi}{\partial x} + \mathbf{j}\frac{\partial \phi}{\partial y} + \mathbf{k}\frac{\partial \phi}{\partial z} \tag{6.4}$$

Note that this is a vector function of the position (a vector field). It is convenient to rewrite (6.4) as

$$\operatorname{grad} \phi \equiv \left(\mathbf{i}\frac{\partial}{\partial x} + \mathbf{j}\frac{\partial}{\partial y} + \mathbf{k}\frac{\partial}{\partial z}\right) \phi(x, y, z) \equiv \nabla \phi \tag{6.5}$$

where the symbol ∇ means a vector-like space-differential operator,

$$\boxed{\nabla \equiv \mathbf{i}\frac{\partial}{\partial x} + \mathbf{j}\frac{\partial}{\partial y} + \mathbf{k}\frac{\partial}{\partial z}} \tag{6.6}$$

which is called the *del* (or *nabla*) *operator*.

By looking at the function ϕ at two neighboring points, (x, y, z) and $(x + dx, y + dy, z + dz)$, we obtain

$$\begin{aligned} d\phi &= \phi(x + dx, y + dy, z + dz) - \phi(x, y, z) \\ &= \frac{\partial \phi}{\partial x} dx + \frac{\partial \phi}{\partial y} dy + \frac{\partial \phi}{\partial z} dz \\ &= \left(\mathbf{i}\frac{\partial \phi}{\partial x} + \mathbf{j}\frac{\partial \phi}{\partial y} + \mathbf{k}\frac{\partial \phi}{\partial z}\right) \cdot (\mathbf{i} dx + \mathbf{j} dy + \mathbf{k} dz) \end{aligned}$$

or

$$\boxed{d\phi = \nabla\phi \cdot d\mathbf{r}} \tag{6.7}$$

where

$$d\mathbf{r} \equiv \mathbf{i} dx + \mathbf{j} dy + \mathbf{k} dz \tag{6.8}$$

is the *infinitesimal displacement vector*. The geometrical meaning of the gradient may be grasped from (6.7). The gradient $\nabla \phi$ points along the direction in which the function ϕ increases most rapidly. This direction is normal to the surface $\phi(x, y, z) = $ constant.

6.2.2
The Divergence

Let us now consider a vector field $\mathbf{A}(x, y, z)$. The *divergence* of the vector $\mathbf{A}$, denoted by div $\mathbf{A}$, is defined by

$$\operatorname{div} \mathbf{A} \equiv \frac{\partial A_x}{\partial x} + \frac{\partial A_y}{\partial y} + \frac{\partial A_z}{\partial z} \tag{6.9}$$

This can be written in terms of the del operator ∇ as follows:

$$\operatorname{div} \mathbf{A} \equiv \left(\mathbf{i}\frac{\partial}{\partial x} + \mathbf{j}\frac{\partial}{\partial y} + \mathbf{k}\frac{\partial}{\partial z}\right) \cdot \left(\mathbf{i} A_x + \mathbf{j} A_y + \mathbf{k} A_z\right)$$

$$\equiv \nabla \cdot \mathbf{A} \tag{6.10}$$

This function $\nabla \cdot \mathbf{A}$ is a scalar field.

6.2.3
The Curl

The *curl* or *rotation* of the vector $\mathbf{A}$ is defined by

$$\operatorname{curl} \mathbf{A} \equiv \operatorname{rot} \mathbf{A}$$

$$\equiv \mathbf{i}\left(\frac{\partial A_z}{\partial y} - \frac{\partial A_y}{\partial z}\right) + \mathbf{j}\left(\frac{\partial A_x}{\partial z} - \frac{\partial A_z}{\partial x}\right) + \mathbf{k}\left(\frac{\partial A_y}{\partial x} - \frac{\partial A_x}{\partial y}\right)$$

$$\equiv \nabla \times \mathbf{A} \tag{6.11}$$

This is a vector field, and it can be written in the form of a determinant:

$$\nabla \times \mathbf{A} = \begin{vmatrix} \mathbf{i} & \mathbf{j} & \mathbf{k} \\ \frac{\partial}{\partial x} & \frac{\partial}{\partial y} & \frac{\partial}{\partial z} \\ A_x & A_y & A_z \end{vmatrix} \tag{6.12}$$

6.2.4
Space Derivatives of Products

The space derivatives of products of two fields satisfy the following properties:

$$\nabla(\phi\psi) = \phi\nabla\psi + \psi\nabla\phi \tag{6.13}$$

$$\nabla \cdot (\phi\mathbf{A}) = \nabla\phi \cdot \mathbf{A} + \phi\nabla \cdot \mathbf{A} \tag{6.14}$$

$$\nabla \cdot (\mathbf{A} \times \mathbf{B}) = \mathbf{B} \cdot (\nabla \times \mathbf{A}) - \mathbf{A} \cdot (\nabla \times \mathbf{B}) \tag{6.15}$$

$$\nabla \times (\phi\mathbf{A}) = \phi\nabla \times \mathbf{A} - \mathbf{A} \times (\nabla\phi) \tag{6.16}$$

The *del* operator ∇ is a differential operator, *and* behaves like a vector. The differentiation acts on each factor separately. By noting these properties we may formally derive (6.13) as follows. We consider the *x*-component of $\nabla(\phi\psi)$:

$$\nabla(\phi\psi)]_x \equiv \frac{\partial}{\partial x}(\phi\psi) = \phi\frac{\partial \psi}{\partial x} + \psi\frac{\partial \phi}{\partial x}$$
$$= [\phi\nabla\psi + \psi\nabla\phi]_x$$

Similar equations hold for the *y*- and *z*-components. Hence, we obtain the desired equation.

We may apply the same technique and derive (6.14)–(6.16). For example, (6.16) may be derived as follows:

$$\nabla \times (\phi\mathbf{A})]_x \equiv \frac{\partial(\phi A_z)}{\partial y} - \frac{\partial(\phi A_y)}{\partial z}$$
$$= \phi\left(\frac{\partial A_z}{\partial y} - \frac{\partial A_y}{\partial z}\right) - \left(A_y\frac{\partial \phi}{\partial z} - A_z\frac{\partial \phi}{\partial y}\right)$$
$$= \phi(\nabla \times \mathbf{A})]_x - (\mathbf{A} \times \nabla\phi)]_x$$

Other components can be written similarly.

The space derivatives of higher orders can be defined, and satisfy the following rules:

$$\nabla \times (\nabla\phi) = 0 \quad [= (\nabla \times \nabla)\phi] \tag{6.17}$$

$$\nabla \cdot (\nabla \times \mathbf{A}) = 0 \quad [= (\nabla \times \nabla) \cdot \mathbf{A}] \tag{6.18}$$

$$\nabla \cdot (\nabla\phi) = \frac{\partial^2 \phi}{\partial x^2} + \frac{\partial^2 \phi}{\partial y^2} + \frac{\partial^2 \phi}{\partial z^2} \equiv \nabla^2\phi \tag{6.19}$$

$$\nabla \times (\nabla \times \mathbf{A}) = \nabla(\nabla \cdot \mathbf{A}) - \nabla^2\mathbf{A} \tag{6.20}$$

The scalar differential operator appearing in (6.19):

$$\nabla^2 \equiv \frac{\partial^2}{\partial x^2} + \frac{\partial^2}{\partial y^2} + \frac{\partial^2}{\partial z^2} \equiv \nabla \cdot \nabla \tag{6.21}$$

is called the *Laplacian*.

The two properties in (6.17) and (6.18) may be thought to arise from the vector character of the del operator. That is, $\nabla \times \nabla = 0$ just as $\mathbf{A} \times \mathbf{A} = 0$. The important identity (6.20) may be derived as follows. Using

$$\mathbf{A} \times (\mathbf{B} \times \mathbf{C}) = \mathbf{B}(\mathbf{A} \cdot \mathbf{C}) - (\mathbf{A} \cdot \mathbf{B})\mathbf{C}$$

with $\mathbf{A} = \mathbf{B} = \nabla$ and $\mathbf{C} = \mathbf{F}$, we obtain

$$\nabla \times (\nabla \times \mathbf{F}) = \nabla(\nabla \cdot \mathbf{F}) - (\nabla \cdot \nabla)\mathbf{F} = \nabla(\nabla \cdot \mathbf{F}) - \nabla^2\mathbf{F}$$

6.3
Space Derivatives in Curvilinear Coordinates

6.3.1
Spherical Coordinates (r, θ, φ)

The gradient of a scalar field ψ, and the divergence and curl of a vector field **A** can be expressed as follows:

$$\nabla \psi = \mathbf{n}\frac{\partial \psi}{\partial r} + \mathbf{l}\frac{1}{r}\frac{\partial \psi}{\partial \theta} + \mathbf{m}\frac{1}{r \sin \theta}\frac{\partial \psi}{\partial \varphi} \tag{6.22}$$

$$\nabla \cdot \mathbf{A} = \frac{1}{r^2}\frac{\partial (r^2 A_r)}{\partial r} + \frac{1}{r \sin \theta}\frac{\partial (\sin \theta\, A_\theta)}{\partial \theta} + \frac{1}{r \sin \theta}\frac{\partial A_\varphi}{\partial \varphi} \tag{6.23}$$

$$\nabla \times \mathbf{A} = \mathbf{n}\frac{1}{r \sin \theta}\left[\frac{\partial (\sin \theta\, A_\varphi)}{\partial \theta} - \frac{\partial A_\theta}{\partial \varphi}\right] + \mathbf{l}\left[\frac{1}{r \sin \theta}\frac{\partial A_r}{\partial \varphi} - \frac{1}{r}\frac{\partial (r A_\varphi)}{\partial r}\right]$$
$$+ \mathbf{m}\frac{1}{r}\left[\frac{\partial (r A_\theta)}{\partial r} - \frac{\partial A_r}{\partial \theta}\right] \tag{6.24}$$

where A_r, A_θ, A_φ are the components of **A** along **n**, **l**, **m**, respectively.

These formulas (6.22)–(6.24) can be obtained in a systematic manner, using the integral theorems. We discuss this method in Section 6.4.

The Laplacian ∇^2 is a scalar differential operator and can act on either a scalar or a vector field. The Laplacian of a field ψ is given by

$$\nabla^2 \psi = \frac{1}{r^2}\frac{\partial}{\partial r}\left(r^2 \frac{\partial \psi}{\partial r}\right) + \frac{1}{r^2 \sin \theta}\frac{\partial}{\partial \theta}\left(\sin \theta \frac{\partial \psi}{\partial \theta}\right) + \frac{1}{r^2 \sin^2 \theta}\frac{\partial^2 \psi}{\partial \varphi^2} \tag{6.25}$$

This formula may be obtained by the combined use of (6.22) and (6.23) with $\mathbf{A} = \nabla \psi$ (Problem 6.3.1).

6.3.2
Cylindrical Coordinates (ρ, φ, z)

Various space derivatives can be expressed in cylindrical coordinates as follows:

$$\nabla \psi = \mathbf{h}\frac{\partial \psi}{\partial \rho} + \mathbf{m}\frac{1}{\rho}\frac{\partial \psi}{\partial \varphi} + \mathbf{k}\frac{\partial \psi}{\partial z} \tag{6.26}$$

$$\nabla \cdot \mathbf{A} = \frac{1}{\rho}\frac{\partial (\rho A_\rho)}{\partial \rho} + \frac{1}{\rho}\frac{\partial A_\varphi}{\partial \varphi} + \frac{\partial A_z}{\partial z} \tag{6.27}$$

$$\nabla \times \mathbf{A} = \mathbf{h}\left[\frac{1}{\rho}\frac{\partial A_z}{\partial \varphi} - \frac{\partial A_\varphi}{\partial z}\right] + \mathbf{m}\left[\frac{\partial A_\rho}{\partial z} - \frac{\partial A_z}{\partial \rho}\right]$$
$$+ \mathbf{k}\frac{1}{\rho}\left[\frac{\partial (\rho A_\varphi)}{\partial \rho} - \frac{\partial A_\rho}{\partial \varphi}\right] \tag{6.28}$$

$$\nabla^2 \psi = \frac{1}{\rho}\frac{\partial}{\partial \rho}\left(\rho \frac{\partial \psi}{\partial \rho}\right) + \frac{1}{\rho^2}\frac{\partial^2 \psi}{\partial \varphi^2} + \frac{\partial^2 \psi}{\partial z^2} \tag{6.29}$$

Problem 6.3.1

By noting that $\nabla^2 \psi = \nabla \cdot \nabla \psi$, derive the formula for $\nabla^2 \psi$ in spherical and cylindrical polar coordinates. (The answers are given in (6.25) and (6.29).)

6.4
Integral Theorems

6.4.1
The Line Integral of $\nabla \phi$

Let ϕ be a scalar field. Then, for any curve C running from $\mathbf{r}_0$ to $\mathbf{r}_1$, as shown in Figure 6.1,

$$\int_{\mathbf{r}_0}^{\mathbf{r}_1} d\mathbf{r} \cdot \nabla \phi(\mathbf{r}) = \phi(\mathbf{r}_1) - \phi(\mathbf{r}_0) \tag{6.30}$$

This can be obtained by integrating $d\mathbf{r} \cdot \nabla \phi(\mathbf{r}) = d\phi$ in (6.7) from $\mathbf{r}_0$ to $\mathbf{r}_1$.

6.4.2
Stokes's Theorem

Consider the surface S bounded by a closed curve C shown in Figure 6.2. If one side of S is chosen to be the "positive" side, then the positive direction along C will be defined by the right-hand screw convention as shown in Figure 6.2. Take a small element of area dS on the surface and let $\mathbf{n}$ be the unit vector normal to the element and pointing toward the positive side. *Stokes's theorem* states that if $\mathbf{A}$ is any vector field, then

$$\boxed{\iint_S dS\,\mathbf{n} \cdot (\nabla \times \mathbf{A}) = \oint_C d\mathbf{r} \cdot \mathbf{A}} \tag{6.31}$$

Let us first prove this theorem for an infinitesimal rectangular area. Introduce Cartesian coordinates such that the four corners are represented by $(x - dx/2, y - dy/2, z)$, $(x + dx/2, y - dy/2, z)$, $(x + dx/2, y + dy/2, z)$, $(x - dx/2, y + dy/2, z)$

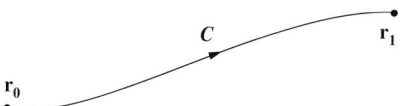

Fig. 6.1 The line integral of $\nabla \phi \cdot d\mathbf{r}$ on the running curve C.

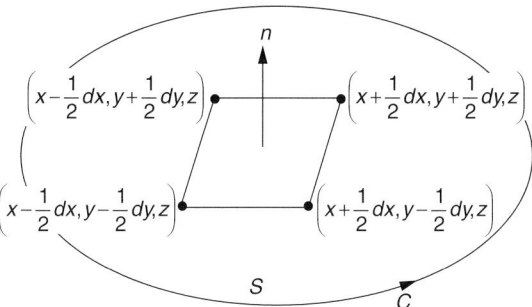

Fig. 6.2 A diagram defining the meaning of Stokes's theorem (6.31). The direction of the outward normal *n* is related to that of the contour *C* by the right-hand screw convention.

and the normal **n** coincides with the unit vector **k** pointing along the *z*-axis. See Figure 6.2. Then, $d\mathbf{S} = \mathbf{k}\,dx\,dy$. The surface integral on the lhs is given by

$$\mathbf{k}\cdot(\nabla\times\mathbf{A})\,dx\,dy = \left(\frac{\partial A_y}{\partial x} - \frac{\partial A_x}{\partial y}\right)dx\,dy \tag{6.32}$$

The line integral on the rhs can be written as the sum of four terms, one from each side. The line integral from $(x-dx/2, y-dy/2, z)$ to $(x+dx/2, y-dy/2, z)$ is given by

$$+dx\,A_x(x, y-dy/2, z)$$

where the argument $(x, y - dy/2, z)$ is chosen to be the midpoint of the segment. Since the path $d\mathbf{r}$ is parallel to the positive *x*-axis, the plus sign + is assigned. By using similar arguments, we can represent the line integrals in the following manner:

$$A_x(x, y-dy/2, z)\,dx + A_y(x+dx/2, y, z)\,dy$$

$$- A_x(x, y+dy/2, z)\,dx - A_y(x-dx/2, y, z)\,dy$$

$$= \left(\frac{\partial A_y}{\partial x} - \frac{\partial A_x}{\partial y}\right)dx\,dy + \text{(terms of higher-order differentials)}$$

We therefore find that the theorem is valid for the infinitesimal element $d\mathbf{S}$. We divide the surface *S* into small elements, and sum over all surface elements in *S*. The sum of the surface terms, $\sum d\mathbf{S}\mathbf{n}\cdot(\nabla\times\mathbf{A})$, becomes, by definition, the surface integral over *S*, and the sum of the line integrals, $\sum \oint d\mathbf{r}\cdot\mathbf{A}$, will be equal to the line integral over the directed curve *C* since the remaining line integrals run on the inner lines (separating the surface elements) in opposite directions to cancel out each other.

6.5
Gauss's Theorem

If V is the volume bounded by a closed surface S as shown in Figure 6.3, then for any vector field $\mathbf{A}$, the following equation connecting the volume integral and the surface integral holds:

$$\boxed{\iiint_V d^3 r \nabla \cdot \mathbf{A} = \iint_S d S \mathbf{n} \cdot \mathbf{A}} \tag{6.33}$$

We may prove this theorem in two steps:
1. Verify the theorem for a small cube shown in Figure 6.4:

$$dx\,dy\,dz\,\nabla \cdot \mathbf{A}(x, y, z)$$
$$= dy\,dz\,[A_x(x + dx/2, y, z) - A_x(x - dx/2, y, z)]$$
$$+ dx\,dz\,[A_y(x, y + dy/2, z) - A_y(x, y - dy/2, z)]$$
$$+ dx\,dy\,[A_z(x, y, z + dz/2) - A_z(x, y, z - dz/2)] \tag{6.34}$$

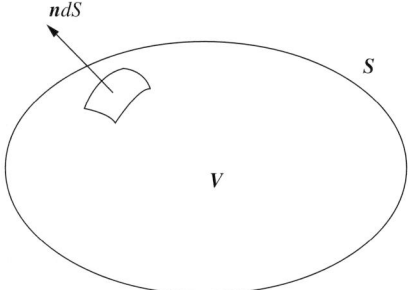

Fig. 6.3 The volume V bounded by a closed surface S.

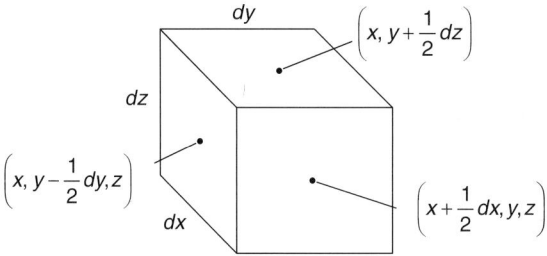

Fig. 6.4 The center of a small cube is located at (x, y, z). The face centers are indicated by dots. The position of the dots are shown in the Cartesian coordinates.

2. Establish a general proof by summing up (6.34) over the cubes, and taking a small-cube limit. The lhs becomes the volume integral indicated in (6.33). The rhs becomes the surface integral since the inner surface terms cancel out each other.

Problem 6.5.1

1. Verify Gauss's theorem for a small cube. Use (6.34).
2. State the validity condition.

6.6
Derivation of the Gradient, Divergence and Curl

One of the uses of the integral theorems is to provide expressions for the gradient, divergence and curl in terms of curvilinear coordinates.

Consider a set of orthogonal curvilinear coordinates q_1, q_2, q_3, and denote the elements of length along the three coordinates curves by $h_1 dq_1, h_2 dq_2, h_3 dq_3$. For example, in cylindrical coordinates,

$$h_\rho = 1, \quad h_\varphi = \rho, \quad h_z = 1 \tag{6.35}$$

and in spherical coordinates

$$h_r = 1, \quad h_\theta = r, \quad h_\varphi = r \sin\theta \tag{6.36}$$

Now consider a scalar field ψ, and two neighboring points (q_1, q_2, q_3) and $(q_1, q_2, q_3 + dq_3)$. Then, the difference between the values of ψ at these points is

$$\frac{\partial \psi}{\partial q_3} dq_3 = d\psi = d\mathbf{r} \cdot \nabla\psi = h_3 dq_3 (\nabla\psi)_3 \tag{6.37}$$

where $(\nabla\psi)_3$ is the component of $\nabla\psi$ in the direction of increasing q_3. From this we obtain

$$(\nabla\psi)_3 = \frac{1}{h_3} \frac{\partial \psi}{\partial q_3} \tag{6.38}$$

Similar expressions hold for other components. Thus, in cylindrical and spherical coordinates we have

$$\nabla\psi = \left(\frac{\partial \psi}{\partial \rho}, \frac{1}{\rho} \frac{\partial \psi}{\partial \varphi}, \frac{\partial \psi}{\partial z} \right) \tag{6.39}$$

and

$$\nabla\psi = \left(\frac{\partial \psi}{\partial r}, \frac{1}{r} \frac{\partial \psi}{\partial \theta}, \frac{1}{r \sin\theta} \frac{\partial \psi}{\partial \varphi} \right) \tag{6.40}$$

To find an expression for the divergence, we use Gauss's theorem, applied to a small volume bounded by the coordinate surfaces. The volume integral is

$$(\nabla \cdot \mathbf{A}) h_1 dq_1 h_2 dq_2 h_3 dq_3$$

In the surface integral, the terms arising from the faces that are surfaces of constant q_3 are of the form $A_3 h_1 dq_1 h_2 dq_2$, evaluated for two different values of q_3. They therefore contribute

$$\frac{\partial}{\partial q_3}(h_1 h_2 A_3) dq_1 dq_2 dq_3$$

Adding the terms from all three pairs of faces, and comparing with the volume integral, we obtain

$$\nabla \cdot \mathbf{A} = \frac{1}{h_1 h_2 h_3}\left\{\frac{\partial}{\partial q_1}(h_2 h_3 A_1) + \frac{\partial}{\partial q_2}(h_3 h_1 A_2) + \frac{\partial}{\partial q_3}(h_1 h_2 A_3)\right\} \qquad (6.41)$$

In particular, in cylindrical and spherical coordinates we have

$$\nabla \cdot \mathbf{A} = \frac{1}{\rho}\frac{\partial(\rho A_\rho)}{\partial \rho} + \frac{1}{\rho}\frac{\partial A_\varphi}{\partial \varphi} + \frac{\partial A_z}{\partial z} \qquad (6.42)$$

and

$$\nabla \cdot \mathbf{A} = \frac{1}{r^2}\frac{\partial(r^2 A_r)}{\partial r} + \frac{1}{r \sin\theta}\frac{\partial(\sin\theta A_\theta)}{\partial \theta} + \frac{1}{r \sin\theta}\frac{\partial A_\varphi}{\partial \varphi} \qquad (6.43)$$

To find the curl, we use Stokes's theorem in a similar manner. If we consider a small element of a surface $q_3 = $ constant, bounded by curves of constant q_1 and q_2, then the surface integral is

$$(\nabla \times \mathbf{A})_3 h_1 dq_1 h_2 dq_2$$

In the line integral round the boundary, the two edges of constant q_2 involve $A_1 h_1 dq_1$ evaluated for different values of q_2, and contribute

$$-\frac{\partial}{\partial q_2}(h_1 A_1) dq_1 dq_2$$

Hence, adding the contribution from the other two edges, we obtain

$$(\nabla \times \mathbf{A})_3 = \frac{1}{h_1 h_2}\left\{\frac{\partial}{\partial q_1}(h_2 A_2) - \frac{\partial}{\partial q_2}(h_1 A_1)\right\} \qquad (6.44)$$

Similar expressions hold for the other components.

Thus, in particular, in cylindrical coordinates we have

$$\nabla \times \mathbf{A} = \left\{\frac{1}{\rho}\frac{\partial A_z}{\partial \varphi} - \frac{\partial A_\varphi}{\partial z}, \frac{\partial A_\rho}{\partial z} - \frac{\partial A_z}{\partial \rho}, \frac{1}{\rho}\left(\frac{\partial(\rho A_\varphi)}{\partial \rho} - \frac{\partial A_\rho}{\partial \varphi}\right)\right\} \qquad (6.45)$$

and in spherical coordinates

$$\nabla \times \mathbf{A} = \left\{ \frac{1}{r\sin\theta} \left(\frac{\partial(\sin\theta\, A_\varphi)}{\partial\theta} - \frac{\partial A_\theta}{\partial\varphi} \right), \right.$$
$$\left. \frac{1}{r\sin\theta} \frac{\partial A_r}{\partial\varphi} - \frac{1}{r} \frac{\partial(rA_\varphi)}{\partial r},\; \frac{1}{r}\left(\frac{\partial(rA_\theta)}{\partial r} - \frac{\partial A_r}{\partial\theta} \right) \right\} \tag{6.46}$$

Finally, combining the expression for the divergence and gradient, we can find the Laplacian of a scalar field. It is

$$\nabla^2 \psi = \frac{1}{h_1 h_2 h_3}$$
$$\times \left\{ \frac{\partial}{\partial q_1}\left(\frac{h_2 h_3}{h_1} \frac{\partial \psi}{\partial q_1} \right) + \frac{\partial}{\partial q_2}\left(\frac{h_3 h_1}{h_2} \frac{\partial \psi}{\partial q_2} \right) + \frac{\partial}{\partial q_3}\left(\frac{h_1 h_2}{h_3} \frac{\partial \psi}{\partial q_3} \right) \right\} \tag{6.47}$$

In cylindrical coordinates we have

$$\nabla^2 \psi = \frac{1}{\rho} \frac{\partial}{\partial \rho}\left(\rho \frac{\partial \psi}{\partial \rho} \right) + \frac{1}{\rho^2} \frac{\partial^2 \psi}{\partial \varphi^2} + \frac{\partial^2 \psi}{\partial z^2} \tag{6.48}$$

and in spherical coordinates we have

$$\nabla^2 \psi = \frac{1}{r^2} \frac{\partial}{\partial r}\left(r^2 \frac{\partial \psi}{\partial r} \right) + \frac{1}{r^2 \sin\theta} \frac{\partial}{\partial \theta}\left(\sin\theta \frac{\partial \psi}{\partial \theta} \right) + \frac{1}{r^2 \sin^2\theta} \frac{\partial^2 \psi}{\partial \varphi^2} \tag{6.49}$$

7
Electromagnetic Waves

Electromagnetic waves are discussed in this chapter. Starting with Maxwell's equations in a vacuum, a wave equation is derived. Its solution in free space is a wave propagating with a speed $c = 3 \times 10^8$ m/s. The wave carries an electric field **E** and a magnetic field **B**, both fields perpendicular to the propagation direction and perpendicular to each other The energy associated with an electromagnetic wave is calculated.

7.1
Electric and Magnetic Fields in a Vacuum

The basic equations governing electromagnetic fields are Maxwell's equations. In a vacuum, the electric field $\mathbf{E}(\mathbf{r}, t)$ and the magnetic field $\mathbf{B}(\mathbf{r}, t)$ obey the two vector equations:

$$\frac{\partial \mathbf{B}}{\partial t} + \nabla \times \mathbf{E} = 0 \tag{7.1}$$

$$\mu_0 \epsilon_0 \frac{\partial \mathbf{E}}{\partial t} - \nabla \times \mathbf{B} = 0 \tag{7.2}$$

and the two scalar equations:

$$\nabla \cdot \mathbf{B} = 0 \tag{7.3}$$

$$\nabla \cdot \mathbf{E} = 0 \tag{7.4}$$

If we introduce a vector potential $\mathbf{A}(\mathbf{r}, t)$ that satisfies

$$\nabla \cdot \mathbf{A} = 0 \quad \text{(Coulomb gauge condition)} \tag{7.5}$$

both **E** and **B** can be expressed in terms of **A** as follows:

$$\mathbf{B} = \nabla \times \mathbf{A}(\mathbf{r}, t) \tag{7.6}$$

$$\mathbf{E} = -\frac{\partial}{\partial t} \mathbf{A}(\mathbf{r}, t) \tag{7.7}$$

Mathematical Physics. Shigeji Fujita and Salvador V. Godoy
Copyright © 2010 WILEY-VCH Verlag GmbH & Co. KGaA, Weinheim
ISBN: 978-3-527-40808-5

It is easy to verify that the fields (**E**, **B**) defined by (7.6) and (7.7) satisfy Maxwell's equations (7.1)–(7.4) (Problem 7.1.1).

Introducing **E** and **B** from (7.6) and (7.7) in (7.2), we obtain

$$-\mu_0\epsilon_0 \frac{\partial^2 \mathbf{A}}{\partial t^2} - \nabla \times (\nabla \times \mathbf{A}) = 0 \tag{7.8}$$

Using the vector identity

$$\nabla \times (\nabla \times \mathbf{C}) = \nabla(\nabla \cdot \mathbf{C}) - \nabla^2 \mathbf{C} \tag{7.9}$$

we obtain

$$\mu_0\epsilon_0 \frac{\partial^2 \mathbf{A}}{\partial t^2} = -\left[\nabla(\nabla \cdot \mathbf{A}) - \nabla^2 \mathbf{A}\right] = \nabla^2 \mathbf{A} \tag{7.10}$$

where we used (7.5). Hence, we have

$$\boxed{\nabla^2 \mathbf{A} - \frac{1}{c^2} \frac{\partial^2 \mathbf{A}}{\partial t^2} = 0} \tag{7.11}$$

where

$$c \equiv (\epsilon_0\mu_0)^{-1/2} = 3 \times 10^8 \text{ m s}^{-1} \tag{7.12}$$

Equation (7.11) is called the *wave equation*, and characterizes the motion of the vector potential **A** in a vacuum. The constant c is the *speed of light*, and has the numerical value indicated.

Applying operators $\partial/\partial t$ and $\nabla \times$ to (7.1) and (7.2), we can verify that both **E** and **B** satisfy the same wave equation:

$$\nabla^2 \mathbf{E} - \frac{1}{c^2} \frac{\partial^2 \mathbf{E}}{\partial t^2} = 0, \quad \nabla^2 \mathbf{B} - \frac{1}{c^2} \frac{\partial^2 \mathbf{B}}{\partial t^2} = 0 \tag{7.13}$$

Earlier in Chapter 5, we set up and solved the wave equation in one dimension, which is associated with the transverse oscillations of a stretched string. We can treat the three-dimensional wave equation (7.11) in an analogous manner.

We assume the *periodic boundary condition*:

$$A(x + L, y, z, t) = A(x, y + L, z, t) = A(x, y, z + L, t) = A(\mathbf{r}, t) \tag{7.14}$$

where L is a length of periodicity. We can show that the exponential functions

$$\exp(-i\omega_k t + i\mathbf{k} \cdot \mathbf{r}) \tag{7.15}$$

where

$$\mathbf{k} = (k_x, k_y, k_z)$$
$$k_x = \frac{2\pi}{L} n_x, \quad k_y = \frac{2\pi}{L} n_y, \quad k_z = \frac{2\pi}{L} n_z \tag{7.16}$$
$$n_x, n_y, n_z = 0, \pm 1, \pm 2, \ldots$$
$$\omega_k \equiv c|\mathbf{k}| = ck \tag{7.17}$$

are solutions of the wave equation (7.11) subject to the periodic boundary condition (7.14) (Problem 7.1.2).

For $\mathbf{k} = (k_x, 0, 0)$, $k_x = k > 0$, the function

$$\exp(-i\omega_k t + i\mathbf{k}\cdot\mathbf{r}) = e^{ik(x-ct)} \tag{7.18}$$

represents a plane wave traveling along the positive x-axis with speed c. For a general $\mathbf{k} = (k_x, k_y, k_z)$, the function $\exp(-i\omega_k t + i\mathbf{k}\cdot\mathbf{r})$ represents a plane wave traveling in the direction of $\mathbf{k}$, as indicated in Figure 7.1.

From (7.14), the vector potential $\mathbf{A}$ is periodic with the period L in the x-, y-, z-directions. We may then expand $\mathbf{A}$ in the following form (Fourier's fundamental theorem):

$$\mathbf{A}(\mathbf{r}, t) = \sum_{n_x=-\infty}^{+\infty} \sum_{n_y=-\infty}^{+\infty} \sum_{n_z=-\infty}^{+\infty}$$
$$\frac{1}{2}\left\{ \mathbf{a}(n_x, n_y, n_z, t) \exp\left[i\frac{2\pi}{L}(xn_x + yn_y + zn_z) \right] + \text{c.c.} \right\} \tag{7.19}$$

By definition, the vector potential $\mathbf{A}$ is real. To insure this property, the complex conjugates (c.c.) is added on the rhs, and the result is divided by two. Introducing the wave vector $\mathbf{k}$ defined in (7.15)–(7.17), we can rewrite (7.19) as

$$\mathbf{A}(\mathbf{r}, t) = \frac{1}{2} \sum_{\mathbf{k}} \left[\mathbf{a}_{\mathbf{k}}(t) e^{i\mathbf{k}\cdot\mathbf{r}} + \text{c.c.} \right] \tag{7.20}$$

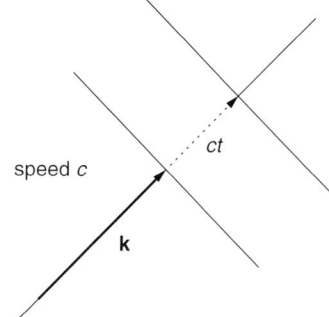

Fig. 7.1 A plane wave traveling in the direction of $\mathbf{k}$ and with speed c.

Substitution of this expression into the wave equation (7.11) yields

$$\ddot{\mathbf{a}}_k(t) + c^2 k^2 \mathbf{a}_k(t) = 0 \tag{7.21}$$

which shows that the vector coefficient $\mathbf{a}_k(t)$ is a periodic function of time with the angular frequency $\omega_k \equiv ck$. We write

$$\mathbf{a}_k(t) = \mathbf{f}_k e^{-i\omega_k t} \tag{7.22}$$

substitute into (7.20), and obtain

$$\mathbf{A}(\mathbf{r}, t) = \frac{1}{2} \sum_k [\mathbf{f}_k \exp(-i\omega_k t + i\mathbf{k} \cdot \mathbf{r}) + \text{c.c.}] \tag{7.23}$$

This expansion can be interpreted as follows: *The general solution for the wave equation with the periodic boundary condition is a superposition of plane-wave solutions.*

Applying the operator $\nabla \cdot$ to (7.20), we obtain

$$\nabla \cdot \mathbf{A} = \frac{1}{2} \sum_k [\mathbf{a}_k \cdot \nabla e^{i\mathbf{k}\cdot\mathbf{r}} + \text{c.c.}] \quad [\mathbf{a}_k \text{ is independent of } \mathbf{r}]$$

$$= \frac{1}{2} \sum_k \left[\left(a_{k,x}\frac{\partial}{\partial x} + a_{k,y}\frac{\partial}{\partial y} + a_{k,z}\frac{\partial}{\partial z} \right) e^{i(xk_x + yk_y + zk_z)} + \text{c.c.} \right]$$

$$= \frac{1}{2} \sum_k \left[i(a_{k,x}k_x + a_{k,y}k_y + a_{k,z}k_z) e^{i(xk_x + yk_y + zk_z)} + \text{c.c.} \right]$$

$$= \frac{1}{2} \sum_k [i\mathbf{k} \cdot \mathbf{a}_k e^{i\mathbf{k}\cdot\mathbf{r}} + \text{c.c.}] = 0 \tag{7.24}$$

Since $e^{i\mathbf{k}\cdot\mathbf{r}}$ does not vanish, we must have

$$\mathbf{k} \cdot \mathbf{a}_k = 0 \tag{7.25}$$

This means that the vector coeficients $\mathbf{a}_k$ are orthogonal to the wave vector $\mathbf{k}$.

Using (7.7) and (7.22) we obtain the electric field $\mathbf{E}$ as follows:

$$\mathbf{E} = -\frac{\partial}{\partial t}\mathbf{A}$$

$$= -\frac{1}{2}\sum_k [\dot{\mathbf{a}}_k(t) \exp(i\mathbf{k}\cdot\mathbf{r}) + \text{c.c.}]$$

$$= \frac{i}{2}\sum_k \omega_k [\mathbf{a}_k \exp(i\mathbf{k}\cdot\mathbf{r}) - \mathbf{a}_k^* \exp(-i\mathbf{k}\cdot\mathbf{r})] \tag{7.26}$$

In a similar manner, we obtain the magnetic field $\mathbf{B}$: (Problem 7.1.3)

$$\mathbf{B} = \frac{i}{2}\sum_k \mathbf{k} \times [\mathbf{a}_k \exp(i\mathbf{k}\cdot\mathbf{r}) - \mathbf{a}_k^* \exp(-i\mathbf{k}\cdot\mathbf{r})] \tag{7.27}$$

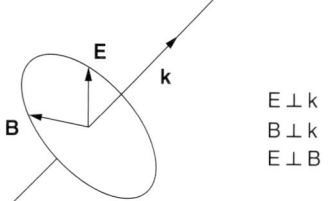

$$E \perp k$$
$$B \perp k$$
$$E \perp B$$

Fig. 7.2 An electromagnetic wave traveling in the direction of **k** carries a pair of fields (**E**, **B**), which are perpendicular to the wave vector **k**, and that are also perpendicular to each other.

The last three equations indicate that the electric and magnetic fields (**E**, **B**) are both perpendicular to the wave vector **k**. In other words, the electromagnetic waves are *transverse waves*.

An electromagnetic traveling wave whose **E**-field oscillates in a fixed plane is called a *plane-polarized wave*. The direction of the electric field **E**, which is perpendicular to the wave vector **k**, is customarily denoted by the *polarization unit* vector $\boldsymbol{\sigma}$. Then, we have by construction

$$\mathbf{k} \cdot \boldsymbol{\sigma} = 0 \tag{7.28}$$

From (7.27) we see that the associated magnetic field **B** oscillates in the direction perpendicular to both **k** and **E**. The relationship between the directions of **E**, **B** and **k** is

$$\frac{\mathbf{E}}{E} \times \frac{\mathbf{B}}{B} = \frac{\mathbf{k}}{k} \tag{7.29}$$

as shown schematically in Figure 7.2.

The electromagnetic wave that is emitted from a single molecule and that has traveled a long distance can be thought of as a plane-polarized wave. In a vacuum, electromagnetic waves of different wave vectors **k**, frequencies ω_k and polarizations $\boldsymbol{\sigma}$ travel with the same speed c and without interaction. Each of these waves carry a certain energy. The sum of these energies represents the energy of the electromagnetic or radiation fields, as we will see in the next section.

Problem 7.1.1

Verify that the fields **E** and **B** given by $\mathbf{B} = \nabla \times \mathbf{A}(\mathbf{r}, t)$, $\mathbf{E} = -\partial \mathbf{A}(\mathbf{r}, t)/\partial t$, where **A** is a vector potential satisfying $\nabla \cdot \mathbf{A} = 0$, satisfy Maxwell's equations in vacuum, (7.1)–(7.4).

Problem 7.1.2

Verify that the exponential functions

$$\exp(-i\omega_k t \pm i\mathbf{k} \cdot \mathbf{r}) \tag{7.30}$$

7 Electromagnetic Waves

with $\mathbf{k}$ and ω_k given by (7.15) and (7.16), satisfy the wave equation (7.11) with the boundary condition (7.14).

Problem 7.1.3

Calculate the magnetic field $\mathbf{B} = \nabla \times \mathbf{A}(\mathbf{r}, t)$, when the vector potential $\mathbf{A}$ is given in the form of (7.20).

$$\mathbf{A}(\mathbf{r}, t) = \frac{1}{2} \sum_{\mathbf{k}} \left[\mathbf{a}_{\mathbf{k}}(t) \exp(i\mathbf{k}\cdot\mathbf{r}) + \mathbf{a}_{\mathbf{k}}^*(t) \exp(-i\mathbf{k}\cdot\mathbf{r}) \right]$$

ANSWER:

$$\mathbf{B} = \frac{i}{2} \sum_{\mathbf{k}} \mathbf{k} \times \left[\mathbf{a}_{\mathbf{k}} \exp(i\mathbf{k}\cdot\mathbf{r}) - \mathbf{a}_{\mathbf{k}}^* \exp(-i\mathbf{k}\cdot\mathbf{r}) \right]$$

7.2
The Electromagnetic Field Theory

In electromagnetism, the energy H of the radiation fields in the volume $V = L^3$ is defined by

$$H = \frac{\epsilon_0}{2} \int d^3 r\, E^2 + \frac{1}{2\mu_0} \int d^3 r\, B^2 \qquad (7.31)$$

In this section we use the symbol H for the energy to avoid confusion with the electric-field magnitude E. We now wish to express this energy in terms of the wave amplitudes $\mathbf{a}_{\mathbf{k}}$.

The absolute square of (7.26) yields

$$E^2 = \left(\frac{1}{2} \sum_{\mathbf{k}} i\omega_k \mathbf{a}_{\mathbf{k}} e^{i\mathbf{k}\cdot\mathbf{r}} + \text{c.c.} \right) \cdot \left(\frac{1}{2} \sum_{\mathbf{k}'} i\omega_{k'} \mathbf{a}_{\mathbf{k}'} e^{i\mathbf{k}'\cdot\mathbf{r}} + \text{c.c.} \right)$$

$$= \frac{1}{4} \sum_{\mathbf{k}} \sum_{\mathbf{k}'} \omega_k \omega_{k'} \Big[-\mathbf{a}_{\mathbf{k}} \cdot \mathbf{a}_{\mathbf{k}'} e^{i(\mathbf{k}-\mathbf{k}')\cdot\mathbf{r}} + \mathbf{a}_{\mathbf{k}}^* \cdot \mathbf{a}_{\mathbf{k}'} e^{-i(\mathbf{k}+\mathbf{k}')\cdot\mathbf{r}}$$

$$+ \mathbf{a}_{\mathbf{k}} \cdot \mathbf{a}_{\mathbf{k}'}^* e^{i(\mathbf{k}-\mathbf{k}')\cdot\mathbf{r}} - \mathbf{a}_{\mathbf{k}}^* \cdot \mathbf{a}_{\mathbf{k}'}^* e^{-i(\mathbf{k}+\mathbf{k}')\cdot\mathbf{r}} \Big] \qquad (7.32)$$

Let us take the integral

$$I(\mathbf{k} + \mathbf{k}') \equiv \int_V d^3 r \exp\left[i(\mathbf{k} + \mathbf{k}') \cdot \mathbf{r} \right]$$

$$= \iiint dx\,dy\,dz \exp\left[i(k_x + k_x')x + i(k_y + k_y')y + i(k_z + k_z')z \right]$$

$$(7.33)$$

which can be factorized into three parts: the x-, y- and z-integrals. The x-integral

$$\int_0^L dx \exp[i(k_x + k'_x)x] = \int_0^L dx \exp[i2\pi x(n_x + n'_x)/L]$$

vanishes unless $n_x + n'_x = 0$, in which case the x-integral equals L. Hence, we obtain

$$\int_0^L dx \exp[i(k_x + k'_x)x] = \begin{cases} L & \text{if } k_x + k'_x = 0 \\ 0 & \text{otherwise} \end{cases} \tag{7.34}$$

similar properties hold for the y- and z-integrals. We therefore obtain

$$I(\mathbf{k} + \mathbf{k}') = \begin{cases} L^3 = V & \text{if } \mathbf{k} + \mathbf{k}' = 0 \\ 0 & \text{otherwise} \end{cases} \tag{7.35}$$

The same expression is valid if we replace $\mathbf{k} + \mathbf{k}'$ by $\mathbf{k} - \mathbf{k}'$ or $-\mathbf{k} - \mathbf{k}'$. In summary, we have:

$$I(\mathbf{q}) \equiv \int_V d^3r \exp[i\mathbf{q} \cdot \mathbf{r}] = V\delta_{\mathbf{q},0}^{(3)} \tag{7.36}$$

Integrating (7.32) over the normalization volume V and using the property (7.36), we obtain

$$\int_V E^2 d^3r = V \sum_k \omega_k^2 \left[2\mathbf{a}_k^* \cdot \mathbf{a}_k - \mathbf{a}_k \cdot \mathbf{a}_{-k} - \mathbf{a}_k^* \cdot \mathbf{a}_{-k}^* \right] \tag{7.37}$$

In a similar manner, we can show that [Problem 7.2.1]

$$\int_V B^2 d^3r = V \sum_k \left[2(\mathbf{k} \times \mathbf{a}_k) \cdot (\mathbf{k} \times \mathbf{a}_k^*) \right.$$

$$\left. + (\mathbf{k} \times \mathbf{a}_k) \cdot (\mathbf{k} \times \mathbf{a}_{-k}) + (\mathbf{k} \times \mathbf{a}_k^*) \cdot (\mathbf{k} \times \mathbf{a}_{-k}^*) \right] \tag{7.38}$$

Applying the general formula

$$(\mathbf{A} \times \mathbf{B}) \cdot (\mathbf{C} \times \mathbf{D}) = \mathbf{A} \cdot [\mathbf{B} \times (\mathbf{C} \times \mathbf{D})]$$
$$= (\mathbf{A} \cdot \mathbf{C})(\mathbf{B} \cdot \mathbf{D}) - (\mathbf{A} \cdot \mathbf{D})(\mathbf{B} \cdot \mathbf{C}) \tag{7.39}$$

to $(\mathbf{k} \times \mathbf{a}_k) \cdot (\mathbf{k} \times \mathbf{a}_k^*)$, we obtain

$$(\mathbf{k} \times \mathbf{a}_k) \cdot (\mathbf{k} \times \mathbf{a}_k^*) = k^2 \mathbf{a}_k \cdot \mathbf{a}_k^* - (\mathbf{k} \cdot \mathbf{a}_k)(\mathbf{k} \cdot \mathbf{a}_k^*)$$
$$= k^2 \mathbf{a}_k \cdot \mathbf{a}_k^* \tag{7.40}$$

In a similar manner, we can get

$$(\mathbf{k} \times \mathbf{a_k}) \cdot (\mathbf{k} \times \mathbf{a_{-k}}) = k^2 \mathbf{a_k} \cdot \mathbf{a_{-k}}$$

$$(\mathbf{k} \times \mathbf{a_k^*}) \cdot (\mathbf{k} \times \mathbf{a_{-k}^*}) = k^2 \mathbf{a_k^*} \cdot \mathbf{a_{-k}^*} \tag{7.41}$$

Using (7.40)–(7.42), we can reduce (7.38) to

$$\int_V B^2 d^3 r = V \sum_k k^2 \left[2\mathbf{a_k} \cdot \mathbf{a_k^*} + \mathbf{a_k} \cdot \mathbf{a_{-k}} + \mathbf{a_k^*} \cdot \mathbf{a_{-k}^*} \right] \tag{7.42}$$

Introducing (7.37) and (7.42) in (7.31), and noting that $c^2 \equiv (\epsilon_0 \mu_0)^{-1}$, we obtain

$$H = 2\epsilon_0 V \sum_k \omega_k^2 (\mathbf{a_k} \cdot \mathbf{a_k^*}) = \epsilon_0 V \sum_k \omega_k^2 |\mathbf{a_k}|^2 \tag{7.43}$$

Here, we observe that the energy of radiation in a vacuum, which is positive by definition, is expressed in terms of the absolute squares of plane-wave amplitudes. Also, note that the radiation energy is proportional to the volume V.

Let us now introduce real vectors:

$$\mathbf{Q_k} \equiv (\epsilon_0 V)^{1/2} (\mathbf{a_k} + \mathbf{a_k^*}) \tag{7.44}$$

$$\mathbf{P_k} \equiv -i(\epsilon_0 V)^{1/2} \omega_k (\mathbf{a_k} - \mathbf{a_k^*}) = \dot{\mathbf{Q}}_k \tag{7.45}$$

The vectors $\mathbf{Q_k}$ clearly obey the same equations of motion, (7.21), as the amplitude vectors $\mathbf{a_k}$:

$$\ddot{\mathbf{Q}}_k = -\omega_k^2 \mathbf{Q}_k \tag{7.46}$$

In terms of $\mathbf{Q_k}$ and $\mathbf{P_k}$, we can rewrite (7.43) in the following form: (Problem 7.2.2)

$$H = \sum_k \frac{1}{2} \left(P_k^2 + \omega_k^2 Q_k^2 \right) \tag{7.47}$$

Since $\mathbf{a_k} \cdot \mathbf{k} = 0$, we obtain

$$\mathbf{Q_k} \cdot \mathbf{k} = 0, \quad \mathbf{P_k} \cdot \mathbf{k} = 0, \quad \dot{\mathbf{Q}}_k \cdot \mathbf{k} = 0 \tag{7.48}$$

Thus, vectors $\mathbf{Q_k}$ and $\mathbf{P_k}$ are both perpendicular to the wave vector $\mathbf{k}$. Let us denote their components in the plane perpendicular to $\mathbf{k}$ by $(Q_{k,1}, Q_{k,2})$ and $(P_{k,1}, P_{k,2})$. We may then express (7.47) in the form

$$H = \sum_k \sum_{\sigma=1,2} \frac{1}{2} \left(P_{k,\sigma}^2 + \omega_k^2 Q_{k,\sigma}^2 \right) \tag{7.49}$$

In this form, the radiation energy is expressed as the sum of the energies of harmonic oscillators, each oscillator corresponding to an electromagnetic plane wave characterized by the wave vector **k**, the angular frequency ω_k and the polarization index σ. Notice the similarity between (7.49) and (5.64), which represents the energy of a vibrating string written in terms of the normal coordinates and momenta.

We can regard the Hamiltonian H with the set of canonical variables $\{Q_{k,\sigma}, P_{k,\sigma}\}$. In fact, one set of the Hamiltonian equations,

$$\frac{d}{dt} Q_{k,\sigma} = \frac{\partial H}{\partial P_{k,\sigma}} = P_{k,\sigma} \tag{7.50}$$

coincides with the definition equation (7.45) for $P_{k,\sigma}$. The other set of equations,

$$\frac{d}{dt} P_{k,\sigma} = -\frac{\partial H}{\partial Q_{k,\sigma}} = -\omega_k^2 Q_{k,\sigma} \tag{7.51}$$

generates the equations of motion.

Problem 7.2.1

Prove (7.38).

HINT: Use (7.27) for **B** in (7.36).

Problem 7.2.2

Derive (7.47) from (7.43).

8
Fluid Dynamics

A fluid motion is represented by the mass density ρ and the velocity field **v**. The continuity equation and the equation of motion are derived and discussed in this chapter. A viscous flow is briefly discussed.

8.1
Continuity Equation

Fluid dynamics deals with the macroscopic motion of a fluid such as air, water, and so on. We disregard the molecular constitution of the fluid, and look at it as a continuously distributed mass moving in space. The basic variables used in this mode of description are the *mass density* ρ and the *fluid velocity* **v**, both of which are regarded as functions of position **r** and a time t:

$$\rho = \rho(\mathbf{r}, t), \quad \mathbf{v} = \mathbf{v}(\mathbf{r}, t) \tag{8.1}$$

The basic evolution equations that govern the evolution of ρ and **v** will be derived in this and the following sections.

Let us consider a continuous fluid with the density $\rho(\mathbf{r}, t)$ and the velocity $\mathbf{v}(\mathbf{r}, t)$. Imagine a volume V_1 fixed in space. The mass of the fluid within V_1 is given by

$$M(t) = \int_{V_1} d^3 r \rho(\mathbf{r}, t) \tag{8.2}$$

Differentiating this with respect to t, we obtain

$$\frac{dM}{dt} = \int_{V_1} d^3 r \frac{\partial}{\partial t} \rho(\mathbf{r}, t) \tag{8.3}$$

Note that the partial derivative of the density $\rho(\mathbf{r}, t)$: $\partial \rho(\mathbf{r}, t)/\partial t$ appears in the integrand. The change in mass is caused by the influx of the fluid through the boundary S_1 of the volume V_1, which may be calculated as follows:

Let us consider the small element dS of the boundary surface as shown in Figure 8.1. The amount of fluid moving out through the element dS in the small time

Mathematical Physics. Shigeji Fujita and Salvador V. Godoy
Copyright © 2010 WILEY-VCH Verlag GmbH & Co. KGaA, Weinheim
ISBN: 978-3-527-40808-5

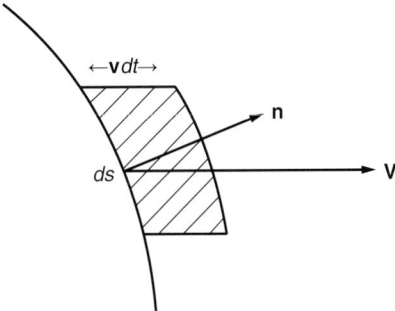

Fig. 8.1 The amount of fluid moving out through dS in the time dt is given by $\rho(\mathbf{v}\cdot\mathbf{n}dt)dS$.

dt will be given by the mass of the fluid within the shaded volume shown in the figure. It is given by

$$\rho(\mathbf{v}\cdot\mathbf{n}dt)dS = \rho\mathbf{v}\cdot\mathbf{n}dS dt \tag{8.4}$$

where $\mathbf{n}$ represents the *outward normal unit vector* at the surface element dS. Integrating this with respect to dS over the surface S_1, we obtain

$$\int_{S_1} d S\mathbf{n}\cdot(\rho\mathbf{v})dt$$

This is the net loss of the mass through the surface S_1 in the time interval dt. Dividing it by dt and changing the sign, we obtain the increment of the mass:

$$-\int_{S_1} d S\mathbf{n}\cdot(\rho\mathbf{v}) = \frac{dM}{dt} \tag{8.5}$$

Using this and (8.3), we obtain

$$\int_{V_1} d^3 r \frac{\partial \rho}{\partial t} = -\int_{S_1} d S\mathbf{n}\cdot(\rho\mathbf{v}) \tag{8.6}$$

The surface integral on the rhs can be converted into a volume integral with the aid of:

$$-\int_{S_1} d S\mathbf{n}\cdot(\rho\mathbf{v}) = -\int_{V_1} d^3 r\, \nabla\cdot(\rho\mathbf{v}) \tag{8.7}$$

Transferring this to the lhs of (8.6), we obtain

$$\int_{V_1} d^3 r \left[\frac{\partial \rho}{\partial t} + \nabla\cdot(\rho\mathbf{v})\right] = 0$$

Since the volume V_1 is arbitrarily chosen, the integrand must vanish. We, therefore, obtain

$$\boxed{\frac{\partial \rho}{\partial t} + \nabla \cdot (\rho \mathbf{v}) = 0} \tag{8.8}$$

This is called the *continuity equation*. Physically, this equation represents the conservation of mass.

8.2
Fluid Equation of Motion

We now turn to the equation of motion for the fluid. Let us suppose that the fluid inside V_1 at time t moves to a new volume V_2 at time $t + dt$, as indicated in Figure 8.2. The momenta of the fluid at t and $t + dt$ are given by

$$\int_{V_1} d^3 r \rho(\mathbf{r}, t) \mathbf{v}(\mathbf{r}, t) \quad \text{and} \quad \int_{V_2} d^3 r \rho(\mathbf{r}, t+dt) \mathbf{v}(\mathbf{r}, t+dt)$$

respectively. We look at the change in the x-component of the momentum of the fluid:

$$\int_{V_2} d^3 r (\rho v_x)_{t+dt} - \int_{V_1} d^3 r (\rho v_x)_t$$

$$= \int_{V_2} d^3 r \left\{ (\rho v_x)_t + \left(\frac{\partial (\rho v_x)}{\partial t}\right)_t dt \right\} - \int_{V_1} d^3 r (\rho v_x)_t$$

$$= \int_{V_2} d^3 r \frac{\partial (\rho v_x)}{\partial t} dt + \left[\int_{V_2} d^3 r \rho v_x - \int_{V_1} d^3 r \rho v_x \right] \tag{8.9}$$

where we used a Taylor expansion and kept terms up to the first order in dt only. The quantity within the brackets represents the change caused by the motion of the fluid through the surface S_1, as indicated in Figure 8.2, and can be represented by the surface integral

$$\int_S d S \mathbf{n} \cdot (\mathbf{v} \rho v_x) dt \tag{8.10}$$

[Note the similarity between this and expression (8.5); the only significant change is that the momentum density ρv_x enters here in place of the mass density ρ.] We can now convert (8.10) into a volume integral by means of the divergence theorem:

$$\int_{S_1} d S \mathbf{n} \cdot (\mathbf{v} \rho v_x) dt = \int_{V_1} d^3 r \nabla \cdot (\mathbf{v} \rho v_x) dt \tag{8.11}$$

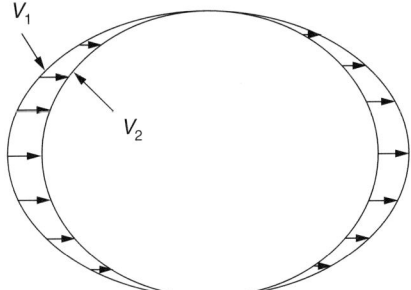

Fig. 8.2 The fluid within the volume V_1 at time t occupies the new volume V_2 at time $t + dt$.

The volume integral $\int_{V_2} d^3 r \partial(\rho v_x)/\partial t \, dt$ in (8.9) is already of the first order in dt, and will be approximated by the integral over the original volume V_1:

$$\int_{V_2} d^3 r \frac{\partial}{\partial t}(\rho v_x) dt \cong \int_{V_1} d^3 r \frac{\partial}{\partial t}(\rho v_x) dt \tag{8.12}$$

We take the sum of (8.11) and (8.12), and divide the result by dt. We obtain

$$\int_{V_1} d^3 r \left[\frac{\partial}{\partial t}(\rho v_x) + \nabla \cdot (\mathbf{v} \rho v_x) \right] \tag{8.13}$$

This integral represents the rate of the change in momentum of the fluid.

This change is caused by the forces acting on the fluid. These forces can be classified into two types: bulk and surface forces. A *bulk force*, such as gravitational force, will act on the fluid everywhere within the volume V_1, and can be represented by

$$\int_{V_1} d^3 r f_x(\mathbf{r}, t) \tag{8.14}$$

where **f** is the force per unit volume; we have taken the x-component of the bulk force here.

An important *surface force* arises from the pressure P that acts on the fluid *inward* (in the direction opposite to the outward normal **n**) at any surface element dS. This force can be written as

$$-\mathbf{n} dS P \tag{8.15}$$

If the fluid flow is not uniform, the fluid is acted upon not only by the pressure P but also by other kinds of surface forces, such as the viscous force (discussed in the following section). In order to treat the general case, we will introduce a *stress tensor* $\mathbb{T}$, and represent the surface force acting on the element dS in the form

$$\mathbf{n} \cdot \mathbb{T} dS \tag{8.16}$$

The definition and principal properties of tensors were collected and discussed in Chapter 3.

If we set

$$\mathbb{T} = -P(\mathbf{ii} + \mathbf{jj} + \mathbf{kk}) \equiv -P\mathbb{E} \tag{8.17}$$

where $\mathbb{E}$ is the *unit tensor*, we obtain

$$\mathbf{n} \cdot \mathbb{T} = -\mathbf{n} \cdot (P\mathbb{E}) = -\mathbf{n}P \tag{8.18}$$

Thus, the force arising from the pressure can be represented in the form (8.16).

Let us integrate (8.16) over the surface S_1:

$$\int_{S_1} dS\,\mathbf{n} \cdot \mathbb{T} \tag{8.19}$$

This represents the net surface force acting on the fluid. By taking the *x*-component of this quantity and applying the divergence theorem, we obtain

$$\int_{S_1} dS(\mathbf{n} \cdot \mathbb{T})_x = \int_{V_1} d^3 r (\nabla \cdot \mathbb{T})_x \tag{8.20}$$

We now apply Newton's equation of motion: momentum change = force. Equation (8.13) is equal to the sum of (8.14) and (8.20). We obtain

$$\int_{V_1} d^3 r \left[\frac{\partial}{\partial t}(\rho v_x) + \nabla \cdot (\mathbf{v}\rho v_x) \right] = \int_{V_1} d^3 r \left[f_x + (\nabla \cdot \mathbb{T})_x \right] \tag{8.21}$$

Since V_1 is arbitrary, we obtain

$$\frac{\partial}{\partial t}(\rho \mathbf{v})_x + (\nabla \cdot \rho \mathbf{v}\mathbf{v})_x = [\mathbf{f} + \nabla \cdot \mathbb{T}]_x \tag{8.22}$$

Similar equations hold for the *y*- and *z*-components. By dropping the component-subscript, we obtain the vector equation:

$$\boxed{\frac{\partial}{\partial t}(\rho \mathbf{v}) + (\nabla \cdot \rho \mathbf{v}\mathbf{v}) = \mathbf{f} + \nabla \cdot \mathbb{T}} \tag{8.23}$$

The terms on the lhs can be reduced as follows:

$$\frac{\partial}{\partial t}(\rho \mathbf{v}) + (\nabla \cdot \rho \mathbf{v}\mathbf{v})$$

$$= \mathbf{v}\left[\frac{\partial \rho}{\partial t} + \nabla \cdot (\rho \mathbf{v})\right] + \rho \frac{\partial \mathbf{v}}{\partial t} + \rho(\mathbf{v} \cdot \nabla)\mathbf{v}$$

$$= \rho \frac{\partial \mathbf{v}}{\partial t} + \rho(\mathbf{v} \cdot \nabla)\mathbf{v}, \quad \text{[use of (8.8)]} \tag{8.24}$$

where

$$\mathbf{v} \cdot \nabla \equiv v_x \frac{\partial}{\partial x} + v_y \frac{\partial}{\partial y} + v_z \frac{\partial}{\partial z} \tag{8.25}$$

The operator $\mathbf{v} \cdot \nabla$ has a meaning only when it is expressed in Cartesian coordinates. The quantity $(\mathbf{v} \cdot \nabla)\mathbf{v}$ is, therefore, not readily expressible in general curvilinear coordinates. One can, however, show that

$$(\mathbf{v} \cdot \nabla)\mathbf{v} = \frac{1}{2}\nabla v^2 - \mathbf{v} \times (\nabla \times \mathbf{v}) \tag{8.26}$$

where the terms on the rhs are both ordinary vectors.

Using (8.23), (8.24) and (8.26), we finally obtain

$$\boxed{\rho \frac{\partial \mathbf{v}}{\partial t} + \frac{1}{2}\rho \nabla v^2 - \rho \mathbf{v} \times (\nabla \times \mathbf{v}) = \mathbf{f} + \nabla \cdot \mathbb{T}} \tag{8.27}$$

This is a general. In practice, this equation can be simplified further. Additional discussion will be presented in the next section.

Problem 8.2.1

Prove (8.10). Hint: follow the derivation of (8.5) as given in the text.

Problem 8.2.2

Verify the relation (8.26) by explicitly writing out the quantities on both sides in Cartesian coordinates.

8.3
Fluid Dynamics and Statistical Mechanics

In the present section, we will discuss the general features of fluid dynamics and Statistical physics, and their interconnection.

Basic evolution equations for the fluid dynamic variables $(\rho, \mathbf{v})$ were derived in previous sections. For illustration, let us take air as a sample fluid. As a first approximation, we will neglect the internal friction and assume that the surface force acting on the air comes from the pressure $P(\mathbf{r})$ alone. We can then write the stress tensor $\mathbb{T}$ in the form $-P\mathbb{E}$. From this we obtain

$$\begin{aligned}
\nabla \cdot \mathbb{T} &= -\nabla \cdot \left[P(\mathbf{r})\mathbb{E} \right] \\
&= -\left(\mathbf{i}\frac{\partial}{\partial x} + \mathbf{j}\frac{\partial}{\partial y} + \mathbf{k}\frac{\partial}{\partial z} \right) \cdot \left[P(\mathbf{ii} + \mathbf{jj} + \mathbf{kk}) \right] \\
&= -\left(\mathbf{i}\frac{\partial P}{\partial x} + \mathbf{j}\frac{\partial P}{\partial y} + \mathbf{k}\frac{\partial P}{\partial z} \right) \equiv -\nabla P
\end{aligned} \tag{8.28}$$

Introducing this result in (8.27), we obtain the equation of motion for the air as follows:

$$\rho \frac{\partial \mathbf{v}}{\partial t} + \frac{1}{2}\rho \nabla \mathbf{v}^2 - \rho \mathbf{v} \times (\nabla \times \mathbf{v}) = \mathbf{f} - \nabla P \tag{8.29}$$

The gravitational force (on Earth) per unit volume, **f** may be represented by

$$\mathbf{f} = -\rho g \mathbf{k} \tag{8.30}$$

where **k** is the unit vector pointing upwards and g the gravity constant.

Counting the three components of the velocity vector field **v**, we have four basic field variables, ρ, v_x, v_y, and v_z, all of which appear in (8.29). The pressure P ordinarily is treated as an unknown. We, therefore, have five unknowns in all. To complete the fluid-dynamic description, we may supplement (8.29) by two other equations, the equation of continuity:

$$\frac{\partial \rho}{\partial t} + \nabla \cdot (\rho \mathbf{v}) = 0 \tag{8.31}$$

and the *equation of state*

$$P = P(\rho) \tag{8.32}$$

which relates the pressure P to the density ρ.

The set of (8.29), (8.31) and (8.32) constitute five equations for five unknowns (ρ, v_x, v_y, v_z, P), and may be solved subject to given boundary and initial conditions.

In fluid dynamics, the equation of state (8.32) must be given. It may be presented as an empirical law, such as *Boyle's law*:

$$P = \text{constant} \times \rho \tag{8.33}$$

In statistical mechanics, we derive this equation, starting with a Hamiltonian of the system under consideration.

An important macroscopic property that distinguishes a liquid from a gas is the fact that the volume occupied by a quantity of liquid does not change much under large variations in pressure. In more precise terms, the (isothermal) compressibility defined by

$$\kappa \equiv -\frac{1}{V}\left(\frac{\partial V}{\partial P}\right)_T \tag{8.34}$$

is much smaller for a liquid than for a gas.

An ideal fluid with zero compressibility is called an *incompressible fluid*. Description of an incompressible fluid is much simpler than that of a real fluid, and will now be reviewed.

For the incompressible fluid ($\kappa = 0$), we obtain

$$\frac{\partial \rho}{\partial P} = \frac{\partial}{\partial P}\left(\frac{M}{V}\right) = -\frac{M}{V^2}\frac{\partial V}{\partial P} = \kappa \rho = 0 \tag{8.35}$$

As a fluid flows, it is subject to varying pressures. Equation (8.35) means that the density ρ of the incompressible fluid does not change with the pressure P. Thus, the mass density ρ does not change with time if we follow the fluid through a small time dt. Since the mass, starting at space-time $(\mathbf{r}, t)$, will have moved to $(\mathbf{r} + \mathbf{v}dt, t + dt)$, this means that

$$\rho(\mathbf{r}, t) = \rho(\mathbf{r} + \mathbf{v}dt, t + dt) \tag{8.36}$$

Expanding the rhs in powers of dt, we obtain

$$\rho(\mathbf{r} + \mathbf{v}dt, t + dt) = \rho(\mathbf{r}, t) + \left[\frac{\partial \rho}{\partial t} + \mathbf{v} \cdot \frac{\partial \rho}{\partial \mathbf{r}}\right] dt + \ldots \tag{8.37}$$

Since dt is arbitrary, the coefficients in front must vanish. We therefore obtain

$$\frac{\partial \rho}{\partial t} + \mathbf{v} \cdot \nabla \rho = 0 \tag{8.38}$$

By comparing this equation with the continuity equation:

$$\frac{\partial \rho}{\partial t} + \nabla \cdot (\rho \mathbf{v}) = \frac{\partial \rho}{\partial t} + \rho \nabla \cdot \mathbf{v} + \mathbf{v} \cdot \nabla \rho = 0$$

We observe that

$$\nabla \cdot \mathbf{v} \equiv \operatorname{div} \mathbf{v} = 0 \tag{8.39}$$

This equation characterizes the incompressible flow.

Because of its very small compressibility, a liquid will move like an incompressible fluid, if it is not subjected to excessively high pressures. (For the treatment of a sound wave propagating in a fluid, however, we cannot regard the fluid as incompressible.)

Real air, or for that matter, any real fluid, suffers a viscous resistance when its motion is not uniform. As illustration, let us suppose that a fluid flows along the x-direction near a wall as shown in Figure 8.3.

If a velocity gradient exists in the y-direction, that is,

$$\frac{\partial v_x}{\partial y} = \text{finite} \tag{8.40}$$

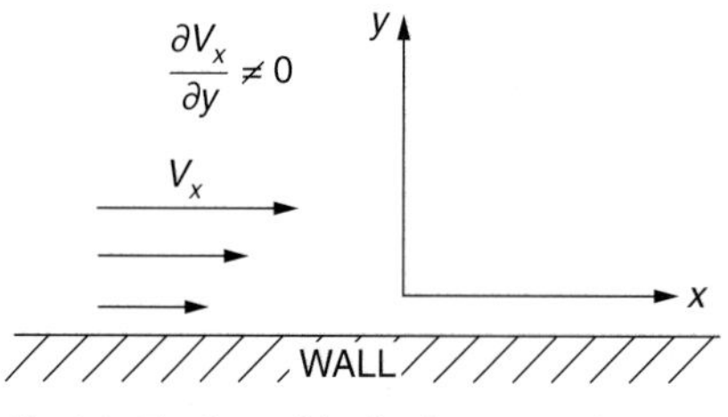

Fig. 8.3 The flow of the fluid is not uniform near the wall, and is subjected to a viscous resistance.

the fluid will experience a viscous stress. Experiments indicate that the viscous force per unit area perpendicular to the y-axis, T_{yx}, is proportional to the velocity gradient $\partial v_x/\partial y$, and directed in such a way as to reduce the gradient. We may write this relation as

$$T_{yx} = -\eta \frac{\partial v_x}{\partial y} \tag{8.41}$$

The constant of proportionality η is called the *viscosity coefficient*. This η depends on the material constituting the fluid, and in general also depends on the density and temperature. In fluid-dynamic problems, the value of η must be given. The microscopic derivation and calculation of a transport coefficient such as the viscosity coefficient as a function of temperature and density is one of the main purposes of nonequilibrium statistical mechanics.

For a viscous fluid, the stress tensor $\mathbb{T}$ generally has nine nonvanishing components:

$$\mathbb{T} = \begin{pmatrix} T_{xx} & T_{xy} & T_{xz} \\ T_{yx} & T_{yy} & T_{yz} \\ T_{zx} & T_{zy} & T_{zz} \end{pmatrix} \tag{8.42}$$

and the equation of motion for the fluid becomes quite complicated.

Problem 8.3.1

From the definition equation (8.41) determine the dimensions of the viscosity coefficient.

9
Irreversible Processes

Viscous flow and diffusion are significant irreversible processes. So is heat conduction. These processes are discussed in this chapter. A simple kinetic theory in terms of a mean free path is developed. Ohm's law connecting the charge current and the electric field is discussed in terms of a mean free time.

9.1
Irreversible Phenomena, Viscous Flow, Diffusion

Let us consider a quantity of water contained in a U-tube. At some initial time the water levels on both sides of the tube are made unequal in height. This state is shown in Figure 9.1 (State A). The water levels will then oscillate up and down under the action of the gravitational force until they come to rest at equal heights as shown in Figure 9.1 (State B).

The phenomenon we have just described is an example of an *irreversible* process, as the water in the state B will never start moving on its own and revert back to state A (unless it is forced by outside means). If we use a macroscopic description we may say that friction between water and glass, and again between water and water, cause the oscillation to dampen out. Any real fluid experiences an internal friction called a *viscous stress* when its motion is not uniform.

Let us consider another irreversible phenomenon: the mixing of water and ink. In Figure 9.2 (State A) a small amount of ink is added to the water in a beaker. We observe that after a sufficient length of time, the ink spreads and mixes with the water, reaching the final state of uniform distribution shown.

We may say that *diffusion* took place, negating the nonuniform concentration of ink. In more quantitative terms, it is experimentally observed that the current density of ink, **j**, is proportional to the gradient of the ink density n, that is,

$$\mathbf{j}(\mathbf{r}, t) = -D \, \nabla n(\mathbf{r}, t) \tag{9.1}$$

where the proportionality constant D is called the *diffusion coefficient*. This quantity D depends on the material (ink) as well as on the temperature and the solvent (water) density. The relation (9.1) is called *Fick's Law*.

Mathematical Physics. Shigeji Fujita and Salvador V. Godoy
Copyright © 2010 WILEY-VCH Verlag GmbH & Co. KGaA, Weinheim
ISBN: 978-3-527-40808-5

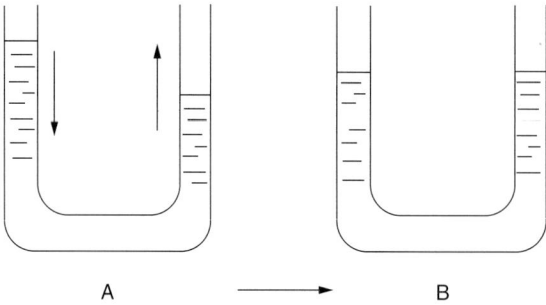

Fig. 9.1 The water in the U-tube moves up and down before coming to rest, eventually reaching the final state B where the water levels are at equal heights.

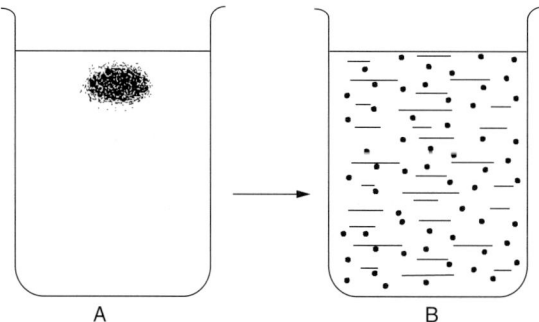

Fig. 9.2 The ink added to the water in a beaker will diffuse, and eventually reach the state of uniform concentration shown in (B).

Let us now look at these two phenomena: viscous damping and diffusion, from the microscopic point of view. In both cases it is clear that the motion of water molecules and ink particles is responsible for the phenomena. Therefore, if we describe this motion properly, we should be able to explain both viscosity and diffusion. We know, however, that the moving particles obey mechanical laws that allow reversible motion. In other words, no irreversible motion at all would seem to be possible. We, thus, face an apparent paradox. How is it possible to explain these irreversible phenomena on a molecular basis? Indeed, this is a fundamental question in nonequilibrium statistical mechanics.

To fully answer this question is quite a difficult task. We first note the important fact that the modes of description used in discussing these phenomena are quite different. In describing irreversible viscous flow, the basic variables used are the stress tensor $\mathbb{T}$ and the velocity field $\mathbf{v}$, both of which are *macroscopic quantities*, that is *quantities that are independent of the coordinates and momenta of the particles*. On the other hand, in speaking of reversible mechanical motion, the basic variables used are the coordinates and momenta of the particles. In fact, irreversible processes are encountered only when we describe them at the macroscopic level.

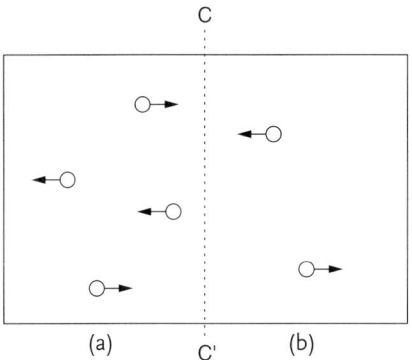

Fig. 9.3 If the particles flow out in all directions with no preference, there will be more particles crossing the imaginary boundary CC' in the (a) to (b) direction than in the opposite direction.

Important steps for the understanding of the fundamental question of irreversibility were undertaken in the last century by Boltzmann (Ludwig Boltzmann 1844–1906), one of the founding fathers of statistical mechanics. The method of the Boltzmann equation, is not only helpful for the understanding of irreversible processes but also provides us with almost exact calculations of transport coefficients for a dilute gas.

In order to clearly understand the phenomenon of diffusion let us look at the following simple situation. Imagine that four ink particles are in space (a), and two ink particles are in space (b), as shown in Figure 9.3. Assuming that both spaces (a) and (b) have the same volume, we may say that the density of ink is higher for that of (b). We assume that half of the particles in each space will be heading toward the boundary CC'. This situation is schematically shown in Figure 9.3. It is then natural to expect that in due time two particles would cross the boundary CC' from (a) to (b), and one particle from (b) to (a). This means that more particles would pass the boundary from (a) to (b), that is, from the side of high density to that of low density. This is, in fact, the cause of diffusion.

The essential points in the above arguments are the (reasonable) assumptions that (a) the particles flow out from a given space in all directions with the same probability, and (b) the rate of this outflow is proportional to the number particles contained in that space. In the present case the condition (a) will be assured by the fact that each ink particle *collides* with water molecules frequently so that it may lose the memory of how it entered the space originally and may leave with no preferred direction. The condition (b) is likely to be met if the density of ink particles is kept low so that the effect of collisions between ink particles may be ignored.

The viscous stress can also be understood in a similar manner. Picture two spaces (a) and (b) with the hypothetical boundary CC' as shown in Figure 9.4. Assume that there is a velocity gradient across CC', such that the molecules in space (a) have greater x-component velocities on the average than those in space (b): $\langle v_X \rangle_a > \langle v_X \rangle_b$. In a short time some molecules near the boundary CC' would cross

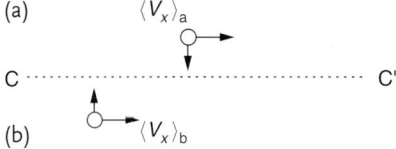

Fig. 9.4 By assumption, the molecules in space a have greater x-component velocities than the molecules in space b: $\langle v_x \rangle_a > \langle v_x \rangle_b$. If molecules are allowed to cross the hypothetical boundary CC′ in both directions with the same probability, a net momentum will be transported across the boundary, generating a viscous force.

it from (a) to (b), and other molecules would do so in the opposite direction; this crossing will take place with the same probability if we assume that the overall current crossing CC′ is zero. Since these molecules carry momenta with them, they will cause transport of the x-component of the momentum across the boundary. Imagine a typical molecule 1, which was in (a) and that carried the momentum component $m \langle v_x \rangle_a$, crossing the boundary while another molecule 2 that was in (b) and that carried the momentum component $m \langle v_x \rangle_b$ crosses in the same time. In this case, the net momentum $m \langle v_x \rangle_a - m \langle v_x \rangle_b$ will be transported across the boundary CC′ from (a) to (b). Looking at this macroscopically, the change in momentum must be caused by a force according to Newton's second law of motion. This force can be identified as the *viscous force*, which acts across the boundary surface CC′.

A heat flow is an important irreversible phenomenon. Let us consider a dilute gas with a nonuniform temperature distribution. It is observed that the heat flow **q** obeys *Fourier's law*:

$$\mathbf{q} = -K \nabla T \tag{9.2}$$

where the proportionality factor K is called the *heat conductivity*. The irreversible nature of the heat flow can be understood by drawing a diagram similar to Figure 9.4, with the assumption that the moving particles carry energies. The energy transfer from the high-temperature to the low-temperature region is regarded as the heat flow.

9.2
Collision Rate and Mean Free Path

Let us consider a particle moving through a medium containing n molecules per unit volume. If the particle proceeds with a speed v, it will sweep the volume $(v \, dt) \times \sigma = v \sigma dt$ during the time interval dt, where σ represents the total cross section. See Figure 9.5. The particle would collide with any molecule if the latter lies within the cylinder. Now, the number of molecules in the cylindrical volume $v \, dt$ is $n(v \sigma dt) = nv\sigma dt$. Dividing this number by dt, we obtain the *number of*

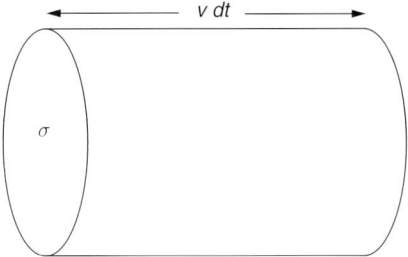

Fig. 9.5 A particle moving with a speed v sweeps the volume $(v\,dt)\sigma$ during the time dt, where σ represents the total cross section.

collisions per unit time

$$\omega_c = nv\sigma \tag{9.3}$$

This quantity ω_c is called the *collision rate*. Note that this collision rate depends linearly on the speed v, the number density n and the cross section σ.

The above consideration may be applied to the molecular collision in a gas. In this case, the particle in question is simply one of the molecules. Let us estimate the average collision rate for a typical gas. For a gas of neon, the interaction potential between two molecules has a range of a few Angstroms,

$$R(\text{range}) \simeq 2 \times 10^{-8}\,\text{cm} = 2\,\text{Å} \tag{9.4}$$

Therefore, the total cross section σ has the following order of magnitude:

$$\sigma = \pi R^2 = 3.14 \times (2 \times 10^{-8})^2\,\text{cm}^2 = 1.2 \times 10^{-15}\,\text{cm}^2 \tag{9.5}$$

A typical molecule will have a kinetic energy of the order of $3k_B T/2$. Therefore, it has the thermal speed

$$(v)_{\text{thermal}} = \left\langle \left(\frac{2\varepsilon}{m}\right)^{1/2} \right\rangle = \left(\frac{2 \cdot \frac{3}{2} k_B T}{m}\right)^{1/2} \tag{9.6}$$

Using the data, $T = 273$ K and $m(\text{neon}) = 20 m_p$ where m_p is the proton mass, we then obtain

$$(v)_{\text{thermal}} = 5.9 \times 10^4\,\text{cm s}^{-1} = 590\,\text{m s}^{-1} \tag{9.7}$$

It is interesting to note that this molecular speed has the same order of magnitude as the sound speed $340\,\text{m s}^{-1}$. At $0\,°\text{C}$ and 1 atmospheric pressure, the number of molecules per cm^3 is

$$n = 2.69 \times 10^{19}\,\text{cm}^{-3} \tag{9.8}$$

If we substitute these values from (9.5)–(9.8) into (9.3), we obtain

$$\omega_c = 1.91 \times 10^9\,\text{s}^{-1} \tag{9.9}$$

Here, we see that molecules collide with each other an enormous number of times per second.

The inverse of the collision rate defined by

$$\tau_c \equiv \frac{1}{\omega_c} = \frac{1}{n \langle v \rangle_{\text{thermal}} \sigma} \tag{9.10}$$

is called the *average time between collisions*. Substituting the numerical value from (9.9), we get

$$\tau_c = \frac{1}{1.91 \times 10^9 \text{ s}^{-1}} = 5.22 \times 10^{-10} \text{ s} \tag{9.11}$$

Let us compare this time τ_c with the $\tau_{\text{duration}} \equiv \tau_d$, that is the average time that the molecule spends within the range R of another particle. This time is defined by

$$\tau_d \equiv \frac{R}{\langle v \rangle_{\text{thermal}}} \tag{9.12}$$

Using the numerical values from (9.4) and (9.7), we have

$$\tau_d = 3.4 \times 10^{-13} \text{ s} \tag{9.13}$$

Comparison between (9.11) and (9.13) shows that

$$\tau_c \gg \tau_d \tag{9.14}$$

This means that in a typical gas the molecules move freely most of the time, and occasionally collide with each other.

By multiplying the thermal speed $\langle v \rangle_{\text{thermal}}$ by the average time between collisions τ_c, we obtain

$$\langle v \rangle_{\text{thermal}} \times \tau_c \equiv l \quad \text{(mean free path)} \tag{9.15}$$

This quantity l, called the *mean free path*, gives a measure of the distance that a typical molecule covers between successive collisions. From (9.10) and (9.15), we obtain

$$l = \langle v \rangle_{\text{thermal}} \times \tau_c = \frac{\langle v \rangle_{\text{thermal}}}{n \langle v \rangle_{\text{thermal}} \sigma} = \frac{1}{n\sigma} \tag{9.16}$$

Note that the mean free path does not depend on the speed of the particle, and therefore has a value independent of temperature. Introducing the numerical values from (9.5) and (9.8), we obtain

$$l = 3.1 \times 10^{-5} \text{ cm} = 3100 \text{ Å} \gg R \sim 2 \text{ Å} \tag{9.17}$$

comparing this and (9.3), we see that the mean free path is about three orders of magnitude greater than the force range.

Problem 9.2.1

Using the numerical values introduced in the present section estimate, for any given particle:
1. The probability of being within the force range of another particle, and
2. The probability of being within the force range of two particles simultaneously.

9.3
Ohm's Law, Conductivity, and Matthiessen's Rule

Let us consider a system of free electrons moving in a potential field of impurities that act as scatterers. The impurities are, by assumption, distributed uniformly in space.

Under the action of an electric field **E** pointed along the positive x-axis, a classical electron will move in accordance with the equation of motion:

$$m\frac{dv_x}{dt} = -eE \tag{9.18}$$

In the absence of the impurity potential this gives rise to a uniform acceleration and therefore a linear change in the velocity along the direction of the field:

$$v_x = -\frac{e}{m}Et + v_x^0 \tag{9.19}$$

where v_x^0 is the x-component of the initial velocity. The velocity increases indefinitely and leads to infinite conductivity. In the presence of the impurities, this uniform acceleration will be interrupted by scattering. When the electron hits a scatterer (an impurity), the velocity will suffer an abrupt change in direction and grow again following (9.19) until the electron hits another scatterer. Let us denote the *average time between successive scatterings* by τ_f, which is also called the *mean free time*. The order of magnitude of the average velocity $\langle v_x \rangle$ is then given by

$$\langle v_x \rangle = -\frac{e}{m}E\tau_f \tag{9.20}$$

In arriving at this expression, we assumed that the electron loses the memory of its preceding motion every time it hits the scatterer, and the average of the velocities just after the collisions vanishes:

$$\langle v_x^0 \rangle = 0 \tag{9.21}$$

The charge current density (average current per unit area), j_x, is given by

$$j_x = \text{(charge)} \times \text{(number density)} \times \text{(average velocity)} = -en\langle v_x \rangle$$

$$= \frac{e^2}{m}n\tau_f E \tag{9.22}$$

where n is the number density of electrons.

According to *Ohm's law* the current density j_x is proportional to the applied field E when this field is small:

$$j_x = \sigma E \tag{9.23}$$

The proportionality factor σ is called the *electrical conductivity*. It represents the facility with which the current is generated in response to the electric field. Comparing the last two equations, we obtain

$$\boxed{\sigma = \frac{e^2}{m} n \tau_f} \tag{9.24}$$

This formula is very useful in the discussion of the charge-transport phenomena. The inverse-law mass dependence means that the ion contribution to the electric conductivity will be smaller by at least three orders of magnitude than the electron contribution. Notice that the conductivity is higher if the number density is greater and/or if the mean free time is greater.

The inverse of the mean free time τ_f:

$$\Gamma \equiv \frac{1}{\tau_f} \tag{9.25}$$

is called the *rate of collision* or the *collision rate*. Roughly speaking, this represents the mean frequency with which the electron is scattered by impurities. The collision rate Γ is given by

$$\Gamma = n_I v A \tag{9.26}$$

where n_I and A are, respectively, the density of scatterers and the scattering cross section. We denote the cross section by A. The symbol σ stands for the conductivity in this section.

If there is more than one kind of scatterer, the rate of collision may be computed by the addition law:

$$\begin{aligned} \Gamma &= n_1 v A_1 + n_2 v A_2 + \cdots \\ &= \Gamma_1 + \Gamma_2 + \cdots \end{aligned} \tag{9.27}$$

This is often called *Matthiessen's rule. The total rate of collision is the sum of collision rates computed separately for each kind of scatterer.*

Historically, and also in practice, the analysis of resistance data for a conductor is done as follows: If the electrons are scattered by impurities and again by phonons, the total resistance will be written as the sum of the resistances due to each separate cause of scattering:

$$R_{\text{total}} = R_{\text{impurity}} + R_{\text{phonon}} \tag{9.28}$$

This is the original statement of Matthiessen's rule. In further detail, the electron–phonon scattering depends on temperature because of the change in the phonon

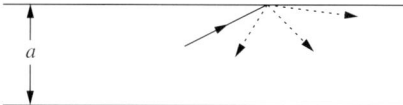

Fig. 9.6 Problem 9.3.1.

population, while the effect of the electron–impurity scattering is temperature independent. By separating the resistance into two parts, one temperature dependent and the other temperature independent, we may apply Matthiessen's rule. Since the resistance R is inversely proportional to the conductivity σ, (9.24) and (9.27) together imply (9.28).

Problem 9.3.1

Free electrons are confined within a long rectangular planar strip shown below. Assume that each electron is *diffusely scattered* at the boundary so that it may move in all allowed directions without preference after the scattering. Find the mean free path along the long strip. Calculate the conductivity σ.

Problem 9.3.2

Do the same as in Problem 9.3.1. for the case in which electrons are confined within a long circular cylinder.

10
The Entropy

After reviewing the foundations of thermodynamics we establish Clausius's theorem. Using this theorem as a base we then introduce the entropy as a function of the thermodynamic state. The entropy plays a central role in many thermodynamic discussions.

10.1
Foundations of Thermodynamics

Let us consider a quantity of a gas of the same atoms as neons in equilibrium. The obvious thermodynamic variables are the volume V, the pressure P and the absolute temperature T. These variables satisfy an *equation of state*:

$$f(P, V, T) = 0 \tag{10.1}$$

For a mole of an ideal gas this equation takes a special form:

$$PV = RT \tag{10.2}$$

where

$$R = 8.314 \text{ J K}^{-1} = 1.986 \text{ cal K}^{-1} \tag{10.3}$$

is the gas constant, and the absolute temperature T is related to the Celsius temperature t by

$$T = 273.16 + t \tag{10.4}$$

The (internal) energy E of the system can be increased by the heat Q absorbed and by the work done W to the system. For a small amount of heat and work, this is represented by

$$dE = dQ + dW \tag{10.5}$$

which is known as the *First Law of Thermodynamics*, or the *energy conservation law*. Joule has established the value of heat equivalent to work as

$$1 \text{ cal} = 4.181 \times 10^7 \text{ ergs} = 4.181 \text{ J} \tag{10.6}$$

Mathematical Physics. Shigeji Fujita and Salvador V. Godoy
Copyright © 2010 WILEY-VCH Verlag GmbH & Co. KGaA, Weinheim
ISBN: 978-3-527-40808-5

For a slow reversible process, the differential work can be expressed as

$$dW = -P\,dV \tag{10.7}$$

If the system absorbs heat, its temperature generally increases. If the volume is kept constant in the process, the heat capacity at constant volume, C_V, is defined by

$$C_V \equiv \left(\frac{dQ}{dT}\right)_V \tag{10.8}$$

Using (10.5) and (10.7), we obtain

$$C_V = \frac{d}{dT}(dE - dW)|_V = \left(\frac{\partial E}{\partial T}\right)_V \tag{10.9}$$

The Pacific Ocean has an enormous heat capacity because of its volume. If a ship could carry an ideal heat-to-work converter (engine) that absorbs heat from the sea and transform it into work, then, the ship could travel without carrying any conventional fuels. *The Second Law of thermodynamics* prohibits such an ideal converter. In more precise terms, Kelvin's statement of the Second Law is:

Kelvin's statement: A thermodynamic transformation whose only final result is to convert heat extracted from a thermal reservoir into positive work is impossible.

Such an imaginary impossible process is schematically shown in Figure 10.1. The Second Law can also be expressed as follows:

Clausius's statement: A thermodynamic transformation whose only final result is to transfer heat from a cold reservoir to a hot reservoir, is impossible.

The impossible ideal transformation is schematically shown in Figure 10.2.

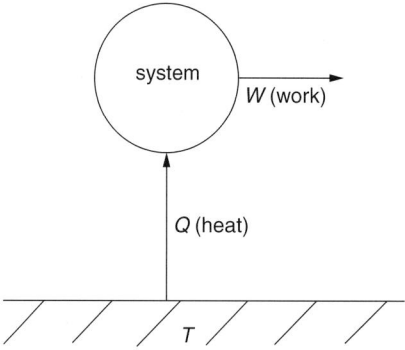

Fig. 10.1 In the imaginary process that violates Kelvin's statement of the Second Law, heat extracted from a reservoir is fully converted into work.

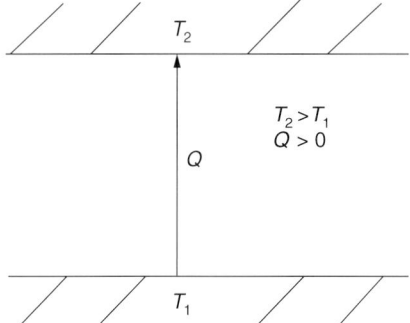

Fig. 10.2 In the imaginary process that violates Clausius' statement of the Second Law, heat is transferred from a cold to a hot thermal reservoir.

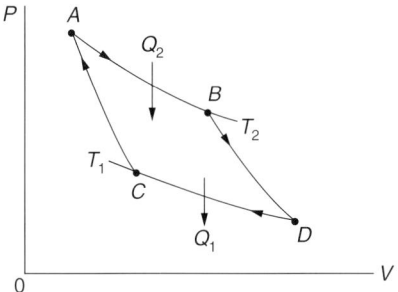

Fig. 10.3 The Carnot cycle is run along isothermal and adiabatic lines. The work done, represented by the area bounded by the four lines, equals the net amount of heat absorbed $Q_2 - Q_1$.

We wish to prove the equivalence of the two statements. We first show that denying Kelvin's statement K means the denial of Clausius's statement C. (According to logic, this implies that the statement C leads to the statement K.) In contradiction to K, let us convert heat taken from a cold reservoir into work, leaving no other traces. This work may be converted into heat by friction, and this heat may be delivered to a hot reservoir. The combined processes can be regarded as a net transfer of heat from the cold to the hot reservoir, which contradicts C. Conversely, the denial of C leads to the denial of K, which will be shown in the next section.

10.2
The Carnot Cycle

Let us consider a gas whose state can be represented in a P–V diagram. In Figure 10.3, we draw two isothermal lines AB and CD corresponding to two temperatures, T_2 and $T_1(<T_2)$, and two adiabatic lines AC and BD. The reversible transformation performed along the cyclic path ABDCA is called the *Carnot cycle*.

A Carnot cycle may be performed with a gas in a cylinder with a movable piston. During the cycle, the system (gas) absorbs heat Q_2 from the T_2-reservoir along the path AB and emits heat Q_1 to the T_1-reservoir along the path DC. The net amount of heat absorbed by the system is $Q_2 - Q_1$. The work W done by the system during the cycle ABDCA is equal to the area bounded by the cycle in Figure 10.3. Applying the First Law, we obtain

$$W = Q_2 - Q_1 \tag{10.10}$$

This equation means that only part of the heat absorbed by the system Q_2 from the high-temperature (T_2) reservoir is transformed into work during the Carnot cycle; the rest of the heat Q_1, instead of being transformed into work, is surrendered to the low-temperature (T_1) reservoir.

A device in which heat is converted into mechanical (or electrical) work is called a *heat engine*. The Carnot cycle just described can be regarded as a cyclic transformation of an idealized heat engine working between two temperatures (T_1, T_2).

We define the *efficiency* η of any heat engine as the ratio of the work W generated to the heat Q_2 absorbed at the high-temperature reservoir:

$$\eta \equiv \frac{\text{work generated}}{\text{heat absorbed}} = \frac{W}{Q_2} \tag{10.11}$$

Substitution from (10.10) yields

$$\eta_C = \frac{Q_2 - Q_1}{Q_2} = 1 - \frac{Q_1}{Q_2} \tag{10.12}$$

which represents the efficiency of the *Carnot engine*.

As the Carnot cycle is reversible, all of the processes described may be performed in the opposite direction. When this is done, the system has an amount of work equal to W done on it, absorbs the heat Q_1 from the low-temperature reservoir, and gives up the heat Q_2 to the high-temperature reservoir.

As a first application of the Carnot cycle, we will complete the proof of the equivalence between Kelvin's and Clausius's statements of the Second Law.

Let us assume, in contradiction to Clausius's statement, that it is possible to transfer a positive amount of heat Q_2 from a low-temperature reservoir to a high-temperature reservoir with no other changes in the state of the system and its environment. With the aid of a Carnot cycle, we could, then, extract the same amount of heat Q_2 from the high-temperature reservoir and produce an amount of work W. Since the high-temperature reservoir receives and gives up the same amount of heat, it undergoes no net change. Thus, the combined processes would have generated the work W from the net heat $Q_2 - Q_1$ extracted from the low-temperature reservoir as the only final result. This is contrary to Kelvin's statement of the Second Law.

10.3
Carnot's Theorem

In the Carnot cyclic engine the work $W = Q_2 - Q_1$ is generated by absorbing heat Q_2 from the T_2-reservoir and emitting heat Q_1 to the T_1-reservoir. The efficiency η_C of the Carnot engine is then $1 - Q_1/Q_2$. It will be shown that this efficiency η_C can be expressed simply by

$$\boxed{\eta_C = 1 - \frac{T_1}{T_2}} \tag{10.13}$$

Note that this expression does not depend on the amount of heat exchanged with the thermal reservoir but only on the temperatures of the two reservoirs.

This remarkable feature of Carnot's engine can be included as part of an even more remarkable theorem stated as follows:

Carnot's theorem:
1. *No engine operating between two temperatures T_2 and $T_1 (< T_2)$ is more efficient than a Carnot engine. That is, the efficiency η of any engine cannot exceed the Carnot efficiency η_C:*

$$\eta \leq \eta_C \tag{10.14}$$

2. *All reversible cyclic engines working between two temperatures T_1 and T_2, have the same efficiency equal to the Carnot efficiency η_C.*
3. *The Carnot efficiency is given by $\eta_C = 1 - T_1/T_2$.*

The theorem may be proved as follows:

Let us assume that an engine A operates during a cycle by absorbing heat Q'_2 from the T_2-reservoir, and emitting heat Q'_1 to the T_1-reservoir, as indicated in Figure 10.4a. This engine generates the work $W' = Q'_2 - Q'_1$, and has the efficiency

$$\eta = 1 - \frac{Q'_1}{Q'_2} \tag{10.15}$$

Imagine now a reversed Carnot cycle that absorbs Q_1 from the T_1-reservoir, delivers Q'_2 to the T_2-reservoir, and generates the negative work $-W = Q_1 - Q'_2$. A Carnot engine that operates in this manner is schematically shown in Figure 10.4b. The efficiency of this Carnot engine is given by

$$\eta_C = 1 - \frac{Q_1}{Q'_2} \tag{10.16}$$

We assume that the reversed Carnot cycle is completed during the same time as the cycle performed by engine A.

We now consider the combination of these two cycles. The T_2-reservoir emits and receives the same amount of heat Q'_2. The total work of the combined cycles

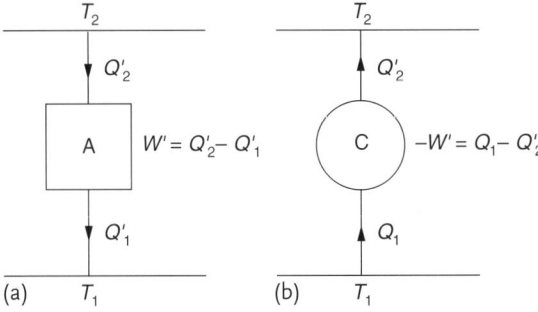

Fig. 10.4 The real engine A absorbs heat Q'_2, emits heat Q'_1, and generates the work equal to $Q'_2 - Q'_1$ in the cycle. The imaginary Carnot engine, run in reverse, returns the heat Q'_2 to the T_2-reservoir.

can be calculated by applying the First Law, and it is given by

$$W' - W = (Q'_2 - Q'_1) + (Q_1 - Q'_2) = Q_1 - Q'_1 \tag{10.17}$$

This work must be negative or equal to zero:

$$Q_1 - Q'_1 \leq 0 \tag{10.18}$$

otherwise it will contradict the Second Law (Kelvin's statement).

Using this inequality, we obtain

$$\eta - \eta_C = 1 - \frac{Q'_1}{Q'_2} - \left(1 - \frac{Q_1}{Q'_2}\right) = \frac{1}{Q'_2}(Q_1 - Q'_1) \leq 0$$

which establishes the proof of (a).

If our engine A can be operated in a reversible manner, we can, then, apply our arguments to the reverse processes, and obtain

$$\eta \geq \eta_C \tag{10.19}$$

Thus, the reversible engine must satisfy (10.13) and (10.19). This is possible only if $\eta = \eta_C$. This completes the proof of (b).

We stress that the proofs of (a) and (b) were carried out on very general grounds. In particular, nothing is specified about the nature of the gas used in the system. We only required that the engine A operates in a cycle between two temperatures.

The last part (c) of Carnot's Theorem can now be proved by considering a Carnot cycle operating with a mole of ideal gas. Let us suppose that the cycle is executed as shown in Figure 10.3. We first look at the isothermal expansion. From the fundamental relation $dE = dQ - PdV$ we obtain

$$dQ = dE + PdV = C_V dT + PdV = 0 + PdV = PdV$$

Integrating this along the path AB we obtain (using $PV = RT$)

$$\int_A^B dQ = \int_A^B P\,dV = \int_A^B \frac{RT_2}{V} dV = RT_2 \ln(V_B/V_A) = Q_2 \qquad (10.20)$$

which equals the heat received from the T_2-reservoir, Q_2.

By applying a similar consideration to the isothermal line DC, we obtain

$$Q_1 = RT_1 \ln(V_D/V_C) \qquad (10.21)$$

for the heat emitted to the T_1-reservoir. Dividing this by (10.20), we obtain

$$\frac{Q_1}{Q_2} = \frac{T_1 \ln(V_D/V_C)}{T_2 \ln(V_B/V_A)} \qquad (10.22)$$

The path BD corresponds to an adiabatic line along which the relation

$$TV^{R/C_V} = TV^{(C_P - C_V)/C_V} = TV^{\gamma - 1} = \text{constant} \qquad (10.23)$$

holds, where $\gamma \equiv C_P/C_V$. We, therefore, have

$$T_2 V_B^{\gamma - 1} = T_1 V_D^{\gamma - 1} \qquad (10.24)$$

The path CA is another adiabatic line. Applying (10.23), we obtain

$$T_1 V_C^{\gamma - 1} = T_2 V_A^{\gamma - 1} \qquad (10.25)$$

From the last two equations, we see that

$$\frac{V_D}{V_C} = \frac{V_B}{V_A} \qquad (10.26)$$

Introducing this result in (10.22), we find that

$$\frac{Q_1}{Q_2} = \frac{T_1}{T_2} \qquad (10.27)$$

Using this, we obtain the desired result,

$$\eta_C = 1 - \frac{Q_1}{Q_2} = 1 - \frac{T_1}{T_2} \qquad \text{Q.E.D.}$$

10.4
Heat Engines and Refrigerating Machines

According to Carnot's theorem, no engine working between two temperatures can have a higher efficiency than a Carnot engine. Thus, the Carnot efficiency $\eta_C =$

T_1/T_2 represents the highest possible efficiency for any engine working between T_1 and $T_2 > T_1$.

For most heat engines the low temperature T_1 is the temperature of the environment, and hence cannot be controlled. It is, therefore, desirable to have the temperature T_2 as high as possible. Hotter engines will work with higher efficiencies! This is in agreement with common experience.

For engines actually in use, such as automobile engines, cycles are far from reversible, and therefore, their efficiencies are considerably lower than the maximum efficiencies theoretically achievable. The efficiency of the usual piston-type engine can only approach about 10 per cent. The diesel engine may achieve as high as a 30 per cent efficiency.

In the Carnot cycle operated in the reverse direction, as shown in Figure 10.4b, and amount of heat Q_1 is extracted from the low-temperature T_1-reservoir. We may look at it as an ideal refrigerator if we regard the high temperature T_2 as the environment temperature. By solving

$$\frac{Q_1}{Q_2} = \frac{T_1}{T_2}, \quad \eta_C = \frac{W}{Q_2} = 1 - \frac{T_1}{T_2}$$

with respect to W, we obtain

$$W = Q_1 \left(\frac{T_1}{T_2} - 1 \right) \tag{10.28}$$

This equation indicates that the work needed to extract a fixed amount of heat Q_1 from a body at the temperature T_1 becomes greater as the temperature T_1 decreases. Accordingly, the operating cost of a refrigerator is higher if the temperature T_1 is set lower. Furthermore, if the temperature T_1 approaches absolute zero, the work W becomes infinitely large. This means that *reaching absolute zero temperature by any means is impossible*.

As in the case of heat engines, all actual refrigerating machines must involve irreversible processes. These machines are then considerably less efficient than the ideal refrigerator.

Problem 10.4.1

1. Derive (10.28).
2. Show that the real work needed is greater than the ideal work W given in (10.28).

10.5
Clausius's Theorem

We saw that a Carnot engine has an efficiency given by $\eta_C \equiv 1 - Q_1/Q_2 = 1 - T_1/T_2$. This means that

$$\frac{Q_1}{T_1} = \frac{Q_2}{T_2} \quad \text{(old notation)} \tag{10.29}$$

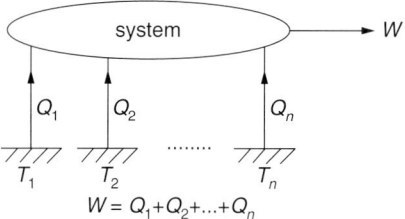

Fig. 10.5 An engine completes a cycle in *n* steps. In the *j*th step it receives (positive or negative) heat Q_j from the thermal reservoir T_j.

Let us recall that Q_2 is the heat absorbed by the system from the high-temperature reservoir T_2, and Q_1 is the heat delivered to the low-temperature reservoir T_1. Both Q_1 and Q_2 are positive.

We now introduce a new notation. We denote by Q_j ($j = 1, 2$ in this case) the heat exchanged between the system and the reservoir T_j, such that Q_j is positive if the heat is absorbed by the system, and negative if emitted. Using the new notation, we may rewrite (10.29) as

$$\frac{Q_1}{T_1} + \frac{Q_2}{T_2} = 0 \quad \text{(new notation)} \tag{10.30}$$

From now on, we will use this new notation. Equation (10.30) holds if the cycle is reversible. We have an inequality for an irreversible cycle:

$$\frac{Q_1}{T_1} + \frac{Q_2}{T_2} < 0 \quad \text{(irreversible cycle)} \tag{10.31}$$

We can express (10.30) and (10.31) together by

$$\frac{Q_1}{T_1} + \frac{Q_2}{T_2} \leq 0 \tag{10.32}$$

where the equality corresponds to the case of a reversible cycle.

Let us now consider an engine that undergoes a more general cycle. We suppose that during the cycle the system exchanges heat $Q_1, Q_2, \ldots, Q_n$ with the thermal reservoirs $T_1, T_2, \ldots, T_n$, respectively. An engine performing such a cycle is schematically shown in Figure 10.5. We note that some of the heats ($Q_1, Q_2, \ldots, Q_n$) must be negative. For example, for $n = 2$, the heat exchanged with the low-temperature reservoir, Q_1 is negative.

Extending (10.32), we have

$$\boxed{\sum_{j=1}^{n} \frac{Q_j}{T_j} \leq 0} \tag{10.33}$$

where *the equality holds if the cycle is reversible*.

This is *Clausius's Theorem*. The proof of the general case may be established as follows:

Let us introduce a heat reservoir with a temperature T_0, called the reservoir T_0, besides the n reservoirs that have different temperatures $(T_1, T_2, \ldots, T_n)$. Let us imagine n Carnot cycles $(C_1, C_2, \ldots, C_n)$ that operate between T_0 and $(T_1, T_2, \ldots, T_n)$, such that during the jth Carnot cycle the heat Q_j is delivered to the T_0-reservoir. (Q_j is the amount of heat received by the original system at the temperature T_j.) The amount of heat received from the reservoir T_0 for the Carnot cycle is definite and given by

$$\bar{Q}_j \equiv \frac{T_0}{T_j} Q_j \tag{10.34}$$

The Carnot engine that performs the cycle C_j is indicated in Figure 10.6.

We now consider a combined cycle consisting of the original cycle S and n Carnot cycles $(C_1, C_2, \ldots, C_n)$. The net exchange of heat at each of the n reservoirs $T_1, T_2, \ldots, T_n$, is zero; the reservoir T_j delivers the heat Q_j to the system in the original cycle S, but it receives the same amount of heat in the cycle C_j. The reservoir T_0, on the other hand, delivers heat equal to the sum of all $\bar{Q}_j$:

$$Q_0 \equiv \sum_{j=1}^{n} \bar{Q}_j = T_0 \sum_{j=1}^{n} \frac{Q_j}{T_j} \tag{10.35}$$

The net result is that our combined cycle receives the heat Q_0 from the reservoir T_0. But we know that in a cyclic transformation the work performed is equal to the total heat received by the system. Thus, the heat Q_0 received from the reservoir T_0 is transformed into work. If this Q_0 were positive, the cycle would have violated Kelvin's statement of the Second Law. Therefore, the heat Q_0 must be nonpositive:

$$Q_0 \equiv T_0 \sum_{j=1}^{n} \frac{Q_j}{T_j} \leq 0 \tag{10.36}$$

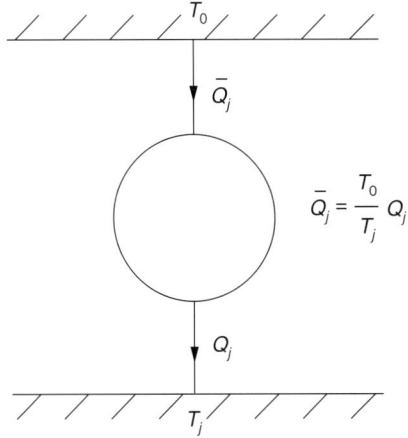

Fig. 10.6 A Carnot engine executes a cycle C_j such that it delivers the heat Q_j to the T_j-reservoir and receives the heat $\bar{Q}_j$ from the T_0-reservoir.

10.5 Clausius's Theorem

Dividing this by T_0 (> 0), we obtain (10.33).

If the original cycle is reversible, we can describe it in the opposite sense, in which case all Q_js will change sign. By considering the reversal of the combined cycle, we deduce that

$$\sum_{j=1}^{n} \frac{(-Q_j)}{T_j} \leq 0$$

or

$$\sum_{j=1}^{n} \frac{Q_j}{T_j} \geq 0 \tag{10.37}$$

Thus, for a reversible cycle, this inequality as well as the inequality (10.33) must be satisfied. This is possible only if the equality sign holds. In other words, we must have

$$\sum_{j=1}^{n} \frac{Q_j}{T_j} = 0 \quad \text{for a reversible cycle} \quad \text{Q. E. D.} \tag{10.38}$$

Clausius's theorem as stated deals with the case in which the system exchanges heat with a finite number of thermal reservoirs. Practically important cases are those in which the system exchanges heat with reservoirs that have a continuously varying temperature. We now formulate the theorem suitable for such cases.

Let us suppose that a system completes a cyclic transformation in a finite time t. We divide the time t into small intervals: $\{\Delta t_j\}$. At the jth interval Δt_j, the system is in contact with the reservoir of temperature T_j and exchanges heat in the amount of ΔQ_j. The sum $\sum_j \Delta Q_j / T_j$ over the cycle in the limit of fine divisions will be denoted by the integral:

$$\oint_{(t)} \frac{dQ}{T} \equiv \lim_{\text{fine time divisions}} \sum_{j}^{\text{cycle}} \frac{\Delta Q_j}{T_j} \tag{10.39}$$

Here, the circle on the integral sign denotes a cyclic transformation, and the symbol (t) denotes the *time-division integral*. Now, Clausius's theorem can be reformulated as follows:

$$\boxed{\oint_{(t)} \frac{dQ}{T} \leq 0} \tag{10.40}$$

For illustration, let us consider a gas that undergoes a reversible cycle. This reversible cycle may be represented by a directed closed curve C in the P–V plane as shown in Figure 10.7. Let us divide this curve into small pieces. The location of the jth piece represents a thermodynamic state, and in particular, should specify the temperature T_j of that state. The heat absorbed by the system along the piece will be denoted by ΔQ_j. We consider the sum $\sum_j \Delta Q_j / T_j$ (on a closed curve C). In the limit of fine divisions, this sum, by definition, becomes a *line integral* along the closed directed path C, and will be denoted by

$$\oint_C \frac{dQ}{T} \equiv \lim_{\text{fine curve divisions}} \sum_{\text{closed curve}} \frac{\Delta Q_j}{T_j} \tag{10.41}$$

For a reversible transformation, the line integral and the time-division integral are clearly equal to each other:

$$\oint_C \frac{dQ}{T} = \oint_{(t)} \frac{dQ}{T} \equiv \oint \frac{dQ}{T} \tag{10.42}$$

We will then drop the symbols C (curve) and (t) (time division) for such a case.

Applying now Clausius's theorem (10.40) to a reversible cycle, we obtain

$$\boxed{\oint \frac{dQ}{T} = 0 \quad \text{(reversible cycle)}} \tag{10.43}$$

If we have an irreversible change from a state A to another state B, then such a charge cannot be represented by a curve connecting A and B. Let us recall that irreversible changes may result from two causes: (1) If we have any dissipative effects, such as friction, viscosity, hysteresis, and so on or (2) if the conditions for quasistatic equilibrium are not satisfied, such as when work is done abruptly or if heat is exchanged with a body whose temperature is much different from the temperature of the gas. Then, the system undergoes an irreversible change of state. Since during such irreversible process the system cannot go through a succession of quasiequilibrium states, the integral defined in (10.39) for an irreversible cycle

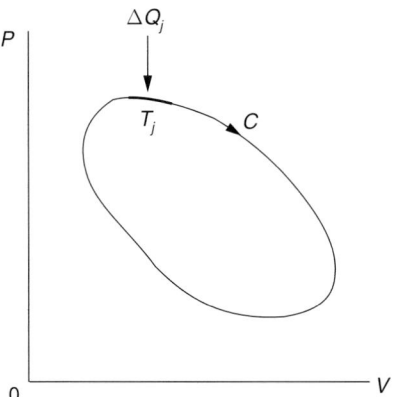

Fig. 10.7 A reversible cycle in the P–V diagram. The system receives the heat ΔQ_j in the small (darkened line) at the temperature T_j.

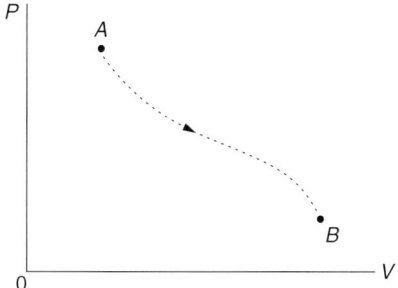

Fig. 10.8 An irreversible change of state from A to B is represented by a directed dotted line.

cannot be expressed in terms of the line integral of the type discussed above. In the text, an irreversible change of state from A to B will be represented by a dotted line directed from A to B as shown in Figure 10.8.

Problem 10.5.1

Consider an engine that completes a cycle in n steps, as indicated in Figure 10.5. Show that the engine must emit heat in some steps. This means that some Q_j must be negative.

HINT: Take the cases of $n = 1, 2, 3$, and extend your arguments for general n.

10.6
The Entropy

We consider reversible processes only in this section. According to Clausius's theorem, we have

$$\oint \frac{dQ}{T} = 0 \quad \text{(reversible cycle)} \tag{10.44}$$

This property can be restated in a different form. Let A and B be two equilibrium states of the gas, which are represented by two points in the P–V plane shown in Figure 10.9. Reversible transformations from the initial state A to the final state B can be represented by directed curves running from A to B. In Figure 10.9 two such paths AIB and AIIB, are shown:

Consider now two line integrals along the paths AIB and AIIB,

$$\int_{AIB} \frac{dQ}{T} \quad \text{and} \quad \int_{AIIB} \frac{dQ}{T}$$

We can show from (10.44) that these two integrals have the same value:

$$\int_{AIB} \frac{dQ}{T} = \int_{AIIB} \frac{dQ}{T} \equiv \int_A^B \frac{dQ}{T} \tag{10.45}$$

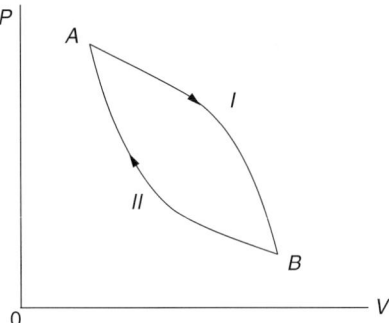

Fig. 10.9 The line integral of dQ/T along the reversible path AIB equals the same quantity along the path AIIB.

This means that *the value of the line integral from A to B depends only on the initial (A) and final (B) states and not on the intermediate states along the path*. The line integral having this property will be denoted by

$$\int_A^B \frac{dQ}{T}$$

To prove (10.45), we consider a cyclic transformation along the closed path AIBI-IA. Since this is a reversible cycle, we have, from (10.44),

$$\int_{\text{AIBIIA}} \frac{dQ}{T} = 0$$

This integral can be split into two integrals:

$$\int_{\text{AIB}} \frac{dQ}{T} + \int_{\text{BIIA}} \frac{dQ}{T} = 0$$

but

$$\int_{\text{BIIA}} \frac{dQ}{T} = -\int_{\text{AIIB}} \frac{dQ}{T}$$

because in the transformation from B to A along BIIA, dQ takes the same value except for the sign. Hence, we obtain (10.45).

The property (10.45) for reversible processes makes it possible to define a new function of state called the *entropy* of the system. The entropy, which plays a most significant role in thermodynamics, will be defined in the following manner.

Let us take a certain equilibrium state 0 of a system and call it the *reference state*. Let A be some other equilibrium state, and consider the integral along a reversible path from 0 to A:

$$S(A; 0) \equiv \int_0^A \frac{dQ}{T} \tag{10.46}$$

10.6 The Entropy

We have seen that such an integral depends only on the end states 0 and A. By looking at the reference state 0 as fixed we may regard $S(A;0)$ as a function of the sate A, and call it the *entropy of the sate A*.

Consider now two equilibrium states A and B, whose entropies are denoted by $S(A;0)$ and $S(B;0)$. Then, we can show that

$$\boxed{S(B;0) - S(A;0) = \int_A^B \frac{dQ}{T}} \tag{10.47}$$

where the integral is taken along any reversible path from A to B. See Figure 10.10.

To prove (10.47), we take a reversible path running from the state A to the reference state 0 and then back to the sate B. For this path, we have

$$\int_A^B \frac{dQ}{T} = \int_A^0 \frac{dQ}{T} + \int_0^B \frac{dQ}{T} \tag{10.48}$$

The second integral is $S(B;0)$. The first integral is $-S(A;0)$ since

$$\int_A^0 \frac{dQ}{T} = -\int_0^A \frac{dQ}{T} = -S(A;0)$$

Substituting these values for the integrals in (10.48) we obtain (10.47). Q.E.D.

It is clear form the proof that (10.47) holds for *any* reference state 0.

The definition (10.46) of the entropy requires the choice of a reference state. We can easily show that if, instead of 0, we choose a different state 0', then the new entropy $\bar{S}(A;0')$ is different from the old one $S(A;0)$ only by an additive constant (Problem 10.6.1):

$$\bar{S}(A;0') = S(A;0) + \text{constant} \tag{10.49}$$

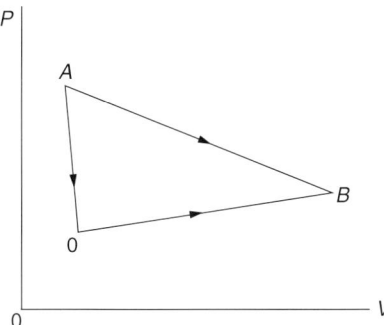

Fig. 10.10 The integral $\int_A^B dQ/T$ along the reversible path A → B equals the integral $\int_A^0 dQ/T$ along the path A → 0 plus the integral $\int_0^B dQ/T$ along the path 0 → B.

The entropy is therefore defined within an additive constant. In most applications, the entropy difference and not the entropy itself is needed for thermodynamic discussion. In such cases, the additive constant plays no role. Hereafter, we will drop designation of the reference state. In certain problems, however, the additive constant does play a significant role. For further discussions of the entropy constant, see the classic book by Fermi [1].

Finally, let us consider an infinitesimal transformation in which the entropy varies by dS and the system receives heat dQ at the temperature T. Then, we obtain

$$\boxed{dS = \frac{dQ}{T}} \tag{10.50}$$

That is, the change in the entropy for an infinitesimal reversible transformation is obtained by dividing the heat absorbed dQ by the temperature T.

Problem 10.6.1

Show that the entropies defined with two arbitrary chosen reference states differ only by a constant.

10.7
The Exact Differential

In the preceding section we have seen that (a) the entropy S is a function of the thermodynamic state, (b) a small change in entropy, dS, is related to a small change of heat dQ by

$$dS = \frac{dQ}{T} \tag{10.51}$$

and (c) the line integral along a reversible path from A to B in the P–V diagram,

$$\int_{ARB} dS \equiv \int_{ARB} \frac{dQ}{T} = S(B) - S(A) \tag{10.52}$$

depends on the end-points A and B only.

On the other hand, the integral of

$$dQ = dE + PdV \tag{10.53}$$

taken along the path ARB,

$$\int_{ARB} dQ = \int_{ARB} (dE + PdV) \tag{10.54}$$

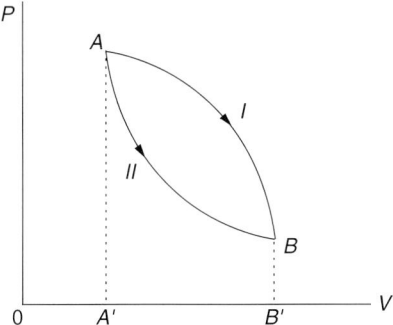

Fig. 10.11 The line integral $\int_{AIB} dQ$ is in general not equal to the line integral $\int_{AIIB} dQ$, while $\int_{AIB} dS = \int_{AIIB} dS = S(B) - S(A)$.

has no simple property like that expressed by (10.52).

In general, if a differential expression of two independent variables x and y of the form

$$df = M(x, y)dx + N(x, y)dy \tag{10.55}$$

is the differential of a certain function of x and y, then it is called an *exact differential*. In such a case, M and N must satisfy the following relation:

$$\frac{\partial M(x, y)}{\partial y} = \frac{\partial N(x, y)}{\partial x} \tag{10.56}$$

Conversely, if (10.56) is satisfied, then it is possible to integrate (10.56) and obtain a function f for which df is an exact differential. If the condition (10.56) is not satisfied, no such function f exists, and df cannot be regarded as the differential of a function of x and y. In such a case, the line integral of $df = Mdx + Ndy$ taken along a path from a point A to another point B on the xy-plane depends not only on the limits A and B but also on the path joining them. As for the two differential expressions (10.51) and (10.53), dS is a perfect differential, while dQ is not.

To see this more clearly let us take two points A and B on the P–V plane and connect them by two different paths as shown in Figure 10.11.

If we integrate dS along the two paths, we obtain the same result $S(B) - S(A)$ in both cases. If, on the other hand, we integrate dQ along these two paths, we obtain two results Q_I and Q_{II}, which generally are not equal.

This can be seen by applying the first Law ($dQ = dE + PdV$) to the two transformations AIB and AIIB. In fact, we find

$$Q_I = E(B) - E(A) + W_I \tag{10.57}$$

$$Q_{II} = E(B) - E(A) + W_{II} \tag{10.58}$$

where W_I and W_{II} are the amounts of work represented by the areas AIBB'A'A and AIIBB'A'A, respectively. Taking the difference, we obtain

$$Q_I - Q_{II} = W_I - W_{II} = \text{area AIBIIA} \tag{10.59}$$

Hence, the difference $Q_I - Q_{II}$ in general does not vanish. Thus, dQ is not a perfect differential. *We cannot regard the heat Q as a function of the thermodynamic state.*

Problem 10.7.1

Check if the following differentials are perfect. If so, find the functions f.
1. $df = y\,dx + x\,dy$
2. $df = y\,dx + 2x\,dy$
3. $df = 2xy\,dx + y^2\,dy$

Reference

1 Fermi, E. (1950) *Thermodynamics*, Dover Publications, New York.

11
Thermodynamic Inequalities

11.1
Irreversible Processes and the Entropy

Let us consider two states A and B of a gas. If the system changes reversibly from A to B, then the difference in entropy is given by the line integral

$$S(B) - S(A) = \int_A^B \frac{dQ}{T} \tag{11.1}$$

We now show that if the system changes from A to B in an arbitrary manner, the following inequality holds:

$$S(B) - S(A) \geq \int_{\substack{A \to B \\ (t)}} \frac{dQ}{T} \tag{11.2}$$

The integral on the rhs represents the integral with fine time divisions as defined in Section 10.5. The equality sign holds if the path A to B is reversible.

To prove (11.2), we imagine the reversible path BRA (solid line) in addition to the original path AIB (dotted line) in Figure 11.1. The two paths together forming an irreversible cycle. We now apply Clausius's theorem, (10.40), and obtain

$$\oint_{(t)} \frac{dQ}{T} = \int_{AIB} \frac{dQ}{T} + \int_{BRA} \frac{dQ}{T}$$

$$= \int_{AIB} \frac{dQ}{T} + S(B) - S(A) \leq 0$$

where we used (11.1). Adding $S(A) - S(B)$ to both sides, we obtain the desired inequality. Q. E. D.

For an isolated system, the inequality (11.2) takes a very simple form. Since the system cannot exchange heat with its surrounding, we have $dQ = 0$. Then, we obtain

$$\boxed{S(B) \geq S(A)} \quad \text{(isolated system)} \tag{11.3}$$

Mathematical Physics. Shigeji Fujita and Salvador V. Godoy
Copyright © 2010 WILEY-VCH Verlag GmbH & Co. KGaA, Weinheim
ISBN: 978-3-527-40808-5

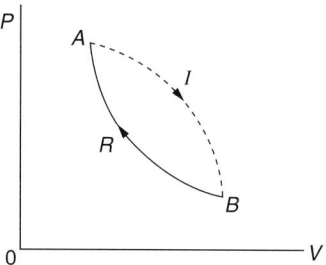

Fig. 11.1 An irreversible path (dotted line) from A to B, and the reversible path (solid line) BRA form an irreversible cycle.

This means that the entropy of the final state B of an isolated system is greater than, or equal to, that of the initial state A. That is, *the entropy of an isolated system cannot decrease with the time.*

An isolated system naturally has a fixed energy and cannot absorb heat from the outside. If such a system has a maximum entropy, it can undergo no further change of state according to (11.3). In other words, *the state of maximum entropy is the most stable state for an isolated system.*

For illustration, let us take a few simple examples.

Consider the free expansion of a dilute gas in Joule's experiment. The gas that was initially confined to one chamber will occupy both chambers with no exchange of heat or work with the surrounding and with no change in temperature. The change in the entropy can be calculated using (11.1) as follows:

From the First Law, $dQ = dE + PdV$. For a mole of ideal gas, we can assume $dE = C_V dT$, which vanishes since $dT = 0$. Now the integral of dQ/T taken along the isothermal path from the initial state A to the final state B can be evaluated as

$$S(B) - S(A) = \int_A^B \frac{dQ}{T} = \int_A^B \frac{PdV}{T}$$

$$= \int_A^B \frac{RdV}{V} = R \ln\left(\frac{V_B}{V_A}\right) \quad (>0) \tag{11.4}$$

which is positive since $V_B > V_A$ in the expansion. Thus, the entropy for the final state B is greater by $R \ln(V_B/V_A)$ than that for the initial state A. We note that *the entropy difference is calculated by postulating a reversible path.*

As a second example, we consider a generation of heat by friction. A system which is heated by friction receives a positive heat $(dQ > 0)$, so the integral on the rhs of (11.2) is positive. That is, the friction will cause the entropy to increase.

As a third example, we consider the exchange of heat by thermal conduction between two parts of a system. To treat this case, we must first study the entropy of a system composed of two or more parts.

Theorem

For a homogeneous substance, the entropy of the whole system is equal to the sum of the entropies of arbitrarily divided parts.

The energy of the system is the sum of the energies of all of the parts, and the work done on the system during a reversible change is the sum of the work done on each of the individual parts. Applying the First Law ($dQ = dE - dW$) and evaluating the resulting integral, we easily establish the proof.

In the case of a system composed of two homogeneous substances such as liquid and saturated vapor in a closed vessel, it is possible to express the total energy of the system as the sum of the energies of the two substances if we neglect the surface energy at the interface. We can neglect the surface energy if the divided subsystems are *macroscopic* so that thermodynamic considerations are applicable to each subsystem individually. Note that this is the definition of *macroscopic* in the present text. We may then express the entropy of the system as the sum of the entropies of the two parts.

We now consider a more general case. We divide a given system into a number of parts $\{j\}$, each of which is in a state of equilibrium. If the energy of the total system E can be expressed as the sum of the energies of the components, $\{E_j\}$:

$$E = \sum_j E_j \tag{11.5}$$

and likewise the work done during a reversible transformation can be summed in the same manner:

$$W = \sum_j W_j \tag{11.6}$$

then the entropy can also be calculated in the same manner:

$$\boxed{S = \sum_j S_j} \tag{11.7}$$

We postulate this *additivity of entropy* even when the system is inhomogeneous or consists of different materials. Note, however, that the entropy must be defined by referring to the equilibrium state and the reversible change of state.

We now go back to the original problem of heat conduction between two parts of an isolated system. Let T_1 and T_2 ($> T_1$) be the temperatures of the two parts 1 and 2. Since heat flows from the hotter body 2 to the colder body 1, the part 2 gives up a positive heat Q that is absorbed by part 1. Then, the entropy of part 1 changes by Q/T_1, while the entropy of part 2 changes by $-Q/T_2$. The entropy change for the total system is

$$\frac{Q}{T_1} - \frac{Q}{T_2} = Q\left(\frac{1}{T_1} - \frac{1}{T_2}\right) \tag{11.8}$$

which is positive since $T_1 < T_2$. That is, the entropy of the total system has increased during the heat conduction.

A mixing of two nonreacting gases occurs in an irreversible manner. This process also generates an increase in entropy.

11.2
The Helmholtz Free Energy

In the case of a purely mechanical system the work L performed by the system is equal to the decrease in its energy. That is,

$$L = -\Delta E \tag{11.9}$$

For a thermodynamic system there is no such simple relationship between the work performed and the change in energy, because energy can be exchanged between the system and its environment in the form of heat. We have, instead, the First Law of thermodynamics, which can be written in the form:

$$L = -W = -\Delta E + Q \tag{11.10}$$

where W is the work done on the system, and Q the heat absorbed by the system.

Many thermodynamic processes occur while the systems are in thermal contact with the environment, allowing an exchange of heat between the system and the environment. Therefore, the work L may be greater or smaller than $-\Delta E$, depending on whether the system absorbs heat from, or gives up heat to, the environment.

Let us now suppose that our system is in thermal contact with an environment of large heat capacity (a thermal reservoir) that is at constant temperature T throughout. We consider a thermodynamic transformation of our system from an initial state A to a final state B. Applying the fundamental inequality (11.2) to this transformation, we have

$$\int_{\substack{A \to B \\ (t)}} \frac{dQ}{T} \leq S(A) - S(B) \tag{11.11}$$

Since the system receives heat only from a reservoir whose temperature is constant, we may remove $1/T$ from under the integral sign, and we obtain

$$Q = \int_{\substack{A \to B \\ (t)}} dQ \leq T[S(A) - S(B)] \tag{11.12}$$

This inequality limits the maximum amount of heat that the system can receive from the reservoir in the transformation from A to B. If the transformation from A to B is reversible, the equality sign holds in (11.12). In this case, the amount of heat absorbed by the system is exactly equal to $T[S(A) - S(B)]$.

From (11.10) and (11.12) we obtain, after putting $\Delta E = E(B) - E(A)$,

$$L \leq E(A) - E(B) + T[S(B) - S(A)] \tag{11.13}$$

This inequality places an upper limit on the amount of work that can be generated in the process from A to B. If the transformation is reversible, the equality sign holds, and the work generated is equal to the upper limit.

Let us now suppose that the temperatures of the initial and final states, A to B, are the same and equal to the temperature T of the environment. (The temperatures of the intermediate states may or may not be equal to T.) We introduce a function F of the state, defined by

$$F \equiv E - TS \tag{11.14}$$

In terms of this function F, we can rewrite (11.13) as

$$L \leq F(A) - F(B) \equiv -\Delta F \tag{11.15}$$

where the equality sign holds if the process is reversible. The function F defined in (11.14) is called the *Helmholtz free energy* and plays an important role in many thermodynamic discussions.

Equation (11.15) may be interpreted as follows: If a system undergoes a reversible transformation from the initial state A to the final state B, both of which have the same temperature T as the environment, and if the system exchanges heat with the environment only, (the isothermal process from A to B is an example), the work performed by the system is equal to the decrease in the Helmholtz free energy. If the change of state is irreversible, the work performed by the system must be less than the decrease in the free energy F.

We now consider a system that is *dynamically insulated* from its environment in the sense that any exchange of energy in the form of work between the system and its environment is prohibited. The system may, however, exchange energy in the form of heat with its environment. For example, a system enclosed in a rigid container may undergo transformations with no exchange of work. Such transformations are called isovolumic or *isochoric* transformations. For any dynamically insulated system, we have, by definition, $L = 0$; we then obtain from (11.15) $0 \leq F(A) - F(B)$ or

$$F(B) \leq F(A) \tag{11.16}$$

That is, if a system is in contact with a thermal reservoir but is dynamically insulated in such a way that it can neither perform nor receive work, its Helmholtz free energy can only decrease or remain constant. If the system attains a minimum free energy, then it can undergo no further transformation. This means that *the stable equilibrium state of the system must have a minimum free energy*. That is, the condition for stable equilibrium of a thermodynamic system enclosed in a rigid container is that its free energy F be at a minimum. This can be contrasted with the case of a mechanical system for which the stable equilibrium is characterized by a minimum potential energy. For this reason, the Helmholtz free energy is also referred to as the *thermodynamic potential at constant volume*.

Let us now consider an infinitesimal, isothermal transformation that generates the change in volume, dV. Because the transformation is reversible, we can apply (11.15) with the equality sign, and obtain

$$dL = -dF, \quad \text{(isothermal transformation.)}$$

Since

$$dL = -dW = PdV$$

and

$$dF = \left(\frac{\partial F}{\partial V}\right)_T dV + \left(\frac{\partial F}{\partial T}\right)_V dT = \left(\frac{\partial F}{\partial V}\right)_T dV + 0$$

we obtain from $dL = -dF$

$$PdV = -\left(\frac{\partial F}{\partial V}\right)_T dV \quad (dT = 0)$$

or

$$\boxed{P = -\left(\frac{\partial F}{\partial V}\right)_T} \tag{11.17}$$

Here, we used a shorthand notation for the partial derivative:

$$\left(\frac{\partial F}{\partial X}\right)_Y \equiv \frac{\partial F(X, Y)}{\partial X} \tag{11.18}$$

Equation (11.17) is an important thermodynamic relation, which will be further expounded in Section 11.4.

11.3
The Gibbs Free Energy

In many thermodynamic transformations, the pressure and the temperature of the system do not change but remain equal to the pressure and the temperature of the environment. In treating such a case, it is convenient to introduce a function of state, G, called the *Gibbs free energy* that has the following property: if the function G is at a minimum for given values of the pressure and temperature, then the system is in a stable equilibrium.

Let us consider an isobaric transformation of a system in contact with a thermal reservoir, which takes the system from state A to another state B. If $V(A)$ and $V(B)$

are the initial and final volumes occupied by the system, then the work performed by the system is

$$L = P[V(B) - V(A)] \tag{11.19}$$

If the temperatures of the initial and final states are equal to that of the thermal reservoir, we may apply (11.15) and obtain

$$PV(B) - PV(A) \leq F(A) - F(B) \tag{11.20}$$

Let us introduce a new function G defined by

$$G \equiv F + PV = E + TS + PV \tag{11.21}$$

In terms of G, the inequality (11.20) can be rewritten as

$$\boxed{G(B) \leq G(A)} \tag{11.22}$$

This means that for an isobaric transformation of a system in contact with a thermal reservoir, the Gibbs free energy can never increase. It follows then that for the same environment, the state of the system for which the Gibbs free energy is at a minimum is a state of stable equilibrium. In this sense, the Gibbs free energy G plays a role analogous to that played by a mechanical potential (when the temperature and pressure are kept constant). Therefore, the Gibbs free energy is sometimes called the *thermodynamic potential at constant pressure*.

The Gibbs free energy is very useful in the discussion of the liquid–vapor coexistence phase, as illustrated in the following example.

We consider a system composed of a liquid and its saturated vapor enclosed in a cylinder and kept at constant temperature and pressure. If E_1, E_2, S_1, S_2, and V_1, V_2 are the energies, entropies, and volume of the liquid and the vapor parts, respectively, and E, S, and V are the corresponding quantities for the total system, then we have

$$E = E_1 + E_2, \quad S = S_1 + S_2, \quad V = V_1 + V_2 \tag{11.23}$$

From (11.21), we also have

$$G = G_1 + G_2 \tag{11.24}$$

where G_1 and G_2 are the Gibbs free energies of the liquid and vapor, respectively.

Let M_1 and M_2 be the masses of the liquid and vapor parts, respectively, and let $\mathcal{E}_1$, $\mathcal{S}_1$, $\mathcal{V}_1$ and $\mathcal{G}_1$, and $\mathcal{E}_2$, $\mathcal{S}_2$, $\mathcal{V}_2$ and $\mathcal{G}_2$ be the *specific* energies, entropies, volumes and Gibbs free energies of the liquid and the vapor. We then have

$$G_1 = M_1 \mathcal{G}_1, \quad G_2 = M_2 \mathcal{G}_2 \tag{11.25}$$

We know from the general properties of saturated vapors that all of the specific quantities $\mathcal{E}_1$, $\mathcal{E}_2$, $\mathcal{S}_1$, $\mathcal{S}_2$, $\mathcal{V}_1$ and $\mathcal{V}_2$, as well as pressure P, are functions of the temperature T only. Therefore, $\mathcal{G}_1$ and $\mathcal{G}_2$ must also depend on T only. We can write

$$G = M_1 \mathcal{G}_1(T) + M_2 \mathcal{G}_2(T) \tag{11.26}$$

We start with the system in equilibrium and perform an isothermal and isobaric transformation such that only M_1 and M_2 may vary. Let M_1 be increased by an amount dM as a result of this transformation. Then, since $M_1 + M_2 =$ total mass = constant, M_2 will decrease by dM. The Gibbs free energy will now be given by

$$(M_1 + dM)\mathcal{G}_1 + (M_2 - dM)\mathcal{G}_2 = G + (\mathcal{G}_1 - \mathcal{G}_2)dM \tag{11.27}$$

According to (11.22), this quantity must be less than, or equal to, G:

$$G + (\mathcal{G}_1 - \mathcal{G}_2)dM \leq G \tag{11.28}$$

Since the transformation is reversible, the equality sign holds here, and we obtain

$$\boxed{\mathcal{G}_1 = \mathcal{G}_2} \tag{11.29}$$

That is, *the specific Gibbs free energy for the liquid and the saturated vapor are equal to each other.*

11.4
Maxwell Relations

The state of a pure gas in equilibrium can be characterized by two independent state variables. These state variables are the pressure P, the volume V, the temperature T and the entropy S. By examining reversible thermodynamic transformations, we can establish important general relations among the thermodynamic quantities. These relations will be discussed in the present section.

According to the First and Second Laws applied to a reversible process, the differential of the internal energy E is given by

$$\boxed{dE = TdS - PdV} \tag{11.30}$$

This indicates that the internal energy E is a natural function of the entropy S and the volume V. Let us express this function by

$$E = E(S, V) \tag{11.31}$$

We obtain from this equation,

$$dE(S, V) = \left(\frac{\partial E}{\partial S}\right)_V dS + \left(\frac{\partial E}{\partial V}\right)_S dV \tag{11.32}$$

Comparing this with (11.30), we find

$$\boxed{T = \left(\frac{\partial E}{\partial S}\right)_V, \quad P = -\left(\frac{\partial E}{\partial V}\right)_S} \tag{11.33}$$

Assuming that E is an analytic function of S and V, we have

$$\frac{\partial^2 E(S,V)}{\partial S \partial V} = \frac{\partial^2 E(S,V)}{\partial V \partial S} \tag{11.34}$$

The lhs can be re-expressed as

$$\frac{\partial}{\partial S}\left(\frac{\partial E}{\partial V}\right)_S = \frac{\partial}{\partial S}(-P) \equiv -\left(\frac{\partial P}{\partial S}\right)_V$$

while the rhs is given by

$$\frac{\partial}{\partial V}\left(\frac{\partial E}{\partial S}\right)_V = \frac{\partial}{\partial V}(T) \equiv \left(\frac{\partial T}{\partial V}\right)_S$$

We obtain from the last three equations

$$\boxed{\left(\frac{\partial P}{\partial S}\right)_V = -\left(\frac{\partial T}{\partial V}\right)_S} \tag{11.35}$$

Equations (11.33) and (11.35) were obtained in a mode of description in which all thermodynamic quantities are regard as functions of the entropy S and the volume V. In this description, the differential equation (11.30) plays an important role. The internal energy $E = E(S, V)$, which is regarded as a function of (S, V), is particularly useful. This function $E(S, V)$ will be referred to as the *characteristic function* for the description in which the entropy S and the volume V are independent variables.

As noted earlier, the thermodynamic state can be specified by any two independent variables. When we choose the temperature T and the volume V as the variables of description, we can obtain the following thermodynamic relations.

Let us recall that the Helmholtz free energy F is defined by

$$F \equiv E - TS \tag{11.36}$$

Its differential is

$$dF = dE - S\,dT - T\,dS$$

Upon substitution from (11.30), we obtain

$$dF = (T\,dS - P\,dV) - S\,dT - T\,dS$$

or

$$\boxed{dF = -S\,dT - P\,dV} \tag{11.37}$$

The last equation indicates that the Helmholtz free energy F is the characteristic function of the temperature T and the volume V. In fact, by noting

$$dF(T,V) = -\left(\frac{\partial F}{\partial T}\right)_V dT + \left(\frac{\partial F}{\partial V}\right)_T dV$$

and comparing it with (11.37), we obtain

$$\boxed{S = -\left(\frac{\partial F}{\partial T}\right)_V, \quad P = -\left(\frac{\partial F}{\partial V}\right)_T} \tag{11.38}$$

Furthermore, from

$$\frac{\partial^2 F}{\partial T \partial V} = \frac{\partial}{\partial T}(-P), \quad \frac{\partial^2 F}{\partial V \partial T} = \frac{\partial}{\partial V}(-S)$$

we can establish the relation

$$\boxed{\left(\frac{\partial P}{\partial T}\right)_V = \left(\frac{\partial S}{\partial V}\right)_T} \tag{11.39}$$

Let us now turn to the mode of description in which the temperature T and the pressure P are independent variables.

The Gibbs free energy G is given by

$$G \equiv F + PV$$

from which we obtain

$$dG = dF + VdP + PdV$$
$$= (-SdT - PdV) + VdP + PdV$$

or

$$dG = -SdT + VdP \tag{11.40}$$

This means that the Gibbs free energy $G = G(T, P)$ is the characteristic function for the thermodynamic description in terms of (T, P).

Comparing

$$dG(T, P) = \left(\frac{\partial G}{\partial T}\right)_P dT + \left(\frac{\partial G}{\partial P}\right)_T dP$$

with (11.38) we obtain

$$\boxed{S = -\left(\frac{\partial G}{\partial T}\right)_P, \quad V = \left(\frac{\partial G}{\partial P}\right)_T} \tag{11.41}$$

By noting that

$$\frac{\partial^2 G}{\partial T \partial P} = \frac{\partial}{\partial T}(V), \quad \frac{\partial^2 G}{\partial P \partial T} = \frac{\partial}{\partial P}(-S)$$

we get

$$\boxed{\left(\frac{\partial V}{\partial T}\right)_P = -\left(\frac{\partial S}{\partial P}\right)_T} \tag{11.42}$$

Finally, we consider the case in which the entropy S and the pressure P are the variables of description.

We define the *enthalpy H* of the system by

$$H \equiv E + PV \tag{11.43}$$

Its differential is given by

$$dH = dE + PdV + VdP$$
$$= (TdS - PdV) + PdV + VdP$$

or

$$\boxed{dH = TdS + VdP} \tag{11.44}$$

This shows that the enthalpy $H = H(S, P)$ is a natural function of the entropy S and the pressure P. Comparing

$$dH(S, P) = \left(\frac{\partial H}{\partial S}\right)_P dS + \left(\frac{\partial H}{\partial P}\right)_S dP$$

with (11.44), we obtain

$$\boxed{T = \left(\frac{\partial H}{\partial S}\right)_P, \quad V = \left(\frac{\partial H}{\partial P}\right)_S} \tag{11.45}$$

By re-expressing

$$\frac{\partial^2 H}{\partial S \partial P} = \frac{\partial^2 H}{\partial P \partial S}$$

we obtain

$$\boxed{\left(\frac{\partial V}{\partial S}\right)_P = \left(\frac{\partial T}{\partial P}\right)_S} \tag{11.46}$$

The four equations (11.34), (11.39), (11.42) and (11.46) are collectively called the *Maxwell relations*. They express rigorous relationships among various thermodynamic variables. These relations will be used later to derive important thermodynamic relations.

We note that the characteristic functions E, F, G and H all have dimensions of an energy. The products PV and ST also have the same dimensions. The pairs of variables (P, V) and (S, T) are referred to as *conjugate variables*.

The set of Maxwell relations were derived in a systematic manner from the differential relations (11.30), (11.37), (11.40) and (11.44). The latter relations may be memorized with the aid of the diagram shown in Figure 11.2.

The energies E, F, G and H are arranged in alphabetic order and counterclockwise at the four corners. The variables P and V are placed vertically, and the variables S and T are laid down horizontally as shown. In addition, two arrows are

11 Thermodynamic Inequalities

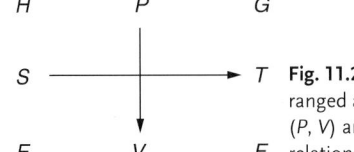

Fig. 11.2 The four characteristic functions (E, F, G, H) are arranged at four corners in alphabetic order. The conjugate pairs (P, V) and (S, T), are placed as shown. The basic differential relations can be read from this figure in a systematic manner.

drawn. Now the diagram may be read as follows: The two variables adjacent to each characteristic function (energy) are those variables preferred. For example, E has S and V as preferred variables. The head and tail of an arrow indicate the negative and positive signs, respectively, of fundamental differential relations. Thus, for example,

$$dE = TdS - PdV$$

where the negative sign of $-PdV$ is read off from the diagram as the variable V is at the head of the arrow. We note that each differential on the rhs should carry the conjugate variable as its coefficient (so that every term may have the dimensions of an energy). In another example,

$$dF = -PdV - SdT$$

where the presence of the two negatives is called for by the fact that both variables V and T are at the arrowheads.

From the fundamental differential relations, we can easily rederive Maxwell relations when desired.

Problem 11.4.1

Using Figure 11.2, derive the four fundamental differentials and the Maxwell relations.

11.5
Heat Capacities

For a gas, the heat capacity at constant volume, C_V, and the heat capacity at constant pressure, C_P, are considerably different from each other. In this section we will investigate the difference $C_P - C_V$ based on thermodynamic arguments.

For an infinitesimal reversible transformation, we have

$$dQ = TdS \tag{11.47}$$

Let us recall the definition equation for the heat capacity at constant volume:

$$dQ = C_V dT, \quad dV = 0 \tag{11.48}$$

This equation indicates that the convenient thermodynamic variables are (T, V). If we regard the entropy S as a function of these variables, we obtain

$$dS = \left(\frac{\partial S}{\partial T}\right)_V dT + \left(\frac{\partial S}{\partial V}\right)_T dV \tag{11.49}$$

Substituting this into (11.47) and comparing with (11.48), we obtain

$$\boxed{C_V = T\left(\frac{\partial S}{\partial T}\right)_V} \tag{11.50}$$

In a similar manner, we can derive

$$\boxed{C_P = T\left(\frac{\partial S}{\partial T}\right)_P} \tag{11.51}$$

These expressions (11.50) and (11.51) indicate that the changes of the entropy with temperature are directly related to the heat capacities.

Using these results, we obtain

$$C_P - C_V = T\left[\left(\frac{\partial S}{\partial T}\right)_P - \left(\frac{\partial S}{\partial T}\right)_V\right] \tag{11.52}$$

Let us go back to (11.49). If we regard the volume V as a function of (T, P), we have

$$dV = \left(\frac{\partial V}{\partial T}\right)_P dT + \left(\frac{\partial V}{\partial P}\right)_T dP \tag{11.53}$$

Substituting this into (11.49), we get

$$\begin{aligned}dS &= \left(\frac{\partial S}{\partial T}\right)_V dT + \left(\frac{\partial S}{\partial V}\right)_T \left[\left(\frac{\partial V}{\partial T}\right)_P dT + \left(\frac{\partial V}{\partial P}\right)_T dP\right] \\ &= \left[\left(\frac{\partial S}{\partial T}\right)_V + \left(\frac{\partial S}{\partial V}\right)_T \left(\frac{\partial V}{\partial T}\right)_P\right] dT + \left(\frac{\partial S}{\partial V}\right)_T \left(\frac{\partial V}{\partial P}\right)_T dP\end{aligned} \tag{11.54}$$

On the other hand, we can regard the entropy S as a function of (T, P) and obtain

$$dS = \left(\frac{\partial S}{\partial T}\right)_P dT + \left(\frac{\partial S}{\partial P}\right)_T dP \tag{11.55}$$

Comparison between (11.54) and (11.55) yields

$$\left(\frac{\partial S}{\partial T}\right)_P = \left(\frac{\partial S}{\partial T}\right)_V + \left(\frac{\partial S}{\partial V}\right)_T \left(\frac{\partial V}{\partial T}\right)_P \tag{11.56}$$

This result can alternatively be obtained as follows. We look at the entropy as a function of (T, V): $S = S(T, V)$. Furthermore, we regard V as a function of (T, P): $V = V(T, P)$. Then, we have $S = S[T, V(T, P)]$. If we differentiate this with respect to T while keeping P constant, we simply obtain (11.56).

Using one of the Maxwell relations: $(\partial S/\partial V)_T = (\partial P/\partial T)_V$, we can rewrite (11.56) as

$$\left(\frac{\partial S}{\partial T}\right)_P = \left(\frac{\partial S}{\partial T}\right)_V + \left(\frac{\partial P}{\partial T}\right)_V \left(\frac{\partial V}{\partial T}\right)_P \tag{11.57}$$

Introducing this result in (11.52), we obtain

$$\boxed{C_P - C_V = T\left(\frac{\partial P}{\partial T}\right)_V \left(\frac{\partial V}{\partial T}\right)_P} \tag{11.58}$$

This is a remarkable relation; the quantities on the rhs can be obtained without calorimetric measurement.

Let us re-express these quantities in terms of the quantities that are normally measured in experiments. the *coefficient of thermal expansion* α is defined by

$$\alpha \equiv \frac{1}{V}\left(\frac{\partial V}{\partial T}\right)_P \tag{11.59}$$

The volume of a substance (gas, liquid, or solid) normally increase with the rise in temperature. Then, the thermal expansion coefficient is positive. A notable exception is the case of water below about $4\,°C$ where the volume decreases as the temperature is raised from the freezing point $0\,°C$ to $4\,°C$. In such a case, the coefficient α is negative.

We further introduce the isothermal compressibility K_T, defined by

$$K_T \equiv -\frac{1}{V}\left(\frac{\partial V}{\partial P}\right)_T \tag{11.60}$$

The volume of a substance must decrease (or remain the same) with an increase in pressure so that

$$\boxed{\left(\frac{\partial V}{\partial P}\right)_T \leq 0}$$

On account of the negative sign in the definition equation (11.60), the compressibility K_T should be nonnegative:

$$K_T \geq 0 \tag{11.61}$$

In terms of the coefficients of thermal expansion α and isothermal compressibility K_T, we can re-express (11.58) as follows:

$$\boxed{C_P - C_V = TV\alpha^2/K_T} \tag{11.62}$$

To show this, we note the following mathematical lemmas: Let X, Y and Z be quantities satisfying a certain functional relation $f(X, Y, Z) = 0$. We then have

$$\left(\frac{\partial X}{\partial Y}\right)_Z = 1 \bigg/ \left(\frac{\partial Y}{\partial X}\right)_Z \tag{11.63}$$

11.5 Heat Capacities

$$\left(\frac{\partial X}{\partial Y}\right)_Z \left(\frac{\partial Y}{\partial Z}\right)_X \left(\frac{\partial Z}{\partial X}\right)_Y = -1 \tag{11.64}$$

which may be proved as follows. We regard X as a function of (Y, Z): $X = X(Y, Z)$. We then have

$$dX = \left(\frac{\partial X}{\partial Y}\right)_Z dY + \left(\frac{\partial X}{\partial Z}\right)_Y dZ \tag{11.65}$$

Consider a change in which Z is kept constant: $dZ = 0$. Dividing both sides by dX, we obtain

$$1 = \left(\frac{\partial X}{\partial Y}\right)_Z \left(\frac{\partial Y}{\partial X}\right)_Z$$

which yields (11.63). Let us consider another change in which X is kept constant: $dX = 0$. Dividing both sides of (11.65) by dZ, we obtain

$$0 = \left(\frac{\partial X}{\partial Y}\right)_Z \left(\frac{\partial Y}{\partial Z}\right)_X + \left(\frac{\partial X}{\partial Z}\right)_Y$$

which yields (11.64) after simple rearrangements. Applying the lemmas (11.63) and (11.64) to $(P = X, T = Y, V = Z)$, we can rewrite the rhs of (11.58):

$$T \left(\frac{\partial P}{\partial T}\right)_V \left(\frac{\partial V}{\partial T}\right)_P = T \left[\frac{-1}{\left(\frac{\partial T}{\partial V}\right)_P \left(\frac{\partial V}{\partial P}\right)_T}\right] \left(\frac{\partial V}{\partial T}\right)_P$$

$$= TV \left[\frac{-1}{\frac{1}{V}\left(\frac{\partial V}{\partial P}\right)_T}\right] \left[\frac{1}{V}\left(\frac{\partial V}{\partial T}\right)_P\right]^2 = TV\alpha^2 K_T^{-1}$$

establishing (11.62).

Since the isothermal compressibility K_T is nonnegative we have, from (11.62),

$$\boxed{C_P \geqq C_V} \tag{11.66}$$

The very significant relation, (11.62), was obtained with the aid of the Maxwell relations. Using these relations we can also obtain the following results:

$$\left(\frac{\partial C_V}{\partial V}\right)_T = T\left(\frac{\partial^2 P}{\partial T^2}\right)_V \tag{11.67}$$

$$\left(\frac{\partial C_P}{\partial P}\right)_T = -T\left(\frac{\partial^2 V}{\partial T^2}\right)_P \tag{11.68}$$

Equation (11.67) may be proved as follows. Notice first that the convenient pair of variables is (V, T). Using (11.50), we obtain

$$\left(\frac{\partial C_V}{\partial V}\right)_T = \frac{\partial}{\partial T}\left[T\left(\frac{\partial S}{\partial T}\right)_V\right]_T = T\frac{\partial^2 S}{\partial V \partial T} = T\frac{\partial^2 S}{\partial T \partial V}$$

We now use one of the Maxwell relations (11.39) and obtain

$$T\frac{\partial}{\partial T}\left[\left(\frac{\partial S}{\partial V}\right)_T\right] = T\frac{\partial}{\partial T}\left[\left(\frac{\partial P}{\partial T}\right)_V\right] = T\left(\frac{\partial^2 P}{\partial T^2}\right)_V \quad \text{Q. E. D.}$$

Equation (11.68) can be proved in a similar manner. (Problem 11.5.3).

Problem 11.5.1

Refer to (11.58). Evaluate the rhs for a mole of ideal gas and obtain the relation $C_P - C_V = R$.

Problem 11.5.2

Assuming an ideal gas, calculate the coefficient of thermal expansion α, the isothermal K_T and the adiabatic K_S compressibilities, where $K_T \equiv -\frac{1}{V}\left(\frac{\partial V}{\partial P}\right)_T$, and $K_S \equiv -\frac{1}{V}\left(\frac{\partial V}{\partial P}\right)_S$.

Problem 11.5.3

Prove that

$$\left(\frac{\partial C_P}{\partial P}\right)_T = -T\left(\frac{\partial^2 V}{\partial T^2}\right)_P$$

HINT: Use the set of independent variables (P, T).

11.6
Nonnegative Heat Capacity and Compressibility

Pauli, in his book [1], based entirely on thermodynamic arguments showed that the heat capacity at constant volume, C_V, and the isothermal compressibility K_T are

11.6 Nonnegative Heat Capacity and Compressibility

nonnegative. His proof is based on the maximum entropy principle for an isolated system. Dividing the system of a monatomic gas in two macroscopic subsystems, for each of which thermodynamic relations hold, and using the stability condition, Pauli established his proof. We show in the present section that the heat capacity C and the compressibility K are nonnegative in each phase (gas, liquid, or solid).

If a material is heated, the temperature T should rise or stay constant. Thus, the heat capacity C for a monatomic gas defined from

$$C\, dT = dQ \tag{11.69}$$

should be nonnegative. In particular, the heat capacity at constant volume, C_V, should be nonnegative:

$$C_V \equiv \left(\frac{dQ}{dT}\right)_V = T\left(\frac{\partial S}{\partial T}\right)_V \geq 0 \tag{11.70}$$

If the system is compressed, the volume V should decrease. Hence, we expect that the isothermal compressibility K_T should be nonnegative:

$$K_T \equiv -\frac{1}{V}\left(\frac{\partial V}{\partial P}\right)_V \geq 0 \tag{11.71}$$

The two inequalities (11.70) and (11.71) will be proved using a more elementary approach based on the principle that the Gibbs free energy G of the system is minimum in the fixed-temperature and the fixed-pressure environment.

We first show that

$$\left(\frac{\partial T}{\partial S}\right)_V \geq 0 \tag{11.72}$$

The entropy S and the volume V are used here. The internal energy E is a characteristic function for the variable set (S, V). The basic differential equation expressing the First and Second Laws for a reversible process is

$$dE = T\, dS - P\, dV \tag{11.73}$$

The Gibbs free energy G is at a stable minimum for the system in the environment with a fixed temperature T_0 and a fixed pressure P_0. Then, we must have

$$\left(\frac{\partial G}{\partial S}\right)_{V,T_0,P_0} = 0 \tag{11.74}$$

$$\left(\frac{\partial^2 G}{\partial S^2}\right)_{V,T_0,P_0} \geq 0 \tag{11.75}$$

We express G in terms of E as $G = E - TS + PV$, calculate the derivatives, $(\partial G/\partial S)_V$ with fixed (T_0, P_0), and obtain

$$\left(\frac{\partial G}{\partial S}\right)_{V,T_0,P_0} = \left(\frac{\partial}{\partial S}(E - T_0 S + P_0 V)\right)\bigg|_{V,T_0,P_0}$$

$$= \left(\frac{\partial E}{\partial S}\right)_V - T_0 = T - T_0 = 0 \tag{11.76}$$

where we used (11.73) and (11.74). Equation (11.76) means that the system temperature T must be equal to the environment temperature T_0.

We calculate the second derivative, use (11.75) and (11.76), to obtain

$$\left(\frac{\partial^2 G}{\partial S^2}\right)_{V,T_0,P_0} = \frac{\partial}{\partial S}(T-T_0)\bigg|_{V,T_0,P_0} = \left(\frac{\partial T}{\partial S}\right)_V \geq 0 \tag{11.77}$$

Using (11.77), we obtain

$$C_V = T\left(\frac{\partial S}{\partial T}\right)_V = T/\left(\frac{\partial T}{\partial S}\right)_V \geq 0 \tag{11.78}$$

establishing the desired inequality (11.70).

From the minimum Gibbs free energy principle we must have

$$\left(\frac{\partial G}{\partial V}\right)_{T,T_0,P_0} = 0 \tag{11.79}$$

$$\left(\frac{\partial^2 G}{\partial V^2}\right)_{T,T_0,P_0} \geq 0 \tag{11.80}$$

The volume V and the temperature T are used here, and the Helmholtz free energy F is the characteristic function of (V, T). The basic differential equation is $dF = -PdV - SdT$. We express G in terms of F: $G = F + PV$. We calculate the derivative $(\partial G/\partial V)_T$ with fixed (T_0, P_0), use (11.79) and obtain

$$\left(\frac{\partial G}{\partial V}\right)_{T,T_0,P_0} = \left(\frac{\partial F}{\partial V}\right)_T + P_0 = -P + P_0 = 0 \tag{11.81}$$

The system pressure P must be equal to the environment pressure P_0.

We calculate the second derivative $(\partial^2 G/\partial V^2)_{T,T_0,P_0}$, use (11.80) and (11.81) to obtain

$$\left(\frac{\partial^2 G}{\partial V^2}\right)_{T,T_0,P_0} = -\left(\frac{\partial P}{\partial V}\right)_T \geq 0 \tag{11.82}$$

Since $(\partial V/\partial P)_T = [(\partial P/\partial V)_T]^{-1}$, the inequality (11.82) means that the isothermal compressibility K_T is nonnegative, as stated in (11.71).

We shall show that the heat capacity at constant pressure, C_P, is nonnegative:

$$C_P = T\left(\frac{\partial S}{\partial T}\right)_P \geq 0 \tag{11.83}$$

We choose the system variables (S, P). The characteristic function is the enthalpy H. The Gibbs free energy G and the enthalpy H are connected by $G =$

$H - TS$. From the minimum free energy principle, we must have

$$\left(\frac{\partial G}{\partial S}\right)_{P,T_0,P_0} = 0 \tag{11.84}$$

$$\left(\frac{\partial^2 G}{\partial S^2}\right)_{P,T_0,P_0} \geq 0 \tag{11.85}$$

We obtain, using (11.84)

$$\left(\frac{\partial G}{\partial S}\right)_{P,T_0,P_0} = \left(\frac{\partial H}{\partial S}\right)_P - T_0 = T - T_0 = 0 \tag{11.86}$$

This confirms that the system temperature T is equal to the environment temperature T_0.

We obtain, after using (11.85) and (11.86),

$$\left(\frac{\partial^2 G}{\partial S^2}\right)_{P,T_0,P_0} = \left(\frac{\partial}{\partial S}\frac{\partial G}{\partial S}\right)_{P,T_0,P_0} = \left(\frac{\partial T}{\partial S}\right)_P \geq 0 \tag{11.87}$$

Hence, we obtain

$$C_P = T\left(\frac{\partial S}{\partial T}\right)_P = T\left[\left(\frac{\partial T}{\partial S}\right)_P\right]^{-1} \geq 0$$

establishing (11.83).

The adiabatic compressibility K_S is defined by

$$K_S \equiv -\frac{1}{V}\left(\frac{\partial V}{\partial P}\right)_S$$

We choose the system variables (V, S). From the stability condition we have

$$\left(\frac{\partial G}{\partial V}\right)_{S,P_0,T_0} = 0 \tag{11.88}$$

$$\left(\frac{\partial^2 G}{\partial V^2}\right)_{S,P_0,T_0} \geq 0 \tag{11.89}$$

Using $G = E - TS + PV$ and (11.88), we obtain

$$\left(\frac{\partial G}{\partial V}\right)_{S,P_0,T_0} = \left(\frac{\partial E}{\partial V}\right)_S + P_0 = -P + P_0 \tag{11.90}$$

confirming that the system pressure P equals the environment pressure P_0.

After further integration, we obtain

$$\left(\frac{\partial^2 G}{\partial V^2}\right)_{S,P_0,T_0} = -\left(\frac{\partial P}{\partial V}\right)_S \geq 0 \tag{11.91}$$

where we used (11.89). Using (11.91) and

$$\left(\frac{\partial P}{\partial V}\right)_S = \left[\left(\frac{\partial V}{\partial P}\right)_S\right]^{-1}$$

we obtain

$$K_S = -\frac{1}{V}\left(\frac{\partial V}{\partial P}\right)_S = -\frac{1}{V}\left[\left(\frac{\partial P}{\partial V}\right)_S\right]^{-1} \geq 0 \qquad (11.92)$$

We have shown that

$$C_V, C_P \geq 0 \qquad (11.93)$$

with the assumption that the Gibbs free energy G is analytic. The set (V, P) can represent a general thermodynamics state for the vapor, the liquid or the solid phase. Hence, the heat capacity C is nonnegative in each phase (vapor, liquid, or solid). In the phase-separation lines, the free energy G is not analytic. In normal experiments the heat capacity at constant pressure, C_P, is measured. Occasionally, the heat capacity C along the phase-separation lines is measured.

We have also shown the two inequalities,

$$K_T, K_S \geq 0 \qquad (11.94)$$

The set (T, S) can also represent a general thermodynamics state. Hence, the compressibility K is nonnegative in each phase.

The heat capacity at constant pressure, C_P, is in general greater than the heat capacity at constant volume, C_V. The difference is from (11.62)

$$C_P - C_V = TV\alpha^2/K_T \geq 0 \qquad (11.95)$$

The ratio K_S/K_T can be shown to be equal to the ratio C_P/C_V: (Problem 11.6.1)

$$\frac{K_S}{K_T} = \frac{C_V}{C_P} \qquad (11.96)$$

Using this relation, we obtain

$$K_T - K_S = K_T\left(1 - \frac{C_V}{C_P}\right) \geq 0 \qquad (11.97)$$

Thus, the isothermal compressibility K_T is generally greater than the adiabatic compressibility K_S.

Equations (11.95) and (11.97) are valid for all temperatures (in each phase). Since the coefficient of thermal expansion α and the isothermal compressibility K_T are both finite, we obtain

$$C_P \to C_V \quad \text{as } T \to 0 \qquad (11.98)$$

Using (11.97) and (11.98), we find that

$$K_T \to K_S \quad \text{as } T \to 0 \tag{11.99}$$

Equations (11.98) and (11.99) are important properties valid in solid and liquids.

Problem 11.6.1

Derive (11.96).

References

1 Pauli, W. (2000) *Thermodynamics and Kinetic Theory of Gases*, Dover Publications, New York, pp. 83–86.

2 Fermi, E. (1956) *Thermodynamics*, Dover Publications, New York, pp. 77–83.

12
Probability, Statistics and Density

The concept of probability is indispensable in statistical physics. Basic elements of probability and statistics are introduced and discussed in the first part of this chapter, Sections 12.1–12.3. The microscopic density is discussed using Dirac's delta-function.

12.1
Probabilities

The concept of probability may be understood simply by flipping a coin. After a large number of throws, we will find that the number of heads divided by the total number of throws is close to one half,

$$\frac{\text{no. of heads}}{\text{total no. of throws}} \simeq \frac{1}{2} \tag{12.1}$$

and the number of tails divided by the total number of throws is also close to one half,

$$\frac{\text{no. of tails}}{\text{total no. of throws}} \simeq \frac{1}{2} \tag{12.2}$$

We may thus expect an even chance that a head or tail will appear in any subsequent throw of the coin. We say that the probability of finding a head, $P(\text{head})$, is equal to the probability of finding a tail, $P(\text{tail})$, and both probabilities are one-half:

$$P(\text{head}) = P(\text{tail}) = \frac{1}{2} \tag{12.3}$$

The number 1/2 need not be obtained by actually flipping a coin, however. The probability $P(\text{head})$ of the event in which head is realized is obtained by dividing the number of possible ways in which a head may be realized by the number of different events possible (in one flip of the coin). In the present case, the number of ways in which a head can occur is one, and the number of different events is two. The probability $P(\text{head})$ is therefore 1/2.

Mathematical Physics. Shigeji Fujita and Salvador V. Godoy
Copyright © 2010 WILEY-VCH Verlag GmbH & Co. KGaA, Weinheim
ISBN: 978-3-527-40808-5

Let us now consider a more complicated example. We throw a die once. We normally expect that the numeral 1 is as likely to appear as any of the other numerals 2, 3, 4, 5, 6. We say that the probability of obtaining the numeral 1, $P(1)$, is equal to the probability of obtaining any other numeral, and this probability is 1/6:

$$P(1) = \frac{\text{no. of possible way to realize event 1}}{\text{no. of all events 1, 2, \ldots, 6}} = \frac{1}{6} \tag{12.4}$$

The definition of probability need not be restricted to single events. For example, the probability of obtaining an odd numeral by a throw of the die can be defined, and calculated as follows:

$$P(\text{odd numeral}) = \frac{\text{no. of events 1, 3, 5}}{\text{no. of all events 1, 2, 3, \ldots, 6}} = \frac{3}{6} = \frac{1}{2} \tag{12.5}$$

We note that

$$P(\text{odd numeral})$$
$$= \frac{(\text{no. of events 1}) + (\text{no. of events 3}) + (\text{no. of events 5})}{\text{no. of all events}}$$
$$= (1) + P(3) + P(5) \tag{12.6}$$

This result may be interpreted as follows: To obtain an odd numeral means to obtain one of the mutually exclusive events 1, 3, or 5. The probability of obtaining an odd numeral is the sum of the probabilities of obtaining 1, 3, or 5. Equation (12.6) is an expression of the general *addition law* in probability theory: *The probability of occurrence of one of a number of mutually exclusive events is equal to the sum of the probabilities for the occurrence of each of the individual events.*

Consider an event in which we obtain the numeral 1 the first time and also the numeral 1 the second time. The probability of obtaining this compound event is

$$P(1,1) = \frac{\text{no. of ways of realizing the event (1,1)}}{\text{total no. of possible events}} = \frac{1}{6 \times 6} = \frac{1}{36} \tag{12.7}$$

This result may be obtained alternatively in the following manner: The probability of obtaining the numeral 1 the first time is $P_I(1) = 1/6$. The probability of obtaining the numeral 1 the second time is $P_{II}(1) = 1/6$. The probability of obtaining the numeral 1 both times is given by the product of the probabilities of obtaining the numeral 1 each time,

$$P(1,1) = P_I(1) \times P_{II}(1) = \frac{1}{6} \times \frac{1}{6} = \frac{1}{36} \tag{12.8}$$

This represents an example of the *multiplication law* of probability. Consider a set of events that concern the result of two operations. If the probability of an event is equal to the product of the probabilities of the subevents, the two operations are said to be *uncorrelated* or *independent*. In the above example, the observation of the numeral 1 the first time and the observation of the numeral 1 the second time are uncorrelated. If the multiplication law does not hold, we say that the two operations are *correlated*.

As we can see from the definition, (12.4), the probability for an event is the number between 0 and 1. If the event is certain to occur, a thrown die must show a numeral. Therefore, the probability of having any numeral at all is unity. Since this numeral must be 1, 2, 3, 4, 5, or 6, we have

$$P(1) + P(2) + \ldots + P(6) \simeq \sum_{n=1}^{6} P(n) = 1 \qquad (12.9)$$

This equation represents the *normalization condition* for the probabilities.

Problem 12.1.1

What are the probabilities of finding different numerals
  1. in a throw of two dice, and
  2. in a throw of three dice?

Problem 12.1.2

What is the probability of finding two numerals whose sum is 10, in a throw of two dice?

Problem 12.1.3

In a card game, the order in which cards are dealt to you does not affect your winning probability. For example, the probability of getting all aces in a deal of four cards must be the same for each player regardless of the order. Prove this statement.

Remark

Consider a simple case first. For example, prove that the chance of getting the ace of spades in a deal of one card is the same for all players, and then generalize your arguments.

Problem 12.1.4

In a deal of three cards, what are the probabilities of having
  1. straights (consecutive numbers)?
  2. flushes (same suits)?
  3. which is the greater of the previous two?
  4. Investigate the same questions in deals of four and five cards.

12.2
Binomial Distribution

Let us consider the square lattice shown in Figure 12.1. An object starts at the lower left-hand corner O and moves one step per unit time either up or to the right. The probabilities of the object moving right and up are denoted by p and q, respectively, with the normalization condition:

$$p + q = 1 \tag{12.10}$$

We may now ask the questions: where will the object reach after N steps, and what is the probability of the object getting there? Before attempting to answer these questions for a general N, let us first examine the problem for specific Ns. For $N = 1$, the object will move to one of the two lattice sites marked by circles. These sites may be labeled by their Cartesian coordinates, that is by (1,0) and (0,1). Clearly, the probabilities of getting to these sites are given, respectively, by p and q:

$$P(1,0) = p, \quad P(0,1) = q \tag{12.11}$$

For $N = 2$, the object will move to one of the three sites (2,0), (1,1) or (0,2), marked by triangles. There are two ways of getting to site (1,1) with the same probability pq. We therefore have

$$P(2,0) = p^2, \quad P(1,1) = 2pq, \quad P(0,2) = q^2 \tag{12.12}$$

For $N = 3$, the object will move to one of the four sites (3,0), (2,1), (1,2) or (0,3), marked by square. The probabilities of reaching these sites are

$$P(3,0) = p^3, \quad P(2,1) = 3p^2 q$$

$$P(1,2) = 3pq^2, \quad P(0,3) = q^2 \tag{12.13}$$

Consider now an arbitrary N. After N steps the object will move to one of the $N + 1$ sites

$$(N,0), (N-1,1), \ldots, (0,N) \tag{12.14}$$

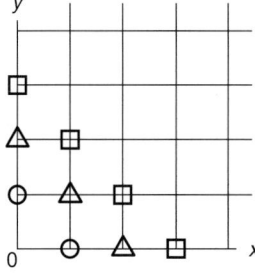

Fig. 12.1 Square lattice.

and the probabilities of reaching these sites are given by

$$P(X, Y) = P(X, N - X) = \binom{N}{X} p^X q^{N-X}, \quad (X = 0, 1, 2, \ldots, N) \quad (12.15)$$

where the quantity

$$\binom{N}{X} \equiv \frac{N(N-1)\ldots(N-X+1)}{X(X-1)\ldots 2 \cdot 1} = \frac{N!}{X!(N-X)!} \quad (12.16)$$

is called the *binomial coefficient* and gives the number of combinations on N objects taken X at a time. The *factorial* (number) $N!$ defined by

$$N! = N(N-1)(N-2)\ldots 2 \cdot 1 \quad (12.17)$$

is a positive integer. By convention 0! is set equal to unity

$$0! \equiv 1 \quad (12.18)$$

The result (12.15) can be obtained without working out the simple cases for small N. In order to reach the point $(X, N - X)$, the object must move X steps to the right. Therefore, the probability $P(X, N - X)$ will have the factor p^X. When the object does not move to the right at a particular unit time, it must move upwards. Therefore, it must have moved $N - X$ steps up, hence, the factor q^{N-X}. The object may move to the final location $(X, N - X)$ in any combination of up and right steps. The number of such combinations is equal to $\binom{N}{X}$. Combining these factors, we obtain the result (12.15). This method is quicker than the first. However, in order to obtain the correct result with the second method, one must somehow know exactly the three factors p^X, q^{N-X} and $\binom{N}{X}$ that are needed for the final result. Since such advance knowledge may not be available in more complicated problems, it is advisable to check any result arrived at by guessing, either by working out simple cases, as was done in the first method, or by other methods of verification (see below). These words of caution apply not only to this particular type of problem in probability but also to many other problems in physics.

It is interesting to observe that terms on the right-hand side (rhs) of (12.15) correspond to terms in the *binomial expansion*:

$$(p + q)^N = q^N + N q^{N-1} p + \ldots + \binom{N}{X} p^X q^{N-X} + \ldots + p^N \quad (12.19)$$

That is, the $(N + 1)$ terms give precisely the probabilities of reaching the $(N + 1)$ sites in (12.14) in N steps. A set of probabilities that correspond to the terms in the binomial expansion is said to form a *binomial distribution*. Thus, our probabilities $P(X, N - X)$ form a binomial distribution. Using (12.15) and (12.19) it then follows that

$$P(0, N) + P(1, N - 1) + \ldots + P(X, N - X) + \ldots + P(N, 0)$$

$$\equiv \sum_{X=0}^{N} P(X, N - X) = (q + p)^N = 1 \quad (12.20)$$

12 Probability, Statistics and Density

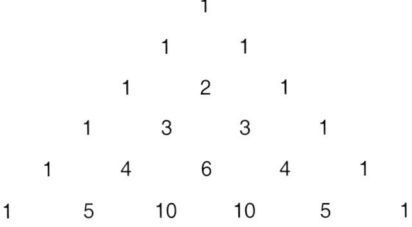

Fig. 12.2 The Pascal triangle.

This shows that the obtained probabilities Ps satisfy the proper normalization condition. We note that checking the normalization is one of the most useful methods of verifying the derived probability distribution.

For the special case

$$p = q = \frac{1}{2} \tag{12.21}$$

that is equal probability for the up and right steps, the probabilities are

$$1, N, \ldots, \binom{N}{X}, \ldots, 1 \tag{12.22}$$

multiplied by the common factor $p^X q^{N-X} = (1/2)^N$. For small N, the set of numbers in (12.22) can be obtained by observing the Pascal triangle as shown in Figure 12.2. (Each number is obtained from the sum of the two neighboring numbers appearing directly above it.) The set of numbers in any row is symmetric about the middle. For $p = q$, the binomial distribution has a single maximum, although then it is not symmetric about $X = N/2$.

Problem 12.2.1

Calculate numerically the binomial distribution

$$P(X, Y) = \binom{N}{X} p^X q^{N-X}$$

for $p = 1/4$ and $N = 3$. Draw a graph of $P(X, Y)$ versus X and verify the existence of a single maximum.

Problem 12.2.2

There are five apples.
1. Find the number of different ways in which these apples can be ordered.
2. Find the number of different ways in which two particular apples can be selected.

12.3
Average and Root-Mean-Square Deviation. Random Walks

In the present section we will discuss general elementary concepts in statistics.

We take the same model introduced in the preceding section. After N steps, the object moves to one of the $N + 1$ possible sites, $(X, N - X)$, $[X = 0, 1, \ldots, N]$ with the arrival possibilities, $P(X, N - X)$.

Let us define the *Average* of the horizontal location X by

$$\langle X \rangle = \sum_{X=0}^{N} X P(X, N - X) \tag{12.23}$$

We note that the rhs represents the sum of the products of the location X and the probability $P(X, N - X)$ over all possible sites. We denote the average either by angular brackets $\langle X \rangle$ or by a super bar $\bar{X}$ in the text. It is convenient to have both notations in practice.

The concept of the average can be applied to a far wider class of events. In general, let us suppose that an event can result in one of many *states* with given probabilities.

The average of a property A, which depends on the states, is defined by

$$\langle A \rangle \equiv \sum_{\text{states}} A(\text{state}) P(\text{state}) \tag{12.24}$$

In the above example, the object can arrive at various states (sites) with assumed probabilities. These states are characterized by the *state variables*, $(X, N - X)$, and the property $A(X, N - X)$ is regarded as a function of the state variables $(X, N - X)$. We may then re-express (12.24) as

$$\langle A \rangle \equiv \sum_{X} A(X, N - X) P(X, N - X) \tag{12.25}$$

By choosing $A = X$ in this expression, and noting that all states are enumerated by varying X from 0 to N, we can simply recover (12.23).

The term "state" is used here in a broader sense than the term "site" or "location". In the present case both terms are identical.

The average $\langle X \rangle$ of the transverse location, defined in (12.23), can be computed as follows:

$$\langle X \rangle \equiv \sum_{X} X P(X, N - X) = \sum_{X} X \binom{N}{X} p^X q^{N-X}$$

$$= \sum_{X} \binom{N}{X} p \left(\frac{d}{dp} p^X \right) q^{N-X} = p \frac{d}{dp} \sum_{X} \binom{N}{X} p^X q^{N-X}$$

$$= p \frac{d}{dp} (p + q)^N = N p (p + q)^{N-1} = N p \quad (p + q = 1) \tag{12.26}$$

This result is in agreement with our expectation since the object moves to the right with probability p and each of the N steps is carried out independently. We note that the average $\langle X \rangle$ is not necessarily an integer, even though the unaveraged X is an integer. For any two quantities A and B,

$$\langle A + B \rangle \equiv \sum_X \left[A(X, N-X) + B(X, N-X) \right] P(X, N-X)$$

$$= \sum_X A(X, N-X) P(X, N-X) + \sum_X B(X, N-X) P(X, N-X)$$

$$= \langle A \rangle + \langle B \rangle \tag{12.27}$$

This means that *the average of the sum is the sum of the averages*. Since no explicit knowledge of the probability distribution is needed in the derivation, the statement is valid for *any* probability distribution.

The quantity

$$\Delta A \equiv A - \langle A \rangle \equiv A - \bar{A} \tag{12.28}$$

is called the *deviation* from the average. From the definition (12.28), the deviation ΔA can be positive, negative or zero. The average of the deviation is zero since

$$\langle A + \bar{A} \rangle = \langle A \rangle - \bar{A} = \bar{A} - \bar{A} = 0 \tag{12.29}$$

The *squared deviation* defined by

$$(\Delta A)^2 \equiv (A - \bar{A})^2 \quad (\geq 0) \tag{12.30}$$

is nonnegative. Its average may be calculated as follows:

$$\langle (\Delta A)^2 \rangle = \langle A^2 - 2A\bar{A} + (\bar{A})^2 \rangle$$
$$= \langle A^2 \rangle - 2\bar{A}\bar{A} + \bar{A}^2 = \langle A^2 \rangle - \bar{A}^2 \tag{12.31}$$

The square root of this number

$$\sqrt{\langle (\Delta A)^2 \rangle} \equiv (\Delta A)_{\text{rms}} \tag{12.32}$$

is called the *root-mean-square deviation* or the *standard deviation*. It gives a good measure of the deviation from the average in a set of trials.

Let us consider a few examples. The mean-square deviation of the transverse location is

$$\langle \Delta X^2 \rangle = \langle X^2 \rangle - \bar{X}^2 \tag{12.33}$$

12.3 Average and Root-Mean-Square Deviation. Random Walks

The average $\langle X^2 \rangle$ may be calculated as follows:

$$\langle X^2 \rangle \equiv \sum_{X=0}^{N} X^2 P(X, N-X) = \sum_{X} X^2 \binom{N}{X} p^X q^{N-X}$$

$$= \sum_{X} \binom{N}{X} \left(p \frac{d}{dp} p \frac{d}{dp} p^X\right) q^{N-X}$$

$$= p \frac{d}{dp} p \frac{d}{dp} \left\{ \sum_{X} \binom{N}{X} p^X q^{N-X} \right\} \quad \text{(exchanging the order)}$$

$$= p \frac{d}{dp} p \frac{d}{dp} (p+q)^N = N(N-1)p^2(p+q)^{N-2} + pN(p+q)^{N-1}$$

$$= N(N-1)p^2 + Np \quad (p+q=1) \tag{12.34}$$

Using this result and (12.26), we obtain

$$\langle X^2 \rangle - \bar{X}^2 = Np(1-p) = Npq \geq 0 \tag{12.35}$$

which is seen to be nonnegative, as it should be. Taking the square root of this quantity, we obtain

$$(\Delta X)_{\text{rms}} = \sqrt{Npq} \tag{12.36}$$

For a particular value, say $p = q = 1/2$, the root mean square deviation $(\Delta X)_{\text{rms}}$ increases monotonically as N increases. Its increase, however, is proportional to $\sqrt{N}$, and thus slower than both the maximum value of X, $(X)_{\text{max}} = N$, and the average $\langle X \rangle = Np$, both of which increase linearly with N. Therefore, the relative deviation

$$\frac{(\Delta X)_{\text{rms}}}{\langle X \rangle} = \frac{\sqrt{Npq}}{Np} = \frac{1}{\sqrt{N}} \sqrt{\frac{q}{p}} \tag{12.37}$$

falls off like $1/\sqrt{N}$.

As an application of the theory of binomial probability distribution, let us take Ehrenfest's model of *random walks*. In this model a walker is allowed to take steps to the right or left on a straight line, with equal step probabilities

$$p = q = \frac{1}{2} \tag{12.38}$$

as indicated in Figure 12.3.

Clearly, the dynamics of the walker can be represented in the two-dimensional lattice shown in Figure 12.1. provided that the walker's moves toward the left are viewed as the object's moves upward in the lattice. The displacement x after N steps, is then given by

$$x = X - Y = X - (N - X) = 2X - N \tag{12.39}$$

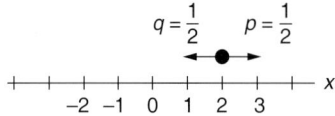

Fig. 12.3 Random walk.

Its average is

$$\langle x \rangle = \langle 2X - N \rangle = 2\langle X \rangle - N$$
$$= 2N\left(\frac{1}{2}\right) - N = 0 \tag{12.40}$$

The mean-square deviation is calculated as follows:

$$\langle (x - \bar{x})^2 \rangle = \langle x^2 \rangle$$
$$= \langle (2X - N)^2 \rangle = 4\langle X^2 \rangle - 4N\langle X \rangle + N^2$$
$$= 4\left[N(N-1)\frac{1}{4} + \frac{1}{2}N\right] - 4N\left[N\left(\frac{1}{2}\right)\right] + N^2$$
$$= N \tag{12.41}$$

Equations (12.40) and (12.41) mean that the walker remains near the origin on the average, but the root-mean-square deviation grows like $\sqrt{N}$.

> **Problem 12.3.1**
>
> Calculate the average and root-mean-square deviation of the number of dots appearing in the throw of two dice.

> **Problem 12.3.2**
>
> In random walks the step probabilities are given at random, so that $p = q = 1/2$. Let us assume that the probability of taking a right step, p, is greater than the probability of taking a left step, q, with the normalization $p + q = 1$. Calculate the mean and mean-square displacement for this model. This model is sometimes called *Bernoulli walks*.

12.4
Microscopic Number Density

Let us consider a cubic box 1 cm on a side containing a gas at *standard temperature and pressure* (STP), that is, $0\,°C$ and 1 atmospheric pressure. The total number N of gas molecules in the box is very large. In fact, a mole of gas should have occupied 22.4 liters at STP. From this the number N can be estimated by multiplying

12.4 Microscopic Number Density

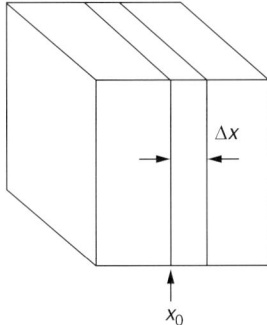

Fig. 12.4 The number of gas molecules within the cubic box of $1.0\,\text{cm}^3$ is very large. The particle number density n is defined by considering the small layer of width Δx.

Avogadro's number $N_0 = 6.02 \times 10^{23}$ by the ratio of volumes $1\,\text{cm}^3/22.4 \times 10^3\,\text{cm}$: $N = (6.02 \times 10^{23}) \times (1/22.4 \times 10^3) = 2.69 \times 10^{19}$. This is indeed a large number.

Let us set up a Cartesian frame of reference along the sides of the cube. Consider a thin layer of gas between $x = x_0$ and $x = x_0 + \Delta x$ as shown in Figure 12.4. The number of molecules in this layer is still very large. For example, when $\Delta x = 0.01$ mm, the number of molecules is of the order 10^{16}. When the gas is in equilibrium, the molecules will be distributed uniformly in the box so that the number of molecules in the layer is proportional to the width Δx. Even if the gas is not in equilibrium, we can expect a linear relation between the number of molecule $N(x_0, \Delta x)$ in the layer and the width of the layer Δx (provided we choose a small width Δx)

$$N(x_0, \Delta x) = n(x_0)\Delta x \quad \text{for small } \Delta x \tag{12.42}$$

The constant of proportionality $n(x_0)$ here is called the *particle-number density* or simply the *number density*.

As the density increases, the probability of finding a particle in the range Δx will be proportionately greater. Therefore, we may also interpret the number density $n(x_0)$ as the *relative probability of finding a particle in the range Δx*. By choosing an infinitesimal width dx at the location x, we then obtain from (12.42)

$$\begin{aligned} n(x,t)dx &= \text{the number of particles in } dx \\ &= \text{relative probability of finding a particle in } dx \end{aligned} \tag{12.43}$$

In equilibrium, the density n is constant. In a nonequilibrium condition, it may vary with the location of the layer and the time. The number density n can in general be regarded as a function of position and time, $n(x, t)$.

By integrating (12.43) over the allowed range, we obtain

$$\int n(x,t)dx = N \tag{12.44}$$

Note that this equation can be regarded as a particular form of the normalization condition for the relative probability distribution function $n(x, t)$. In other words, the distribution function $n(x, t)$ is normalized to N rather than to unity. The two interpretations of the function $n(x, t)$ are both convenient and useful as we will see later in many occasions.

If we know the x-coordinates of all particles $(x_1, x_2, \ldots, x_N)$, we can express the number of the particles within Δx, $N(x_0, \Delta x)$, in terms of the following integrals:

$$N(x_0, \Delta x) = \int_{x_0}^{x_0+\Delta x} dx \sum_{j=1}^{N} \delta(x - x_j) \tag{12.45}$$

Here, $\delta(y)$ is defined by

$$\delta(y) = 0 \quad \text{if } y \neq 0$$

$$\int_{-\infty}^{\infty} dy\, F(y)\delta(y) = F(0) \tag{12.46}$$

where $F(y)$ is an arbitrary function of y that is continuous at $y = 0$.

The properties of the delta function may be understood in the following manner. Let us take a "block" function centered at the origin as shown in Figure 12.5. This function $\Delta(x)$ is given by

$$\Delta(x) = \begin{cases} 1/a & \text{if } -a/2 \leqslant x \leqslant a/2 \quad (a > 0) \\ 0 & \text{otherwise} \end{cases} \tag{12.47}$$

Note that $\Delta(x)$ is positive and symmetric about the origin. The area enclosed by it and the x-axis, is unity. If we let the number a approach zero, we obtain the delta function $\delta(x)$ whose properties are given by (12.46). We stress that the delta function is meaningful only when it is a factor in an integrand.

Equation (12.45) can be proved as follows. If the x-coordinate of a chosen molecule, say the jth molecule, is within the range $(x_0, x_0 + \Delta x)$ so that $x_0 < x_j < x_0 + \Delta x$, then

$$\int_{x_0}^{x_0+\Delta x} dx\, \delta(x - x_j) = \int_{x_0-x_j<0}^{x_0+\Delta x-x_j>0} dy\, \delta(y) = 1 \quad (y = x - x_j)$$

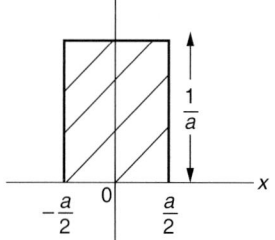

Fig. 12.5 A block function centered at the origin has an area equal to one. This function approaches $\delta(x)$ in the limit of small a.

12.4 Microscopic Number Density

If the x-coordinate is outside the range $(x_0, x_0 + \Delta x)$, the delta function remains zero throughout. Therefore, the molecule is counted only when it is located within the range $(x_0, x_0 + \Delta x)$. By knowing the x-coordinates of the molecules, we can calculate the integral on the rhs of (12.45), and its value clearly is equal to the number of molecules within the range. Q. E. D.

Dividing (12.45) by Δx and taking the small-width limit, we obtain

$$\hat{n}(x_0) = \lim_{\Delta x \to 0} \frac{1}{\Delta x} \int_{x_0}^{x_0 + \Delta x} dx \sum_{j=1}^{N} \delta(x - x_j) = \sum_{j=1}^{N} \delta(x_0 - x_j)$$

or

$$\boxed{\hat{n}(x_0) = \sum_{j=1}^{N} \delta(x_j - x_0)} \qquad (12.48)$$

The expression obtained here for $\hat{n}(x_0)$ depends on the coordinates of the particles, $\{x_j\}$, and will be referred to as the *microscopic number density*. Its connection with the ordinary number density $n(x_0)$, which is independent of the dynamical variables of the particles, will be discussed later. The microscopic number density $\hat{n}(x)$ is normalized to N:

$$\int dx \, \hat{n}(x) = N \qquad (12.49)$$

just as the normal number density $n(x)$. To see this let us substitute $\hat{n}(x)$ from (12.48) into (12.49):

$$\int dx \sum_{j=1}^{N} \delta(x_j - x)$$

We look at the jth term: $\int dx \, \delta(x_j - x)$. Since the particle j must be within the cubic box, the argument $x - x_j$ changes sign in the domain of integration and therefore, the integral equals unity:

$$\int dx \, \delta(x_j - x) = 1$$

We can apply the same argument to each term in the sum and obtain

$$\int dx \sum_{j=1}^{N} \delta(x_j - x) = \sum_{j=1}^{N} \int dx \, \delta(x_j - x) = \sum_{j=1}^{N} (1) = N$$

12.5
Dirac's Delta Function

Some useful properties of delta functions that can be obtained from the definition equations (12.46), are

$$\delta(-x) = \delta(x) \tag{12.50}$$

$$F(x)\delta(x - x_1) = F(x_1)\delta(x - x_1) \tag{12.51}$$

$$\delta(cx) = \frac{1}{|c|}\delta(x), \quad c \neq 0 \tag{12.52}$$

$$\delta[(x-a)(x-b)] = \frac{\delta(x-a) + \delta(x-b)}{|a-b|}, \quad a \neq b \tag{12.53}$$

$$\frac{\partial}{\partial x}\delta(x-y) = -\frac{\partial}{\partial y}\delta(x-y) \tag{12.54}$$

Equation (12.50) means that $\delta(x)$ is symmetric. Clearly, $\delta(-x)$ and $\delta(x)$ are singular at $x = 0$. The integration property can be checked as follows. Multiply (12.50) by a well-behaved function $F(x)$ continuous at $x = 0$, and integrate from $-\infty$ to ∞, and obtain

$$\int_{-\infty}^{\infty} dx\, F(x)\delta(-x) = \int_{-\infty}^{\infty} dx\, F(x)\delta(x) \tag{12.55}$$

The rhs is $F(0)$. The lhs is calculated by a change of variable: $y = -x$ as

$$\text{lhs} \equiv \int_{\infty}^{-\infty} d(-y) F(-y)\delta(y) = F(0)$$

which equals the rhs. Hence, the integration property is established.

The integration property of (12.52) can be checked as follows:

$$\int_{-\infty}^{\infty} dx\, F(x)\delta(cx) = \int_{-\infty}^{\infty} dx\, \frac{1}{|c|} F(x)\delta(x) \tag{12.56}$$

The rhs is clearly $F(0)/|c|$. The lhs can be calculated to yield the same result by considering the two cases, where $c > 0$ and $c < 0$, and using the change of variable $cx = y$.

The proof of (12.51)–(12.54) are left as the reader's exercises.

Problem 12.5.1

Prove (12.51).

Problem 12.5.2

Prove (12.53) and (12.54). In each case, check the singular points first. The integration property can be checked by multiplying by a well-behaved function $F(x)$ and integrating from $-\infty$ to ∞.

12.6
The Three-Dimensional Delta Function

We can extend our theory to the 3-D density as follows. In analogy with (12.43) we express the definition for the density n as follows:

$n(x, y, z, t)\, dx\, dy\, dz$
= the number of particles in the volume element $dx\, dy\, dz$ at time t
= the relative probability of finding a particle in $dx\, dy\, dz$

(12.57)

Figure 12.6 shows the 3-D element $dx\, dy\, dz$. We may denote the position (x, y, z) by the position vector $\mathbf{r}$ and the volume element $dx\, dy\, dz$ by $d^3 r$. In a shorthand notation, we can then express (12.57) as

$n(\mathbf{r}, t)\, d^3 r$
= the number of particles in $d^3 r$ at time t
= the relative probability of finding a particle in $d^3 r$ (12.58)

We can also define the number density, using a set of non-Cartesian coordinates. For example, by taking spherical coordinates (r, θ, ϕ) as shown in Figure 12.7, we may choose the volume element such that

$$d^3 r = (dr)(r\, d\theta)(r \sin \theta\, d\phi) = r^2 \sin \theta\, d\theta\, d\phi \tag{12.59}$$

and apply (12.58).

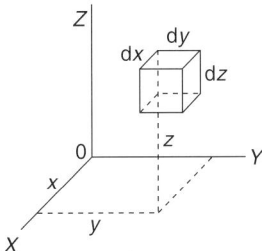

Fig. 12.6 The volume element in Cartesian coordinates.

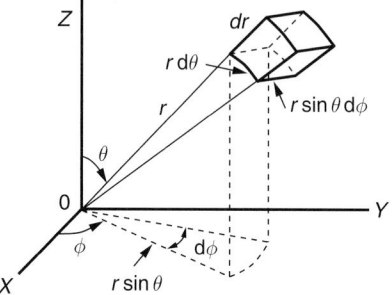

Fig. 12.7 The volume element in the spherical polar coordinates.

Integrating (12.58) over the allowed volume, we obtain

$$\iiint n(x,y,z,t)dxdydz \equiv \int d^3r\, n(\mathbf{r},t) = N \tag{12.60}$$

which represents the density's normalization condition.

If the position of all N particles $(x_1, y_1, z_1, \ldots, x_N, y_N, z_N) \equiv (\mathbf{r}_1, \ldots, \mathbf{r}_N)$ are known, we can express the number of particles within the volume $\Delta x \Delta y \Delta z$, $N(x, y, z, \Delta x \Delta y \Delta z)$, in the form

$$N(x,y,z,\Delta x \Delta y \Delta z)$$

$$= \int_x^{x+\Delta x} d\xi \int_y^{y+\Delta y} d\eta \int_z^{z+\Delta z} d\zeta \sum_j \delta(\xi - x_j)\delta(\eta - y_j)\delta(\zeta - z_j) \tag{12.61}$$

Dividing this by $\Delta x \Delta y \Delta z$ and taking the small-volume limit, we obtain the expression for the microscopic number density as follows:

$$\boxed{\hat{n}(x,y,z) = \sum_j \delta(x-x_j)\delta(y-y_j)\delta(z-z_j)} \tag{12.62}$$

In a shorthand notation, we may write this as

$$\boxed{\hat{n}(\mathbf{r}) = \sum_j \delta^{(3)}(\mathbf{r}-\mathbf{r}_j)} \tag{12.63}$$

where

$$\delta^{(3)}(\mathbf{r}) \equiv \delta(x)\delta(y)\delta(z) \tag{12.64}$$

is a three-dimensional delta function.

To express $\delta^{(3)}(\mathbf{x}-\mathbf{x}')$ in terms of the coordinates (ξ_1, ξ_2, ξ_3) related to (x_1, x_2, x_3), the Jacobian

$$J \equiv \frac{\partial(x_1, x_2, x_3)}{\partial(\xi_1, \xi_2, \xi_3)} \tag{12.65}$$

is useful. We note that the meaningful quantity is $\delta^{(3)}(\mathbf{x}-\mathbf{x}')d^3x$. Hence

$$\delta^{(3)}(\mathbf{x}-\mathbf{x}') = \frac{1}{|J|}\delta(\xi_1-\xi_1')\delta(\xi_2-\xi_2')\delta(\xi_3-\xi_3') \qquad (12.66)$$

So, in spherical coordinates (r, θ, ϕ) we have

$$\delta^{(3)}(\mathbf{x}-\mathbf{x}') = \frac{1}{r^2 \sin\theta}\delta(r-r')\delta(\phi-\phi')\delta(\theta-\theta') \qquad (12.67)$$

13
Liouville Equation

13.1
Liouville's Theorem

Let us consider a particle moving in one dimension, acted upon by a conservative force $F = -dV/dx$. The Hamiltonian H of this system is given by

$$H(q, p) = \frac{p^2}{2m} + V(q) \tag{13.1}$$

In order to emphasize the fact that the coordinate and momentum are a pair of canonical variables, we have denoted these variables by (q, p) rather than (x, p). The dynamical state of the particle specified by (q, p) changes with time, and its evolution is determined from Hamilton's equations of motion:

$$\dot{q} = \frac{\partial H}{\partial p} = \frac{p}{m}, \qquad \dot{p} = -\frac{\partial H}{\partial q} = -\frac{dV}{dq} \tag{13.2}$$

The dynamical state can be represented by a point in phase space, see Figure 13.1. This point will move along the constant-energy curve defined by

$$\frac{p^2}{2m} + V(q) = E \qquad \text{or} \qquad p = \pm\sqrt{2m[E - V(q)]} \tag{13.3}$$

where E is the constant total energy.

The curve given by (13.3) does not branch out into two or more lines. In other words, the point must move along the curve of a constant total energy with velocities $(\dot{q}, \dot{p})$. For different values of E, the curves are different and never intersect each other.

Let us now choose an arbitrary closed curve C, imagine that a particle starts at $t = 0$ with the initial condition represented by a point P on the curve C, and this point arrives at P' at a later time t. This natural evolution of the point from P to P' is schematically represented by a dotted line in Figure 13.1. If the particle started with an initial condition infinitesimally close to that represented by P, then its final state will also be infinitesimally close to that represented by P'. It is then clear that the infinite set of points located on the closed curve C in phase space generates the

Mathematical Physics. Shigeji Fujita and Salvador V. Godoy
Copyright © 2010 WILEY-VCH Verlag GmbH & Co. KGaA, Weinheim
ISBN: 978-3-527-40808-5

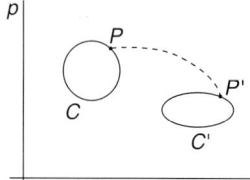

Fig. 13.1 The area in phase space enclosed by the curve C′ equals that enclosed by the curve C (Liouville Theorem).

infinite set of points forming another closed curve C′. In general, the shape of C′ will be different from that of C. We now state an important theorem:

Theorem

The area A′ enclosed by C′ is equal to the area A enclosed by C (for any time t).

This is one form of the Liouville theorem. We will prove it in two steps.

First, let us consider a small rectangle ABCD with sides parallel to the q- and p-axes. See Figure 13.2. The corners A, B, C, D have coordinates (q, p), $(q + \Delta q, p)$, $(q + \Delta q, p + \Delta p)$, $(q, p + \Delta p)$, respectively. A short time Δt later, these points move to new positions A′, B′, C′, D′, which again form the four corners of a new quasirectangle. The deviation of A′B′C′D′ from a true rectangular shape can be made arbitrarily small by choosing sufficiently small Δt. We now wish to show that the area $\omega(t + \Delta t)$ of the new rectangle A′B′C′D′ is equal to the area $\omega(t)$ of the old rectangle ABCD. The distance between A and B, denoted by $\overline{AB}$, equals Δq. Similarly $\overline{AD} = \Delta p$. We therefore obtain

$$\omega(t) = \overline{AB}\,\overline{AD} = \Delta q \Delta p \tag{13.4}$$

The new area $\omega(t + \Delta t)$ is given by

$$\omega(t + \Delta t) = \overline{A'B'}\,\overline{A'D'} \tag{13.5}$$

The difference in the area, $\Delta \omega$, is therefore

$$\Delta \omega \equiv \omega(t + \Delta t) - \omega(t) = \overline{A'B'}\,\overline{A'D'} - \overline{AB}\,\overline{AD}$$
$$= (\overline{A'B'} - \overline{AB})\overline{A'D'} + \overline{AB}(\overline{A'D'} - \overline{AD}) \tag{13.6}$$

The line $\overline{A'B'}$ runs nearly parallel to the q-axis, and therefore the length $\overline{A'B'}$ can be approximately equal to the difference in the q-coordinates of B′, $q(B')$, and the

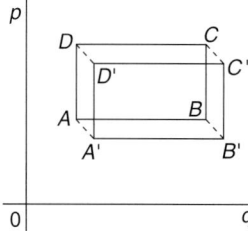

Fig. 13.2 The rectangle ABCD will form the "rectangle" of the same position-momentum area at a later time.

q-coordinate of A', $q(A')$:

$$\overline{A'B'} \cong q(B') - q(A') \tag{13.7}$$

The point A' is reached by the system from the point A in the time Δt. The q-coordinates of A', $q(A')$, can therefore be computed as

$$q(A') \cong q(A) + \dot{q}(A)\Delta t = q + \dot{q}(q, p)\Delta t \tag{13.8}$$

In the same manner, we obtain

$$q(B') \cong q(B) + \dot{q}(B)\Delta t = q + \Delta q + \dot{q}(q + \Delta q, p)\Delta t \tag{13.9}$$

Using the last three equations we obtain

$$\begin{aligned}\overline{A'B'} &\cong q(B') - q(A') \\ &= [q + \Delta q + \dot{q}(q + \Delta q, p)\Delta t] - [q + \dot{q}(q, p)\Delta t] \\ &= \Delta q + [\dot{q}(q + \Delta q, p) - \dot{q}(q, p)]\Delta t \\ &= \Delta q + \left[\frac{\partial}{\partial q}\dot{q}(q, p)\right]\Delta q \Delta t \end{aligned} \tag{13.10}$$

Since $\overline{AB} = \Delta q$, we find that

$$\overline{A'B'} - \overline{AB} = \left[\frac{\partial}{\partial q}\dot{q}(q, p)\right]\Delta q \Delta t \tag{13.11}$$

We also find that

$$\overline{A'D'} - \overline{AD} = \left[\frac{\partial}{\partial p}\dot{p}(q, p)\right]\Delta p \Delta t \tag{13.12}$$

Introducing (13.11) and (13.12) in (13.6), we obtain

$$\begin{aligned}\Delta\omega &= (\overline{A'B'} - \overline{AB})\overline{A'D'} + \overline{AB}(\overline{A'D'} - \overline{AD}) \\ &= \left[\frac{\partial}{\partial q}\dot{q}(q, p)\right]\Delta q \Delta t \Delta p + \left[\frac{\partial}{\partial p}\dot{p}(q, p)\right]\Delta p \Delta t \Delta q \\ &= \left[\frac{\partial}{\partial q}\dot{q}(q, p) + \frac{\partial}{\partial p}\dot{p}(q, p)\right]\Delta q \Delta p \Delta t + O(\Delta t^2) \end{aligned} \tag{13.13}$$

In arriving at this expression, we retained the leading term of the order Δt explicitly; the rest is represented by $O(\Delta t^2)$. Dividing (13.13) by Δt and taking the small-time limit, we obtain

$$\begin{aligned}\frac{d\omega}{dt} &= \lim_{\Delta t \to 0}\left[\left(\frac{\partial \dot{q}}{\partial q} + \frac{\partial \dot{p}}{\partial p}\right)\Delta q \Delta p + O(\Delta t)\right] \\ &= \left(\frac{\partial^2 H}{\partial q \partial p} - \frac{\partial^2 H}{\partial p \partial q}\right)\omega(t) = 0\end{aligned}$$

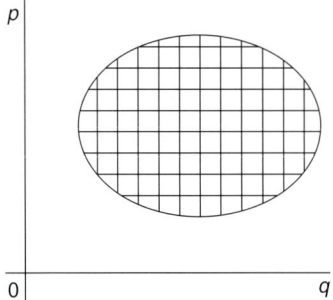

Fig. 13.3 A division of an area into small cells in the phase space.

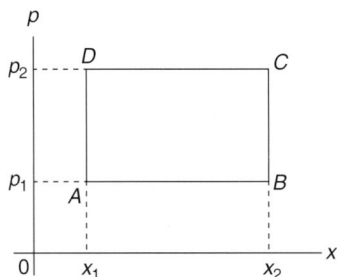

Fig. 13.4 See Problem 13.1.1.

or

$$\boxed{\frac{d\omega}{dt} = 0} \qquad (13.14)$$

where we used (13.2). Equation (13.14) means that the infinitesimal phase-space area does not change with the time.

Next, consider a finite area enclosed by a curve. Divide the area into infinitesimal rectangular cells as shown in Figure 13.3. According to (13.14), each cell will evolve into a new cell of the same area. Summing up over all the cells, we can deduce that this finite area will also evolve into a new area of the same magnitude. Q. E. D.

The present form of Liouville's theorem was given in geometrical terms. The content of Liouville's theorem can be rephrased such that its physical meaning can be grasped in a more substantial manner. This will be discussed in the following section.

Problem 13.1.1

Consider a free particle in one dimension. The Hamiltonian is $p^2/2m$.
1. Write down Hamilton's equation of motion. The state of motion can be represented by a point in phase plane (x, p). Hamilton's equations describe how such a point moves in time.

2. Rectangle ABCD is drawn in the phase plane. Where do the corner points move after a time t? Give analytical answers. Exhibit the movement in the figure.
3. Verify that the area enclosed by the quadrangle A'B'C'D', where A', B', C', D' are the end-points of A, B, C and D, respectively, is equal to the area enclosed by the rectangle ABCD.

13.2
Probability Distribution Function. The Liouville Equation

Let us consider a particle moving in one dimension and characterized by the Hamiltonian.

$$H = \frac{p^2}{2m} + V(x)$$

The dynamical state of this system can be specified by position x and momentum p. These variables change in time, following Hamilton's equations of motion in (13.2). Assume that the particle starts with a total energy equal to E. This energy is conserved subsequently. The change in the state can be represented by the point in phase space, moving on the constant-energy curve specified by the equation

$$\frac{p^2}{2m} + V(x) = E \tag{13.15}$$

A typical constant-energy line is indicated in Figure 13.5.

Suppose that we do not know from what state the particle started initially, but we are given the initial probability distribution in the phase space (x–p). In the present section, we discuss a general manner of calculating the dynamical properties of a system.

Let us take a small element $dx\,dp$ surrounding a point (x, p) in the phase space as shown in Figure 13.6.

The probability that the particle has a position within the range $(x, x + dx)$ and a momentum within the range $(p, p + dp)$ will be proportional to the element $dx\,dp$. We express this relationship in the following manner:

$$\rho(x, p)\frac{dx\,dp}{2\pi\hbar} = \text{the probability of finding a particle in } dx\,dp \tag{13.16}$$

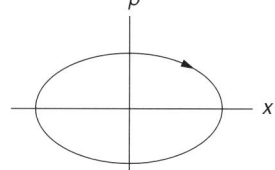

Fig. 13.5 The point representing the dynamical state of a particle moves along a constant-energy line in the phase space.

13 Liouville Equation

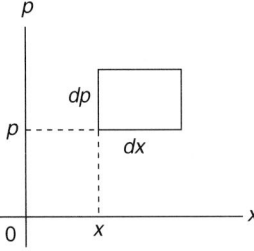

Fig. 13.6 A small surface element $dxdp$ in the phase space.

The proportionality factor ρ is called the *probability distribution function* in phase space. This ρ in general depends on the point (x, p). The element $dxdp$ has the dimension of angular momentum. By choosing the factor $(2\pi\hbar)^{-1}$ ($\hbar$: Planck's constant), the distribution function ρ will become a dimensionless number. The convenience of this choice will become clear later. Integrating (13.16) over the entire phase space, we obtain the *normalization condition*:

$$\frac{1}{2\pi\hbar} \iint dxdp\,\rho(x, p, t) = 1 \tag{13.17}$$

A dynamical property ξ such as position, momentum, kinetic and potential energies, is a function of x and p. We denote this property by $\xi(x, p, t)$. If the system is distributed over the phase space with the probability distribution function $\rho(x, p, t)$, the average value of ξ will be given by

$$\langle \xi \rangle_t = \frac{1}{2\pi\hbar} \iint dxdp\,\xi(x, p, t)\rho(x, p, t) \tag{13.18}$$

Note that the rhs can be obtained in the standard manner: If the particle is found in $dxdp$, then the value of the dynamical quantity is $\xi(x, p, t)$. Multiplying $\xi(x, p, t)$ by the probability $\rho(x, p, t)dxdp/2\pi\hbar$ and integrating the product over the phase space, we obtain the average $\langle \xi \rangle_t$.

In Section 13.1 we saw that the point representing the dynamical state moves with the definite velocities $(\dot{x}, \dot{p})$, given by Hamilton's equations of motion (13.2). This means that the probability distribution will change with time. We now wish to find the evolution equation of the distribution function $\rho(x, p, t)$. This can be achieved with the aid of Liouville's theorem, which was proved in Section 13.1.

Let us suppose that a typical point P within the element $dxdp \equiv d\omega$ at the time t moves to a new point $P'(x', p')$ at the time t', as indicated in Figure 13.7. At the same time, the boundary of the phase-space volume $d\omega$ will move and form a new boundary. The point P' must be within the new boundary, which encloses a phase-space volume $d\omega'$. In fact, if the point P had crossed the boundary of the moving boundary at some intermediate time, it would have occupied the same point in the phase space as one of the points forming the boundary. But this cannot happen since the motion of the particle started with different initial states must be different forever since the evolution lines cannot intersect each other. From this consideration, we can deduce that the probability of finding the particle in the volume $d\omega$ at

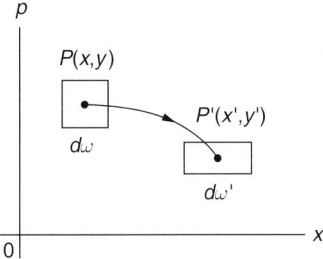

Fig. 13.7 A point P within the phase-space element $d\omega$ at time t moves to a new point P' within the phase-space element $d\omega'$ at time t'.

the time t is equal to the probability of finding it in the new volume $d\omega'$ at time t':

$$\rho(x, p, t)\frac{d\omega}{2\pi\hbar} = \rho(x', p', t')\frac{d\omega'}{2\pi\hbar} \tag{13.19}$$

Applying Liouville's theorem:

$$d\omega = d\omega' \tag{13.20}$$

we obtain from (13.19)

$$\rho(x, p, t) = \rho(x', p', t') \tag{13.21}$$

This means that the probability distribution function does not change if we follow the motion of the definition volume element.

If we choose an infinitesimal time interval dt so that $t' = t+dt$, then $x' = x+\dot{x}\,dt$ and $p' = p + \dot{p}\,dt$. We can then obtain from (13.21)

$$\rho(x, p, t) = \rho(x + \dot{x}\,dt, p + \dot{p}\,dt, t + dt) \tag{13.22}$$

This equation is similar to (8.36), which represents the fact that the mass density for an incompressible fluid does not change along the material flow. In the text, we happen to use the same symbol ρ for the probability distribution function in the phase space and the mass density in the ordinary space. Accidental coincidence such as this is not desirable but practically unavoidable because there are not enough letters to distinguish physical quantities by separate symbols. A list of symbols for important physical quantities is given at the end of the text to help avoid confusion.

By expanding the rhs of (13.22) in powers of dt and comparing terms of the first order in dt, we obtain (Problem 13.2.1)

$$\boxed{\frac{\partial \rho}{\partial t} + \dot{x}\frac{\partial \rho}{\partial x} + \dot{p}\frac{\partial \rho}{\partial p} = 0} \tag{13.23}$$

This is the *Liouville equation*. It is a fundamental equation that governs the flow of the probability distribution in the phase space. The velocities $(\dot{x}, \dot{p})$ appearing in the equation can be eliminated with the aid of Hamilton's equation of motion, (13.2). After rearrangements, we obtain

$$\boxed{\frac{\partial \rho}{\partial t} = \frac{\partial H}{\partial x}\frac{\partial \rho}{\partial p} - \frac{\partial H}{\partial p}\frac{\partial \rho}{\partial x}}$$
(13.24)

which is also called the *Liouville equation*.

Note that the Liouville equation is a *linear* equation with the time-, position- and momentum-derivatives. It may be solved with given initial and boundary conditions. Some discussions of the boundary condition will be given below.

We have just observed that the distribution function ρ in general changes with time and must be regarded as a function of the time t. Let us go back and look at (13.17), which represents the normalization condition for the function $\rho(x, p, t)$. This equation should hold for all time. This must be so by a general probability argument that the probability of finding the particle in *any* dynamical state is unity. The integration with respect to x and p, precludes the x and p dependence of the lhs. But since no t integration is prescribed, the time independence of the integral on the lhs,

$$\iint dx\,dp\,\rho(x, p, t) \equiv A(t)$$
(13.25)

is not obvious. In the following, we will prove that this integral $A(t)$ is independent of time.

To do this, we compute the time derivative of (13.25):

$$\frac{dA}{dt} = \frac{d}{dt}\iint dx\,dp\,\rho(x, p, t) = \iint dx\,dp\,\frac{\partial}{\partial t}\rho(x, p, t)$$

$$= \iint dx\,dp \left[\frac{\partial H}{\partial x}\frac{\partial \rho}{\partial p} - \frac{\partial H}{\partial p}\frac{\partial \rho}{\partial x}\right] \quad \text{[use of (13.24)]}$$
(13.26)

Let us take the first double integral. Integrating it by parts, we obtain

$$\int dx \int_{-\infty}^{\infty} dp\,\frac{\partial H}{\partial x}\frac{\partial \rho}{\partial p}$$

$$= \int dx \left[\frac{\partial H}{\partial x}\rho\right]_{p=-\infty}^{p=\infty} - \int dx \int dp\,\frac{\partial^2 H}{\partial p\,\partial x}\rho$$

The first term can be dropped since the probability of finding the particle with an infinitely large momentum (magnitude) is zero:

$$\rho(x, +\infty, t) = \rho(x, -\infty, t) = 0$$
(13.27)

We, therefore, obtain

$$\int dx \int_{-\infty}^{\infty} dp\,\frac{\partial H}{\partial x}\frac{\partial \rho}{\partial p} = -\int dx \int dp\,\frac{\partial^2 H}{\partial p\,\partial x}\rho$$
(13.28)

Next consider the second double integral in (13.26). By integrating by parts in x, we obtain,

$$\int dp \int dx \, \frac{\partial H}{\partial p} \frac{\partial \rho}{\partial x}$$
$$= \int dp \left[\frac{\partial H}{\partial p} \rho \right]_{\text{boundary}} - \int dp \int dx \, \frac{\partial^2 H}{\partial x \partial p} \rho \qquad (13.29)$$

The boundary term can be dropped if we impose the fixed-ends boundary condition:

$$\rho(0, p, t) = \rho(L, p, t) = 0 \qquad (13.30)$$

where L is the distance between the fixed points. We then obtain

$$\int dp \int dx \, \frac{\partial H}{\partial p} \frac{\partial \rho}{\partial x} = -\int dp \int dx \, \frac{\partial^2 H}{\partial p \partial x} \rho \qquad (13.31)$$

Using the last few equations, we obtain

$$\frac{dA}{dt} = \iint dx \, dp \left(-\frac{\partial^2 H}{\partial x \partial p} + \frac{\partial^2 H}{\partial p \partial x} \right) \rho = 0 \qquad (13.32)$$

This means that the normalization condition holds for all time.

Note that this result was obtained without introducing any explicit form of the Hamiltonian. Thus, the normalization is maintained for a general system.

Problem 13.2.1

Derive the Liouville equation (13.23) from (13.22).

HINT: Expand the rhs of (13.22) in powers of dt and compare terms of the first order in dt.

Problem 13.2.2

Derive the Liouville equation for
1. a free particle ($V = 0$), and
2. a simple harmonic oscillator ($V = kx^2/2$).
3. Start from (13.23) or (13.24)

13.3 The Gibbs Ensemble

In the preceding section, we introduced a statistical method of calculating average dynamical properties of a system in terms of the probability distribution function ρ in the phase space. As we will see later, it is convenient to reinterpret the statistical processes from a somewhat different angle. This reinterpretation, due to Gibbs, (J. Willard Gibbs, American, 1839–1903), will be presented in the present section.

Let us recall that the density field $n(x, t)$ can be regarded as either the number density or the relative probability distribution function in the x-space. In practice, however, the interpretation of $n(x, t)$ as the particle number density is almost always preferred. For one thing, the particle number density is one of the most basic concepts familiar to us physicists. Let us now apply a similar interpretation to the probability distribution function $\rho(x, p.t)$ in the phase space.

Let us take a collection of a large number N_G of dynamical systems, each representing a single particle in one-dimensional motion, characterized by the same Hamiltonian, but, in general, having a different initial condition. The dynamical state of each system can be represented by a point in the phase space. The distribution of these system points in the phase space can be specified by the *system-points density* $\rho'(x, p, t)$, which is defined by

$$\rho'(x, p, t) \frac{dx\,dp}{2\pi\hbar} = \text{the number of system points in } dx\,dp \text{ at time } t \quad (13.33)$$

Integrating this over the phase space, we obtain

$$\frac{1}{2\pi\hbar} \iint dx\,dp\, \rho'(x, p, t) = N_G \quad (13.34)$$

Since each system point moves with the velocities $(\dot{x}, \dot{p})$ as given by Hamilton's equations of motion, the density $\rho'(x, p, t)$ will obey the same Liouville equation

$$\frac{\partial \rho'}{\partial t} = \frac{\partial H}{\partial x}\frac{\partial \rho'}{\partial p} - \frac{\partial H}{\partial p}\frac{\partial \rho'}{\partial x} \quad (13.35)$$

as for the probability distribution function ρ.

If we now postulate that at the initial time the probability distribution function ρ and the system-point density ρ' are related by

$$\rho' = N_G \rho \quad (13.36)$$

then this relation will be maintained for all time since both ρ and ρ' obey the same evolution equation. That is, the distribution functions ρ and ρ' will be different only by a constant factor N_G. The collection of dynamical systems so constructed, will be called the *Gibbs ensemble*.

The main advantage of the Gibbs ensemble lies in the conceptual ease in the calculations of the averages. The average $\langle \xi \rangle_t$ of a dynamical quantity ξ of the system, given by (13.18), can be re-expressed as follows:

$$\boxed{\langle \xi \rangle_t = \frac{(2\pi\hbar)^{-1} \iint dx\,dp\, \xi(x, p)\rho'(x, p, t)}{(2\pi\hbar)^{-1} \iint dx\,dp\, \rho'(x, p, t)}} \quad (13.37)$$

where we used (13.18) and (13.36). This allows us to interpret the average $\langle \xi \rangle_t$ as the *average of ξ over the Gibbs ensemble*.

In the language of mathematical statistics, the system-point density ρ' merely represents a *relative probability distribution function* normalized to a large number N_G. Since all statistical calculations can in principle be formulated in terms of either function ρ or ρ', introduction of a Gibbs ensemble is purely a matter of convenience. But the Gibbs ensemble is a useful physical concept. In the development of statistical mechanics, we use the ensemble language and/or the probability language without stressing the distinction. We will often drop the prime on ρ.

We have introduced the system-point density in the phase space in analogy with the particle-number density in the x-space. This analogy, however, should not be extended too far. The *member systems in the Gibbs ensemble*, by definition, *do not interact* with each other, while the molecules in the real world in general interact with each other. In fact, the system-point density ρ changes following the Liouville equation. On the other hand, the evolution of the particle-number density obeys the continuity equation, which is quite different in character from the Liouville equation.

13.4
Many Particles Moving in Three Dimensions

The foregoing theory developed for a single particle in linear motion can simply be extended to a system of many particles moving in the ordinary three-dimensional (3-D) space.

Such a system can in general be described in terms of a set of canonical variables $(q_1, p_1, q_2, p_2, \ldots, q_{3N}, p_{3N}) \equiv (q, p)$, where N represents the total number of particles. These variables change, following Hamilton's equations of motion:

$$\dot{q}_k = \frac{\partial H(q, p)}{\partial p_k} \quad \dot{p}_k = -\frac{\partial H(q, p)}{\partial q_k}, \quad k = 1, 2, \ldots, 3N \tag{13.38}$$

where H represents the Hamiltonian of the system.

The dynamical state, characterized by the variables (q, p), can be represented by a point in the $6N$-dimensional phase space, called the Γ-space. Since the state changes with the velocities $(\dot{q}_1, \dot{p}_1, \ldots, \dot{q}_{3N}, \dot{p}_{3N}) \equiv (\dot{q}, \dot{p})$ the point representing the state will move in the Γ-space: *The "volume" in the Γ-space bounded by a closed surface will remain the same if the moving surface is followed.*

We can recast this theorem in the form of the Liouville equation as follows: Let us introduce the system-point density ρ in the Γ-space by

$$\rho(q_1, p_1, \ldots, q_{3N}, p_{3N}, t) dq_1 dp_1 \cdots dq_{3N} dp_{3N} (2\pi\hbar)^{-3N}$$
$$\equiv \rho(q, p, t) d^{3N}q \, d^{3N}p \, (2\pi\hbar)^{-3N}$$
$$= \text{the number of system points in the phase-space volume } d^{3N}q \, d^{3N}p$$
at time t
$$= \text{the relative probability of finding the system in } d^{3N}q \, d^{3N}p$$

$$\tag{13.39}$$

From the interpretation given in the last line, the function $\rho(q, p, t)$ is also called the *distribution function* in the Γ-space.

The function $\rho(q, p, t)$ obeys the Liouville equation:

$$\boxed{\frac{\partial \rho}{\partial t} + \sum_{k=1}^{3N} \left(\dot{q}_k \frac{\partial \rho}{\partial q_k} + \dot{p}_k \frac{\partial \rho}{\partial p_k} \right) = 0} \tag{13.40}$$

or alternatively, using Hamilton's equations,

$$\boxed{\frac{\partial \rho}{\partial t} + \sum_{k=1}^{3N} \left(\frac{\partial H}{\partial q_k} \frac{\partial \rho}{\partial p_k} - \frac{\partial H}{\partial p_k} \frac{\partial \rho}{\partial q_k} \right) = 0} \tag{13.41}$$

These equations are straightforward extensions of (13.23) and (13.24).

The Liouville equation plays a fundamental role in statistical mechanics. It deals with the time evolution of the density of system points in the $6N$-dimensional phase space. It is noted that the transition from classical to quantum mechanics, can be carried out most directly with the aid of the Hamiltonian mechanics.

All statistical properties of a system can be expressed in terms of the distribution function ρ. If the system ever reaches equilibrium, its properties do not change with time, and will be described in terms of a stationary (time-independent) distribution function. In seeking such a stationary distribution function, the following theorem is very important.

Theorem

If the distribution function ρ depends on (q, p) only through a time-independent Hamiltonian $H(q, p)$, then $\partial \rho / \partial t = 0$. That is, this function ρ is stationary.

This theorem may be proved as follows. First, let us consider the case of a single particle in linear motion. From premises $\rho = \rho[H(x, p)]$. Using the chain rule of differentiation, we obtain

$$\frac{\partial \rho}{\partial x} = \frac{d\rho}{dH} \frac{\partial H}{\partial x}, \quad \frac{\partial \rho}{\partial p} = \frac{d\rho}{dH} \frac{\partial H}{\partial p}$$

Introducing these in the Liouville equation (13.24) we obtain

$$\frac{\partial \rho}{\partial t} = \frac{\partial H}{\partial x} \frac{\partial \rho}{\partial p} - \frac{\partial H}{\partial p} \frac{\partial \rho}{\partial x}$$

$$= \frac{\partial H}{\partial x} \left(\frac{d\rho}{dH} \frac{\partial H}{\partial p} \right) - \frac{\partial H}{\partial p} \left(\frac{d\rho}{dH} \frac{\partial H}{\partial x} \right) = 0$$

This proof can simply be extended to the multidimensional case.

It is cautioned that this theorem holds only for systems characterized by Hamiltonians that do not depend on time explicitly. When the Hamiltonian H depends on time, the distribution function $\rho(q, p, t)$ still obeys the Liouville equation. However, the equation does not have a stationary solution.

Problem 13.4.1

Consider a particle moving in three dimensions under a conservative force $F = -\nabla V$.

1. Write down the Hamiltonian H in Cartesian coordinates and momenta (x, y, z, p_x, p_y, p_z).
2. Using (13.40) and (13.41) write down the Liouville equation in two alternative forms.
3. Verify that if the distribution function is given in the form: $\rho = c \exp(-\beta H)$, the distribution is stationary.

13.5 More about the Liouville Equation

The Liouville equation governs the time evolution of the system-point density ρ in the Γ-space. In the present section, we will show that the density ρ has the same value for any canonical representation in which the Γ-space is constructed. Also, we will re-express the Liouville equation in a compact form, using Poisson brackets. The discussions presented here are important in Dirac's formulation of quantum mechanics.

Let us take a general system of N particles moving in the three-dimensional space. The dynamical state of this system can be specified by a set of $6N$ canonical variables $(q_1, \ldots, q_{3N}, p_1, \ldots, p_{3N}) \equiv (q, p)$ or again by another set of variables $(Q_1, \ldots, Q_{3N}, P_1, \ldots, P_{3N}) \equiv (Q, P)$. The Jacobian of the transformation from (q, p) to (Q, P) is unity. The phase-space volume elements, therefore, satisfy

$$dQ_1 \ldots dQ_{3N} dP_1 \ldots dP_{3N} = dq_1 \ldots dq_{3N} dp_1 \ldots dp_{3N}$$

or in a shorthand notation

$$d^{3N}Q d^{3N}P = d^{3N}q d^{3N}p \tag{13.42}$$

An important consequence of this relation is that the system-point density defined at a given point in the (q, p) space and that corresponding to the same state in the (Q, P) space have the same value:

$$\rho'(Q, P) = \rho(q, p) \quad \text{for the same state} \tag{13.43}$$

This may be demonstrated in the following manner. Let us take a point A in the (q, p) space as shown in Figure 13.8a. The dynamical state represented by the point A can also be represented by the point A' in the (Q, P) space, which is indicated in Figure 13.8b. Let us consider a phase-space volume element $d^{3N}q d^{3N}p$ surrounding the point A in the (q, p) space. By recalling the definition equations (13.39), the number of system points in $d^{3N}q d^{3N}p$ is given by

$$\rho_A d^{3N}q d^{3N}p (2\pi\hbar)^{-3N}$$

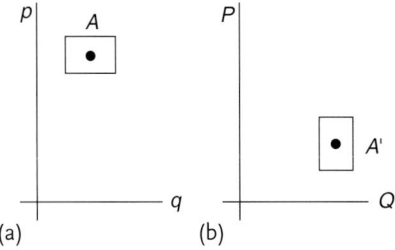

Fig. 13.8 The points A and A' correspond to the same state. The system-point density at A is found to be equal to the system-point density at A'.

By (13.42), this number is equal to

$$\rho_A d^{3N}Q d^{3N}P (2\pi\hbar)^{-3N}$$

where the element $d^{3N}Q d^{3N}P$ should surround the point A' (corresponding to the same state) in the (Q, P) space. Each system point within $d^{3N}q d^{3N}p$ will have a corresponding point within $d^{3N}Q d^{3N}P$ and vice versa. Therefore, the number of system points $\rho_A d^{3N}Q d^{3N}P (2\pi\hbar)^{-3N}$ should be equal to the number of system points in $d^{3N}Q d^{3N}P$, that is, the number that, by definition, is given by $\rho'_{A'} d^{3N}Q d^{3N}P (2\pi\hbar)^{-3N}$. In other words, we must have

$$\rho_A = \rho'_{A'}$$

By representing A and A' by (q, p) and (Q, P) respectively, we then obtain (13.43). Q. E. D.

The Liouville equation, given in (13.41), can be transformed as follows:

$$\frac{\partial \rho}{\partial t} = \sum_{k=1}^{3N} \left(\frac{\partial H}{\partial q_k} \frac{\partial \rho}{\partial p_k} - \frac{\partial \rho}{\partial q_k} \frac{\partial H}{\partial p_k} \right) \equiv \{H, \rho\}$$

or

$$\boxed{\frac{\partial \rho}{\partial t} = \{H, \rho\}} \qquad (13.44)$$

where the curly brackets denote the *Poisson brackets* defined by

$$\{u, v\} \equiv \sum_k \left(\frac{\partial u}{\partial q_k} \frac{\partial v}{\partial p_k} - \frac{\partial v}{\partial q_k} \frac{\partial u}{\partial p_k} \right) \qquad (13.45)$$

The Poisson brackets are representation independent, that is, they can be computed with an arbitrary choice of canonical variables. The evolution equation (13.44) is similar to the equation of motion for a dynamical function ξ. Only the sign is different. Equations of general nature that are independent of representation, are likely to survive (in a modified form) when a "new" (quantum) mechanics is introduced.

13.6
Symmetries of Hamiltonians and Stationary States

A system of a monatomic gas is characterized by the Hamiltonian

$$H = \sum_j p_j^2/2m + \sum_{j>k}\sum_k V\left(|\mathbf{r}_j - \mathbf{r}_k|\right) \tag{13.46}$$

Let us choose a Cartesian frame of reference and express this Hamiltonian in Cartesian coordinates and momenta,

$$\sum_j \left(p_{jx}^2 + p_{jy}^2 + p_{jz}^2\right)/2m$$

$$+ \sum_{j>k}\sum_k V\left(\sqrt{(x_j - x_k)^2 + (y_j - y_k)^2 + (z_j - z_k)^2}\right)$$

$$\equiv H(x_1, y_1, z_1, \ldots, p_{1x}, p_{1y}, p_{1z}, \ldots) \tag{13.47}$$

Take another Cartesian frame that has the same orthogonal unit vector as the first frame but that has the origin *displaced* by a constant vector **a** as shown in Figure 13.9. If we denote the Cartesian coordinates and momenta referring to the second frame by capital letters, we then have the following relations:

$$X_j + a_x \equiv x_j, \quad Y_j + a_y \equiv y_j, \quad Z_j + a_z \equiv z_j \tag{13.48}$$

$$P_{jx} \equiv m\frac{dX_j}{dt} = m\frac{dx_j}{dt} \equiv p_{jx}, \quad P_{jy} = p_{jy}, \quad P_{jz} = p_{jz} \tag{13.49}$$

Using these relations, we can easily show that the Hamiltonian function has the same form in both frames, that is,

$$\sum_j \left(P_{jX}^2 + P_{jY}^2 + P_{jZ}^2\right)/2m$$

$$+ \sum_{j>k}\sum_k V\left(\sqrt{(X_j - X_k)^2 + (Y_j - Y_k)^2 + (Z_j - Z_k)^2}\right)$$

$$= \sum_j \left(p_{jx}^2 + p_{jy}^2 + p_{jz}^2\right)/2m$$

$$+ \sum_{j>k}\sum_k V\left(\sqrt{(x_j - x_k)^2 + (y_j - y_k)^2 + (z_j - z_k)^2}\right)$$

or

$$H(X_1, Y_1, Z_1, \ldots, P_{1X}, P_{1Y}, P_{1Z}, \ldots) = H(x_1, y_1, z_1, \ldots, p_{1x}, p_{1y}, p_{1z}, \ldots) \tag{13.50}$$

In fact, this is true for *any* two frames of coordinates that are mutually generated by a *parallel translation* of the Cartesian bases. We will say that our Hamiltonian is

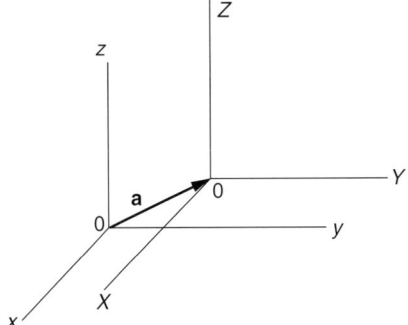

Fig. 13.9 A parallel translation of the Cartesian axes (x, y, z) by the vector **a**.

invariant under translation. A dynamical function is said to be invariant under a certain operation if it can be expressed in the same functional form. For example, the total momentum $\sum_j \mathbf{p}_j$ clearly is invariant under translation, but the total angular momentum $\sum_j \mathbf{r}_j \times \mathbf{p}_j$ is not since

$$\sum_j \mathbf{R}_j \times \mathbf{P}_j = \sum_j (\mathbf{r}_j - \mathbf{a}) \times \mathbf{p}_j = \sum_j \mathbf{r}_j \times \mathbf{p}_j - \mathbf{a} \times \sum_j \mathbf{p}_j \tag{13.51}$$

A many-body system whose Hamiltonian is translation invariant can have a *homogeneous stationary state*. This may be argued as follows. Suppose that such a system reaches, at a certain time, a *homogeneous* state, that is, a state in which the local properties (densities) are the same in the entire space. Now, the dynamics of particles, which is characterized by the Hamiltonian, does not favor one particular place over any other place. Then, the system should stay homogeneous. On the other hand, if the Hamiltonian for a system is not invariant under translation, the motion of the molecules depends on the location. Then, the system can hardly maintain a homogeneous state.

The foregoing arguments imply that *for a system to have a homogeneous stationary state, its Hamiltonian must be invariant under translation*. The inverse of this statement, however, is not true. That is, the translation invariance of the Hamiltonian does not necessarily lead to the existence of a homogeneous stationary state. For example, a system of Ne atoms has a homogeneous gaseous state at STP; it has a homogeneous liquid state below 27.26 K but freezes to form a nonhomogeneous crystalline (solid) state at 24.56 K and below.

Our Hamiltonian H in (13.48) possesses a further property: it is *invariant under rotation*. This means the following: assume that a Cartesian frame of reference was chosen, and in terms of Cartesian coordinates and momenta, the Hamiltonian function was obtained in the form of (13.47); let us now rotate the orthogonal bases (without changing the origin) and obtain a second Cartesian frame; when we construct the new Hamiltonian in the second frame, we observe that this function has the *same* form as the original Hamiltonian.

13.6 Symmetries of Hamiltonians and Stationary States

In general, the Cartesian coordinates in the first and second frames are connected by linear relations:

$$X_j = a_{11}x_j + a_{12}y_j + a_{13}z_j$$
$$Y_j = a_{21}x_j + a_{22}y_j + a_{23}z_j$$
$$Z_j = a_{31}x_j + a_{32}y_j + a_{33}z_j \quad (13.52)$$

or in matrix notation,

$$\begin{pmatrix} X_j \\ Y_j \\ Z_j \end{pmatrix} = \begin{pmatrix} a_{11} & a_{12} & a_{13} \\ a_{21} & a_{22} & a_{23} \\ a_{31} & a_{32} & a_{33} \end{pmatrix} \begin{pmatrix} x_j \\ y_j \\ z_j \end{pmatrix} \quad (13.53)$$

where the coefficients a_{ij} ($i, j = 1, 2, 3$) are real numbers (independent of the particle coordinates) elements of an orthogonal matrix: $A^{-1} = \tilde{A}$.

Here, we are interested in seeing how the Hamiltonian function transforms under such a rotation. As an illustration let us take a particular rotation about the z-axis by 90° counterclockwise; by this rotation, the new positive X-axis points towards the old y-axis and the new positive Y-axis towards the old negative x-axis. It is easy to verify the following relations between the new (capital) and old (lower case) coordinates and momenta

$$X_j = y_j, \quad Y_j = -x_j, \quad Z_j = z_j \quad (13.54)$$

$$P_{jX} \equiv m\frac{dX_j}{dt} = m\frac{dy_j}{dt} \equiv p_{jy}$$
$$P_{jY} \equiv m\frac{dY_j}{dt} = -m\frac{dx_j}{dt} \equiv -p_{jx}$$
$$P_{jZ} \equiv m\frac{dZ_j}{dt} = m\frac{dz_j}{dt} \equiv p_{jz} \quad (13.55)$$

Using these relations, we obtain

$$p_{jx}^2 + p_{jy}^2 + p_{jz}^2 = P_{jX}^2 + P_{jY}^2 + P_{jZ}^2 \quad (13.56)$$

$$(x_j - x_k)^2 + (y_j - y_k)^2 + (z_j - z_k)^2$$
$$= (X_j - X_k)^2 + (Y_j - Y_k)^2 + (Z_j - Z_k)^2 \quad (13.57)$$

These equations clearly ensure the fact that our Hamiltonian retains the same form under this rotation.

The demonstration of the invariance can easily be extended to the rotation about the z-axis by an arbitrary angle.

The rotational invariance of the Hamiltonian in (13.46) means that the dynamics of the gas has no preference in direction. It is, therefore, likely that the system approaches a stationary state in which all the local properties are without directional preference. Such a state is called an *isotropic* stationary state.

In summary, our Hamiltonian H in (13.46) is invariant under translation *and* rotation. We may therefore expect that the stationary state of our system is likely to be homogeneous *and* isotropic. In fact, this is true for the gas and liquid phase of the system. It is, however, not true for the solid phase. In the solid phase, the equilibrium state of the system is neither homogeneous nor isotropic. It forms a crystal lattice of a special structure. The crystalline state is sometimes called a *symmetry-breaking state* since the translation and rotation symmetry is broken.

Problem 13.6.1

Study the invariance properties with respect to the space translation of the following Hamiltonians:

1. $H = \frac{1}{2m} p^2 + mgx$
2. $H = \frac{1}{2m} p_1^2 + \frac{1}{2M} p_2^2 + \frac{k_0 e^2}{|r_1 - r_2|}$
3. $H = \frac{1}{2m} p^2 + \frac{1}{2} k x^2 + x(eE)$

14
Generalized Vectors and Linear Operators

Generalized vectors and linear operators are discussed in this chapter. The eigenvalue problem for a Hermitean operator is set up, and its solution is discussed.

14.1
Generalized Vectors. Matrices

We wish to construct a class of mathematical quantities that have the same algebraic properties as vectors. Such quantities, called *generalized vectors*, can be generated by matrix algebra, which will be discussed in the present section.

A *vector* may be defined geometrically as a quantity having *magnitude and direction*. Various properties of vectors are enumerated in Chapter 2. Among the basic properties of vectors are: (a) any two vectors **A** and **B** can be added together, yielding the vector **A** + **B**; (b) any vector **A** can be multiplied by a real number p, yielding the vector $p\mathbf{A}$; (c) from two vector **A** and **B**, a scalar product can be constructed, yielding a real number $\mathbf{A} \cdot \mathbf{B}$; (d) every vector **A** has a nonnegative magnitude (or length) equal to $\sqrt{\mathbf{A} \cdot \mathbf{A}} \equiv A$; in addition, the following algebraic properties are satisfied by vectors.

$$\mathbf{A} + \mathbf{B} = \mathbf{B} + \mathbf{A} \quad \text{(commutative)}$$

$$(\mathbf{A} + \mathbf{B}) + \mathbf{C} = \mathbf{A} + (\mathbf{B} + \mathbf{C}) = \mathbf{A} + \mathbf{B} + \mathbf{C} \quad \text{(associative)} \tag{14.1}$$

$$p\mathbf{A} = \mathbf{A}p$$
$$p(q\mathbf{A}) = q(p\mathbf{A}) = (pq)\mathbf{A} = pq\mathbf{A}$$
$$p(\mathbf{A} + \mathbf{B}) = p\mathbf{A} + p\mathbf{B} \quad \text{(distributive)}$$
$$(p + q)\mathbf{A} = p\mathbf{A} + q\mathbf{A} \tag{14.2}$$

$$\mathbf{A} \cdot (p\mathbf{B}) = p(\mathbf{A} \cdot \mathbf{B}) = (p\mathbf{A}) \cdot \mathbf{B}$$

$$\mathbf{A} \cdot (\mathbf{B} + \mathbf{C}) = \mathbf{A} \cdot \mathbf{B} + \mathbf{A} \cdot \mathbf{C} \tag{14.3}$$

$$\mathbf{A} \cdot \mathbf{A} \equiv A^2 \geq 0 \tag{14.4}$$

Mathematical Physics. Shigeji Fujita and Salvador V. Godoy
Copyright © 2010 WILEY-VCH Verlag GmbH & Co. KGaA, Weinheim
ISBN: 978-3-527-40808-5

In general, a rectangular array of numbers with m rows and n columns will be called an $m \times n$ *matrix*, for example

$$\begin{pmatrix} a_{11} & a_{12} & \cdots & a_{1n} \\ a_{21} & a_{22} & \cdots & a_{2n} \\ \cdots & \cdots & \cdots & \cdots \\ a_{m1} & a_{m2} & \cdots & a_{mn} \end{pmatrix} \equiv (a_{ij})_{(m,n)} = \mathsf{A} \tag{14.5}$$

where a_{ij} is a general complex number. Two matrices A and B are equal if and only if they have the same elements:

$$\mathsf{A} = \mathsf{B} \quad \text{if} \quad a_{ij} = b_{ij} \tag{14.6}$$

The following operations are assumed:
1. The sum of two matrices $(a_{ij})_{(m,n)}$ and $(a_{ij})_{(m,n)}$ is a matrix whose elements are the sum of the corresponding elements. For example, for $m = n = 2$,

$$\begin{pmatrix} a_{11} & a_{12} \\ a_{21} & a_{22} \end{pmatrix} + \begin{pmatrix} b_{11} & b_{12} \\ b_{21} & b_{22} \end{pmatrix} \equiv \begin{pmatrix} a_{11}+b_{11} & a_{12}+b_{12} \\ a_{21}+b_{21} & a_{22}+b_{22} \end{pmatrix} \tag{14.7}$$

2. The product of a matrix A and a complex number c is a matrix whose elements are c times the corresponding elements of A. For example,

$$c\mathsf{A} \equiv c \begin{pmatrix} a_{11} & a_{12} \\ a_{21} & a_{22} \end{pmatrix} = \begin{pmatrix} ca_{11} & ca_{12} \\ ca_{21} & ca_{22} \end{pmatrix} \tag{14.8}$$

3. The ordered product of a $m \times n$ matrix $(a_{ij})_{(m,n)}$ an a $n \times p$ matrix $(b_{ij})_{(n,p)}$ is a $m \times p$ matrix whose (i, j) elements are given by

$$\sum_{k=1}^{n} a_{ik} b_{kj} \tag{14.9}$$

In particular, the product of two $m \times m$ matrices A and B is an $m \times m$ matrix. For example, for $m = 2$,

$$\begin{aligned} \mathsf{AB} &= \begin{pmatrix} a_{11} & a_{12} \\ a_{21} & a_{22} \end{pmatrix} \begin{pmatrix} b_{11} & b_{12} \\ b_{21} & b_{22} \end{pmatrix} \\ &= \begin{pmatrix} a_{11}b_{11}+a_{12}b_{21} & a_{11}b_{12}+a_{12}b_{22} \\ a_{21}b_{11}+a_{22}b_{21} & a_{21}b_{12}+a_{22}b_{22} \end{pmatrix} \end{aligned} \tag{14.10}$$

The ordered product of a $1 \times n$ matrix and a $n \times 1$ matrix is a number. For example,

$$\begin{pmatrix} a_1 & a_2 \end{pmatrix} \begin{pmatrix} b_1 \\ b_2 \end{pmatrix} = a_1 b_1 + a_2 b_2 \tag{14.11}$$

Consider the $m \times n$ matrix $(a_{ij})_{(m,n)}$ in (14.5). Interchanging rows and columns, we obtain a $n \times m$ matrix

$$\begin{pmatrix} a_{11} & a_{21} & \cdots & a_{m1} \\ a_{12} & a_{22} & \cdots & a_{m2} \\ \cdots & \cdots & \cdots & \cdots \\ a_{1n} & a_{2n} & \cdots & a_{nm} \end{pmatrix} \equiv \mathsf{A}^\mathsf{T} \qquad (14.12)$$

This matrix is called the *transpose* of matrix A, and is denoted by A^T.

A complex number c can, in general, be written as the sum of real and imaginary numbers:

$$c = p + iq \qquad (14.13)$$

where

$$i = \sqrt{-1} \qquad (14.14)$$

is the imaginary unit, and p and q are real numbers. The *complex conjugate* of c is defined by $p - iq$, and will be denoted by c^*:

$$c^* = (p + iq)^* \equiv p - iq \qquad (14.15)$$

The *Hermitean conjugate* of a matrix A is obtained by applying transposition *and* complex-conjugation, and will be denoted by $\mathsf{A}^\dagger$, read as A dagger. Thus, for the matrix A in (14.5),

$$\mathsf{A}^\dagger \equiv (\mathsf{A}^*)^\mathsf{T} = (\mathsf{A}^\mathsf{T})^* = \begin{pmatrix} a_{11}^* & a_{21}^* & \cdots & a_{m1}^* \\ a_{12}^* & a_{22}^* & \cdots & a_{m2}^* \\ \cdots & \cdots & \cdots & \cdots \\ a_{1n}^* & a_{2n}^* & \cdots & a_{nm}^* \end{pmatrix} \qquad (14.16)$$

From the definition, we can show that

$$(c^*)^* = c, \quad (\mathsf{A}^\mathsf{T})^\mathsf{T} = (\mathsf{A}^\dagger)^\dagger = \mathsf{A} \qquad (14.17)$$

A matrix having the same number of rows and columns, that is a $n \times n$ matrix is called a *square matrix*. A $1 \times n$ matrix, that is a matrix with a single row, is called a *row matrix*. A $n \times 1$, that is a matrix with a single column, is called a *column matrix* (vector). The transpose (and the Hermitean conjugate) of a square matrix is a square matrix. The transpose of a row matrix (vector) is a column matrix (vector) and vice versa.

A matrix whose elements are all real: $a_{ij}^* = a_{ij}$, is called a real matrix. For real matrices, $\mathsf{A}^\dagger = \mathsf{A}^\mathsf{T}$. That is, the distinction between transposition and Hermitean conjugation disappears for real matrices.

We can easily see from (14.9) and (14.10) that addition of matrices is commutative and associative, just as for ordinary vectors (see (14.1)). Multiplication of matrices by a number is also commutative, associative and distributive.

Let us take two column matrices,

$$\begin{pmatrix} a_1 \\ a_2 \\ a_3 \end{pmatrix} \equiv |A\rangle \;, \quad \begin{pmatrix} b_1 \\ b_2 \\ b_3 \end{pmatrix} \equiv |B\rangle \tag{14.18}$$

Their Hermitean conjugates are row matrices given by

$$(|A\rangle)^\dagger = \begin{pmatrix} a_1^* & a_2^* & a_3^* \end{pmatrix} \equiv \langle A|$$

$$(|B\rangle)^\dagger = \begin{pmatrix} b_1^* & b_2^* & b_3^* \end{pmatrix} \equiv \langle B| \tag{14.19}$$

Column and row matrices are different mathematical entities. For example, they cannot be added together. In the above, we distinguished column and row matrices by "ket" $|\rangle$ and "bra" $\langle|$ symbols, respectively. The names "bra" and "ket" (due to Dirac) are derived from the first and second halves of the word "braket".

The scalar product $\langle B||A\rangle$, the row matrix on the left and the column matrix on the right, can be calculated as follows:

$$\langle B||A\rangle \equiv \langle B|A\rangle$$

$$\equiv \begin{pmatrix} b_1^* & b_2^* & b_3^* \end{pmatrix} \begin{pmatrix} a_1 \\ a_2 \\ a_3 \end{pmatrix} = b_1^* a_1 + b_2^* a_2 + b_3^* a_3 \tag{14.20}$$

That is, the product written in the form of a complete bracket $\langle B|A\rangle$, is a (complex) number.

Since

$$\langle A|B\rangle = \begin{pmatrix} a_1^* & a_2^* & a_3^* \end{pmatrix} \begin{pmatrix} b_1 \\ b_2 \\ b_3 \end{pmatrix} = a_1^* b_1 + a_2^* b_2 + a_3^* b_3 \tag{14.21}$$

The numbers $\langle B|A\rangle$ and $\langle A|B\rangle$ are complex conjugates to each other:

$$\langle B|A\rangle = (\langle A|B\rangle)^* \tag{14.22}$$

In particular, if $|B\rangle = |A\rangle$, then

$$\langle A|A\rangle = a_1^* a_1 + a_2^* a_2 + a_3^* a_3 \geq 0 \tag{14.23}$$

which is nonnegative.

It is noted that ordered products such as $|A\rangle\langle B|$ and $|B\rangle\langle A|$ do not generate numbers, but instead they are more complicated mathematical quantities (called linear operators, see Section 14.2).

Let us now consider an ordinary vector in three dimensions. In the Cartesian representation, such a vector **A** can be specified by the set of three components

(A_x, A_y, A_z), which are *all real*. We can represent the vector **A** by a real column matrix:

$$\mathbf{A} = \begin{pmatrix} A_x \\ A_y \\ A_z \end{pmatrix} \equiv |A\rangle \tag{14.24}$$

With the provisions that: (a) the magnitude of the vector is given by the square root of

$$\begin{aligned} \langle A|A\rangle &\equiv A_x^* A_x + A_y^* A_y + A_z^* A_z \\ &= A_x^2 + A_y^2 + A_z^2 = |\mathbf{A}|^2 \end{aligned} \tag{14.25}$$

and (b) the scalar product $\mathbf{B} \cdot \mathbf{A}$ with a real vector **B** is given by

$$\begin{aligned} \langle B|A\rangle &\equiv B_x^* A_x + B_y^* A_y + B_z^* A_z \\ &= B_x A_x + B_y A_y + B_z A_z \end{aligned} \tag{14.26}$$

We can alternatively represent ordinary vectors in terms of real row matrices. The resulting algebras for row matrices are not much different from those for column matrices. Essentially, the two algebras are Hermitean conjugates to each other.

The above representation of vectors in terms of column or row matrices suggests a natural generalization. Let us consider $n \times 1$ column matrices with *complex elements*. Clearly the collection of such $n \times 1$ column matrices satisfies the same algebraic laws as ordinary vectors: [compare with (14.1)–(14.4)]

$$|A\rangle + |B\rangle = |B\rangle + |A\rangle$$

$$(|A\rangle + |B\rangle) + |C\rangle = |A\rangle + (|B\rangle + |C\rangle) \equiv |A\rangle + |B\rangle + |C\rangle \tag{14.27}$$

$$c|A\rangle = |A\rangle c$$

$$c(d|A\rangle) = d(c|A\rangle) = (dc)|A\rangle$$

$$c(|A\rangle + |B\rangle) = c|A\rangle + c|B\rangle$$

$$(c+d)|A\rangle = c|A\rangle + d|A\rangle \tag{14.28}$$

$$\langle B|(c|A\rangle) = c\langle B|A\rangle = (\langle B|c)|A\rangle = \langle B|c|A\rangle$$

$$\langle B|(|A\rangle + |C\rangle) = \langle B|A\rangle + \langle B|C\rangle \tag{14.29}$$

$$\langle B|A\rangle^* = \langle A|B\rangle$$

$$\langle A|A\rangle \geq 0 \tag{14.30}$$

where c and d are complex numbers.

With the rule that the column matrix $|A\rangle$ has the *magnitude* given by

$$\sqrt{\langle A|A\rangle} = \text{magnitude of } |A\rangle \tag{14.31}$$

we may say that $n \times 1$ column matrices represent *generalized vectors*. The generalization is two-fold. First, n can be any positive integer; second, the elements are complex numbers. Because of the latter condition, we have introduced minor changes in algebraic properties between (14.1)–(14.4) and (14.27)–(14.30). In order to assure that $\langle A|A\rangle \geq 0$, we chose $\langle A|$ to be the Hermitean conjugate (rather than the transpose) of $|A\rangle$ [see (14.27) and (14.31)].

In summary, the column matrices (and again row matrices) satisfy the same algebraic properties as ordinary vectors. Because of this, these column and row matrices are also called *column and row vectors*.

Problem 14.1.1

Assuming that a, b, x and y are real numbers, show that
1. $[(x+iy)(a+ib)]^* = (x-iy)(a-ib)$
2. $[(x+iy)^n]^* = (x-iy)^n$
  HINT: Use mathematical induction
3. $[e^{(a+ib)}]^* = e^{(a-ib)}$
4. $[e^{(a+ib)(x+iy)}]^* = e^{(a-ib)(x-iy)}$

Problem 14.1.2

Calculate the lengths of the following column vectors

$$\begin{pmatrix} 1 \\ 2 \\ -3 \end{pmatrix}, \begin{pmatrix} 1 \\ i \\ -1 \\ -i \end{pmatrix}$$

Problem 14.1.3

Show that
1. $\langle A|(c|B\rangle) = c\langle A|B\rangle = (\langle A|c)|B\rangle$
2. $\langle A|B\rangle^* = \langle B|A\rangle$. Use a 2×1 column vector for $|A\rangle$.

14.2
Linear Operators

Let us consider a $n \times 1$ column vector $|A\rangle$. We multiply it from the left by a $n \times n$ matrix α. According to the matrix multiplication rule, we then obtain a $n \times 1$ column vector, which will be denoted by $|B\rangle$. In mathematical terms,

$$|B\rangle = \alpha|A\rangle \tag{14.32}$$

We may look at this equation from a few different angles. First, we may regard it as the definition equation for the vector $|B\rangle$ in terms of the vector $|A\rangle$ and the matrix α. Second, we may say that the two vectors $|B\rangle$ and $|A\rangle$ are related by the *mapping* operation α. Third, we may say that the matrix α, upon *acting* on the vector $|A\rangle$ *from the left*, generates the vector $|B\rangle$ with the action being the operation of matrix multiplication. In general, a quantity capable of performing a mathematical operation is called an *operator*. Thus, in the last view, the $n \times n$ matrix α, which upon acting on a column vector generates another column vector, is an operator.

An operator is said to be known (or defined) if all of the results of the operation are given. In the case of (14.32), if the results $\alpha|A\rangle$ for all possible vectors $|A\rangle$ are known, then the operator α is defined.

We now take the special case of (14.1):

$$|A\rangle = I|A\rangle \text{ for all } |A\rangle \tag{14.33}$$

The operator I is well defined, and clearly is given by

$$I = \begin{pmatrix} 1 & 0 & 0 \\ 0 & 1 & 0 \\ 0 & 0 & 1 \end{pmatrix} \quad \text{for } n = 3 \tag{14.34}$$

A matrix having nonzero elements only along the diagonal is called a *diagonal matrix*. If these diagonal elements are all equal to unity, the diagonal matrix is called a *unit matrix*. The matrix I is a unit matrix.

Using the operational rules of matrices, we can show that

$$\begin{aligned} \alpha(|A\rangle + |B\rangle) &= \alpha|A\rangle + \alpha|B\rangle \\ \alpha(c|A\rangle) &= c(\alpha|A\rangle) \end{aligned} \tag{14.35}$$

An operator that satisfies the above algebraic relations is called a *linear operator*. Thus, the matrix α is a linear operator.

Consider now a complex number a. We can multiply any $n \times 1$ column vector by this number and obtain a column vector. We can also obtain relations like (14.35) with a replacing α throughout. The number a can therefore be regarded as a linear operator. Let us denote this operator by $\hat{a}$. This operator may be defined by

$$\hat{a}|A\rangle = a|A\rangle \quad \text{for all } |A\rangle \tag{14.36}$$

It is clear that $\hat{a}$ can be represented by the matrix

$$\hat{a} \equiv \begin{pmatrix} a & 0 & 0 \\ 0 & a & 0 \\ 0 & 0 & a \end{pmatrix} = a \begin{pmatrix} 1 & 0 & 0 \\ 0 & 1 & 0 \\ 0 & 0 & 1 \end{pmatrix} = aI \tag{14.37}$$

In order words, the number a can be represented by the matrix aI.

In general, linear operators relating any two $n \times 1$ column vectors in the form of (14.1) can be represented by $n \times n$ matrices, see below. Conversely, as we have said before, $n \times n$ matrices are linear operators. Because of this close connection, the same symbol will be used to denote both a linear operator and its corresponding $n \times n$ matrix. An advantage of referring to such a quantity as a linear operator is that the name reminds us of its principal operational properties. The term "matrix" can be used to describe a quantity less restricted in operational characters. A tensor (whose definition and properties are given Chapter 3) can be regarded as a linear operator connecting two vectors in ordinary space.

Any two linear operators α and γ can be added together, yielding a third linear operator, denoted by $\alpha + \gamma$, which is defined by

$$(\alpha + \gamma)|A\rangle \equiv \alpha|A\rangle + \gamma|A\rangle \quad \text{for any } |A\rangle \tag{14.38}$$

From this definition, we can obtain

$$\alpha + \gamma = \gamma + \alpha \quad \text{[commutative]}$$

$$(\alpha + \gamma) + \delta = \alpha - (\gamma + \delta) \equiv \alpha + \gamma + \delta \quad \text{[associative]} \tag{14.39}$$

The product of two linear operators α and γ is defined by

$$(\alpha\gamma)|A\rangle \equiv \alpha(\gamma|A\rangle) \quad \text{for any } |A\rangle \tag{14.40}$$

From this, we may deduce that

$$(\alpha\gamma)\delta = \alpha(\gamma\delta) \equiv \alpha\gamma\delta \tag{14.41}$$

However, the product $\alpha\gamma$ and the product of the reversed order, $\gamma\alpha$, are not in general equal to each other:

$$\alpha\gamma \neq \gamma\alpha \quad \text{in general} \tag{14.42}$$

For illustration we take the following two products:

$$\begin{pmatrix} a_{11} & a_{12} \\ a_{21} & a_{22} \end{pmatrix} \begin{pmatrix} b_{11} & b_{12} \\ b_{21} & b_{22} \end{pmatrix} = \begin{pmatrix} a_{11}b_{11} + a_{12}b_{21} & a_{11}b_{12} + a_{12}b_{22} \\ a_{21}b_{11} + a_{22}b_{21} & a_{21}b_{12} + a_{22}b_{22} \end{pmatrix}$$

$$\begin{pmatrix} b_{11} & b_{12} \\ b_{21} & b_{22} \end{pmatrix} \begin{pmatrix} a_{11} & a_{12} \\ a_{21} & a_{22} \end{pmatrix} = \begin{pmatrix} a_{11}b_{11} + a_{21}b_{12} & a_{12}b_{11} + a_{12}b_{12} \\ a_{11}b_{21} + a_{21}b_{22} & a_{12}b_{21} + a_{22}b_{22} \end{pmatrix}$$

Clearly, the two matrices are different.

We often refer to (14.42) by saying that *two linear operators do not necessarily commute*. This is a significant property distinct from that of the product of two numbers since any two numbers commute with each other. It is observed from the second of (14.35) that any number commutes with a linear operator.

Consider now a row vector $\langle C|$. If we multiply it from the right by matrix α, we obtain a row vector $\langle C|\alpha$. The scalar product of this row vector $\langle C|\alpha$ and a column vector $|B\rangle$ is by definition a number, denoted by $(\langle C|\,\boldsymbol{\alpha})|B\rangle$. It is easy to show that the same number can be obtained if we first operate $\boldsymbol{\alpha}$ on $|B\rangle$ and then make a scalar product from $\langle C|$ and $\boldsymbol{\alpha}\,|B\rangle$:

$$(\langle C|\,\boldsymbol{\alpha})|B\rangle = \langle C|\,(\boldsymbol{\alpha}\,|B\rangle) \equiv \langle C|\,\boldsymbol{\alpha}\,|B\rangle \tag{14.43}$$

For $n = 2$, this equality may be exhibited as follows:

$$\langle C| \equiv \begin{pmatrix} c_1^* & c_2^* \end{pmatrix}, \quad \boldsymbol{\alpha} \equiv \begin{pmatrix} a_{11} & a_{12} \\ a_{21} & a_{22} \end{pmatrix}, \quad |B\rangle \equiv \begin{pmatrix} b_1 \\ b_2 \end{pmatrix}$$

$$(\langle C|\,\boldsymbol{\alpha})|B\rangle = \left[\begin{pmatrix} c_1^* & c_2^* \end{pmatrix} \begin{pmatrix} a_{11} & a_{12} \\ a_{21} & a_{22} \end{pmatrix}\right] \begin{pmatrix} b_1 \\ b_2 \end{pmatrix}$$

$$= \begin{pmatrix} c_1^* a_{11} + c_2^* a_{21} & c_1^* a_{12} + c_2^* a_{22} \end{pmatrix} \begin{pmatrix} b_1 \\ b_2 \end{pmatrix}$$

$$= \left(c_1^* a_{11} + c_2^* a_{21}\right) b_1 + \left(c_1^* a_{12} + c_2^* a_{22}\right) b_2 \tag{14.44}$$

$$\langle C|\,(\boldsymbol{\alpha}\,|B\rangle) = \begin{pmatrix} c_1^* & c_2^* \end{pmatrix} \left[\begin{pmatrix} a_{11} & a_{12} \\ a_{21} & a_{22} \end{pmatrix} \begin{pmatrix} b_1 \\ b_2 \end{pmatrix}\right]$$

$$= \begin{pmatrix} c_1^* & c_2^* \end{pmatrix} \begin{pmatrix} a_{11} b_1 + a_{12} b_2 \\ a_{21} b_1 + a_{22} b_2 \end{pmatrix}$$

$$= c_1^* (a_{11} b_1 + a_{12} b_2) + c_2^* (a_{21} b_1 + a_{22} b_2) \tag{14.45}$$

We see that the last lines of (14.44) and (14.45) are equal.

The *Hermitean conjugate or adjoint* of a linear operator $\boldsymbol{\alpha}$, denoted by $\boldsymbol{\alpha}^\dagger$, is defined by

$$\langle B|\,\boldsymbol{\alpha}^\dagger\,|C\rangle \equiv (\langle C|\,\boldsymbol{\alpha}\,|B\rangle)^* \tag{14.46}$$

for every $|B\rangle$ and $|C\rangle$.

If $\boldsymbol{\alpha}$ is represented by a matrix, the matrix representation of $\boldsymbol{\alpha}^\dagger$ is the Hermitean conjugate of that matrix. This may be seen as follows. Using the same example used earlier, we have

$$(\langle C|\,\boldsymbol{\alpha}\,|B\rangle)^* = \left[c_1^* (a_{11} b_1 + a_{12} b_2) + c_2^* (a_{21} b_1 + a_{22} b_2)\right]^*$$
$$= c_1 \left(a_{11}^* b_1^* + a_{12}^* b_2^*\right) + c_2 \left(a_{21}^* b_1^* + a_{22}^* b_2^*\right) \tag{14.47}$$

Let $\boldsymbol{\alpha}^\dagger$ be represented by the matrix (x_{ij}). Then,

$$\langle B|\,\boldsymbol{\alpha}^\dagger\,|C\rangle = \begin{pmatrix} b_1^* & b_2^* \end{pmatrix} \begin{pmatrix} x_{11} & x_{12} \\ x_{21} & x_{22} \end{pmatrix} \begin{pmatrix} c_1 \\ c_2 \end{pmatrix}$$

$$= \left(b_1^* x_{11} + b_2^* x_{21}\right) c_1 + \left(b_1^* x_{12} + b_2^* x_{22}\right) c_2 \tag{14.48}$$

Comparing the last lines of (14.47) and (14.48) and noticing that c_1, c_2, b_1^* and b_2^* are chosen arbitrarily, we obtain

$$\begin{pmatrix} x_{11} & x_{12} \\ x_{21} & x_{22} \end{pmatrix} = \begin{pmatrix} a_{11}^* & a_{12}^* \\ a_{21}^* & a_{22}^* \end{pmatrix} \tag{14.49}$$

This means that the matrix (x_{ij}) representing $\alpha^\dagger$ is the Hermitean conjugate of the matrix (a_{ij}) representing α:

$$x_{ij} = a^*_{ji} \tag{14.50}$$

A linear operator that equals its adjoint

$$\alpha^\dagger = \alpha \tag{14.51}$$

is called a *self-adjoint operator* or a *Hermitean operator*. Hermitean operators play very important roles in quantum mechanics.

Problem 14.2.1

Pauli's spin matrices $(\sigma_x, \sigma_y, \sigma_z)$ are defined by

$$\sigma_x \equiv \begin{pmatrix} 0 & 1 \\ 1 & 0 \end{pmatrix}, \quad \sigma_y \equiv \begin{pmatrix} 0 & -i \\ i & 0 \end{pmatrix}, \quad \sigma_z \equiv \begin{pmatrix} 1 & 0 \\ 0 & -1 \end{pmatrix}$$

Show that
1. $\sigma_x^2 = \sigma_y^2 = \sigma_z^2 = 1$
2. $[\sigma_y, \sigma_z] = 2i\sigma_x$, $[\sigma_z, \sigma_x] = 2i\sigma_y$, $[\sigma_x, \sigma_y] = 2i\sigma_z$
3. $\sigma_y \sigma_z + \sigma_z \sigma_y = 0$, $\sigma_z \sigma_x + \sigma_x \sigma_z = 0$, $\sigma_x \sigma_y + \sigma_y \sigma_x = 0$
4. $\sigma_x \sigma_y \sigma_z = i$

14.3
The Eigenvalue Problem

A square matrix (a_{ij}) whose elements satisfy

$$a_{kj} = a_{jk} \tag{14.52}$$

is called a *symmetric matrix*. A real, symmetric matrix is Hermitean since

$$a_{kj} = a_{jk} = a^*_{jk} \tag{14.53}$$

Let us take a general Hermitean matrix α, which satisfies the relation

$$\alpha^\dagger = \alpha \quad \text{or} \quad a^*_{ij} = a_{ji} \tag{14.54}$$

The nonzero column vector $|B\rangle$, $[\langle B|B\rangle \neq 0]$, which satisfies

$$\alpha |B\rangle = \lambda |B\rangle \tag{14.55}$$

is called the *eigenvector* of α belonging to the *eigenvalue* λ. Equivalently, this equation may be written in the form:

$$(\alpha - \lambda 1)|B\rangle = 0 \tag{14.56}$$

For $n = 3$, this means that

$$(a_{11} - \lambda)b_1 + a_{12}b_2 + a_{13}b_3 = 0$$
$$a_{21}b_1 + (a_{22} - \lambda)b_2 + a_{23}b_3 = 0$$
$$a_{31}b_1 + a_{32}b_2 + (a_{33} - \lambda)b_3 = 0 \tag{14.57}$$

The necessary and sufficient condition for the existence of a nonzero column vector $|B\rangle$ is that

$$\det(\boldsymbol{\alpha} - \lambda \mathsf{I}) = 0$$

that is,

$$\begin{vmatrix} a_{11} - \lambda & a_{12} & a_{13} \\ a_{21} & a_{22} - \lambda & a_{23} \\ a_{31} & a_{32} & a_{33} - \lambda \end{vmatrix} = 0 \tag{14.58}$$

When expanded, this determinant is a cubic equation in λ. There are therefore three roots for λ. We will now show that if $\boldsymbol{\alpha}$ is Hermitean *all roots λ are real*.

Multiplying in (14.55) from the left by $\langle B|$, we obtain

$$\langle B|\boldsymbol{\alpha}|B\rangle = \lambda \langle B|B\rangle \tag{14.59}$$

The complex conjugate of the left-hand term can be calculated as follows:

$$[\langle B|\boldsymbol{\alpha}|B\rangle]^* = \langle B|\boldsymbol{\alpha}^\dagger|B\rangle \quad \text{[use of (14.46)]}$$
$$= \langle B|\boldsymbol{\alpha}|B\rangle \quad \text{[use of (14.54)]}$$
$$= \lambda \langle B|B\rangle \quad \text{[use of (14.55)]}$$

Complex conjugation of the right-hand term yields

$$|\lambda \langle B|B\rangle|^* = \lambda^* \langle B|B\rangle^*$$
$$= \lambda^* \langle B|B\rangle \quad \text{[use of (14.46)]}$$

Therefore, we obtain

$$\lambda \langle B|B\rangle = \lambda^* \langle B|B\rangle \quad \text{or} \quad (\lambda - \lambda^*) \langle B|B\rangle = 0$$

Since $\langle B|B\rangle > 0$ (by assumption), we must have

$$\lambda = \lambda^* \tag{14.60}$$

Thus, the eigenvalue λ is real. It should be noted that the above proof does not depend on the dimension of the matrix. The theorem, therefore, is valid for an arbitrary dimension.

Corresponding to different eigenvalues, there will be different eigenvectors. Let $|B_1\rangle$ and $|B_2\rangle$ be eigenvectors corresponding to two different eigenvalues λ_1

and λ_2,

$$\boldsymbol{\alpha} |B_1\rangle = \lambda_1 |B_1\rangle \tag{14.61}$$

$$\boldsymbol{\alpha} |B_2\rangle = \lambda_2 |B_2\rangle \tag{14.62}$$

$$\lambda_1 \neq \lambda_2$$

We now show that *the eigenvectors corresponding to two different eigenvalues are orthogonal*, that is their scalar product vanishes:

$$\langle B_1 | B_2\rangle = 0 \tag{14.63}$$

Multiplying of (14.61) by $\langle B_2|$ from the left, we obtain

$$\langle B_2| (\boldsymbol{\alpha} |B_1\rangle) = \langle B_2| \lambda_1 |B_1\rangle = \lambda_1 \langle B_2|B_1\rangle$$

The first member of these equations can also be calculated as follows:

$$\langle B_2| \boldsymbol{\alpha} |B_1\rangle = (\langle B_1| \boldsymbol{\alpha}^\dagger |B_2\rangle)^* = (\langle B_1| \boldsymbol{\alpha} |B_2\rangle)^*$$
$$= (\lambda_2 \langle B_1|B_2\rangle)^* = \lambda_2^* \langle B_1|B_2\rangle^*$$
$$= \lambda_2 \langle B_2|B_1\rangle$$

Comparing the two results, we obtain

$$(\lambda_1 - \lambda_2) \langle B_2|B_1\rangle = 0$$

Dividing this by $\lambda_1 - \lambda_2 \,(\neq 0)$, we obtain

$$\langle B_2|B_1\rangle = 0 \quad \text{Q. E. D.}$$

If $|B\rangle$ is an eigenvector associated with a certain eigenvalue λ, then a complex number c times the vector $|B\rangle$, $c|B\rangle$ is also an eigenvector with the eigenvalue $c\lambda$. The two vectors, $|B\rangle$ and $c|B\rangle$, have the same algebraic properties except for their magnitudes (and phases, see below). The set of such similar eigenvectors can all be generated from the eigenvector

$$\frac{|B\rangle}{\sqrt{\langle B|B\rangle}} \equiv |\lambda\rangle \tag{14.64}$$

which is normalized to unity,

$$\langle \lambda|\lambda\rangle = 1 \tag{14.65}$$

We note that the normalization condition (14.65) alone does not determine the eigenvector $|\lambda\rangle$ uniquely. In fact, if $|\lambda\rangle$ is the vector defined by (14.64), then $e^{i\phi} |\lambda\rangle$, where ϕ is a real number called a *phase angle*, also satisfies the same normalization condition since

$$\langle \lambda| e^{-i\phi} \cdot e^{i\phi} |\lambda\rangle = \langle \lambda|\lambda\rangle = 1 \tag{14.66}$$

14.3 The Eigenvalue Problem

If $\boldsymbol{a}$ is a $n \times n$ matrix, the order of the determinant equation is n. There will, therefore, be n real roots. If the n eigenvalues are all distinct, then we can construct a set of n orthogonal and normalized, or *orthonormal*, eigenvectors.

Even if the eigenvalues are not distinct and some of them are accidentally equal, we can still find n orthonormal eigenvectors. This may be understood by studying the following example. If the eigenvalue equation (14.58) has a repeated root, we can find two (or more) independent eigenvectors corresponding to this root. (By "independent" we mean that if $|B\rangle$ and $|C\rangle$ are two vectors, then $|B\rangle$ cannot be obtained from $|C\rangle$ by merely multiplying $|C\rangle$ by a number, and vice versa.) Let $|\lambda_a\rangle$ and $|\lambda_b\rangle$ be such eigenvectors belonging to the same eigenvalue λ:

$$\boldsymbol{a}|\lambda_a\rangle = \lambda|\lambda_a\rangle, \quad \boldsymbol{a}|\lambda_b\rangle = \lambda|\lambda_b\rangle \tag{14.67}$$

Then, any linear combination of these vectors,

$$a|\lambda_a\rangle + b|\lambda_b\rangle \tag{14.68}$$

is also an eigenvector. We can always find a pair of orthonormal vectors among the linear combinations; for example, $|\lambda_a\rangle$ and

$$|\lambda_{b'}\rangle \equiv |\lambda_b\rangle - |\lambda_a\rangle\langle\lambda_a|\lambda_b\rangle \tag{14.69}$$

are orthonormal, as we can verify easily that $\langle\lambda_{b'}|\lambda_a\rangle = 0$ (Problem 14.3.3).

In summary, for any $n \times n$ Hermitean matrix, we can in principle find n eigenvalues $\lambda_1, \lambda_2, \lambda_3, \ldots, \lambda_n$ and n orthonormal eigenvectors $|\lambda_1\rangle, |\lambda_2\rangle, \ldots, |\lambda_n\rangle$. The eigenvalues λ_i may be coincidentally equal.

The eigenvalue problem is very important in quantum mechanics. In fact as we will see later, it is inseparably connected with the fundamental postulates of quantum mechanics. The eigenvalue problem is also quite important in some branches of classical mechanics. For example, in Chapter 4 we discussed the vibration of particles on a stretched string in terms of normal modes. Finding the characteristic frequencies and the associated amplitude relations can be regarded as the solution of the eigenvalue problem of a particular type. Another significant example of a classical eigenvalue problem is the principal-axis transformation associated with the inertia tensor of a rigid body.

Problem 14.3.1

Find the eigenvalues of Pauli's spin matrices $(\sigma_x, \sigma_y, \sigma_z)$ defined in Problem 14.2.1.

Problem 14.3.2

Find the eigenvalues of the matrices $(\alpha_x, \alpha_y, \alpha_z)$ defined by

$$\alpha_x = \begin{pmatrix} 0 & 0 & 0 & 1 \\ 0 & 0 & 1 & 0 \\ 0 & 1 & 0 & 0 \\ 1 & 0 & 0 & 0 \end{pmatrix}, \quad \alpha_y = \begin{pmatrix} 0 & 0 & 0 & -i \\ 0 & 0 & i & 0 \\ 0 & -i & 0 & 0 \\ i & 0 & 0 & 0 \end{pmatrix}, \quad \alpha_z = \begin{pmatrix} 0 & 0 & 1 & 0 \\ 0 & 0 & 0 & -1 \\ 1 & 0 & 0 & 0 \\ 0 & -1 & 0 & 0 \end{pmatrix}$$

Problem 14.3.3

Verify that the ket $|\lambda_{b'}\rangle$ given in (14.69)
1. is orthogonal to the ket $|\lambda_a\rangle$, and
2. it is normalized to unity $\langle \lambda_{b'} | \lambda_{b'} \rangle = 1$.

14.4
Orthogonal Representation

In Section 14.1 we introduced generalized vectors by means of matrix algebra. A vector in the ordinary space of three dimensions, called an *ordinary vector*, can be represented by a 3×1 [or by a 1×3] real matrix. The principal algebraic properties satisfied by 3×1 real matrices are also satisfied by $n \times 1$ column matrices with complex elements. The latter may then be thought to represent quantities that we call *generalized vectors*. The generalization is two-fold. First, n can be any positive integer. Secondly, the elements of column matrices can be complex numbers.

Let us take the case $n = 1$, that is the 1×1 matrix with a complex element, which is just a complex number. A complex number z can always be written as

$$z = x + iy \tag{14.70}$$

where x and y are real numbers. The length of z is, by definition,

$$|z| \equiv \sqrt{z^*z} = \sqrt{x^2 + y^2} \tag{14.71}$$

Any complex number $z = x+iy$ can be represented by a point on the *Gaussian plane* as show in Figure 14.1. It is customary to choose the real axis along the horizontal direction and the imaginary axis along the vertical direction. The complex number z now can be regarded as a vector in two dimensions; the two independent real variables (x, y) that represent the real and imaginary parts of z, correspond to the two Cartesian variables.

On the other hand, by analogy with ordinary vectors, $n \times 1$ real matrices can be considered as vectors in an n-dimensional space. We may, therefore, say that a column vector with n complex elements can be looked at as a vector in the $2n$-dimensional space, called the *Hilbert space*. We note that by allowing the matrix elements to be complex, we could increase the number of independent variables by a factor two.

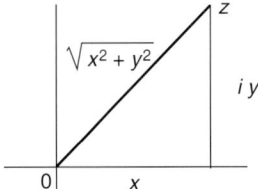

Fig. 14.1 The Gaussian representation of a complex number $z = x + iy$.

14.4 Orthogonal Representation

Any ordinary vector **A** can be expressed in terms of a set of three orthogonal vectors bases $\{\mathbf{i}, \mathbf{j}, \mathbf{k}\}$:

$$\mathbf{A} = \mathbf{i} A_x + \mathbf{j} A_y + \mathbf{k} A_z \tag{14.72}$$

where the set of three vectors $\{\mathbf{i}, \mathbf{j}, \mathbf{k}\}$, satisfy the *orthonormality* condition

$$\mathbf{i} \cdot \mathbf{i} = \mathbf{j} \cdot \mathbf{j} = \mathbf{k} \cdot \mathbf{k} = 1$$

$$\mathbf{i} \cdot \mathbf{j} = \mathbf{j} \cdot \mathbf{k} = \mathbf{k} \cdot \mathbf{i} = 0 \tag{14.73}$$

the components A_x, A_y, A_z are projections of **A** along $\mathbf{i}, \mathbf{j}, \mathbf{k}$, respectively, and can be written as

$$A_x = \mathbf{i} \cdot \mathbf{A}, \quad A_y = \mathbf{j} \cdot \mathbf{A}, \quad A_z = \mathbf{k} \cdot \mathbf{A} \tag{14.74}$$

Using these expressions, we can rewrite (14.72) as

$$\mathbf{A} = \mathbf{i}(\mathbf{i} \cdot \mathbf{A}) + \mathbf{j}(\mathbf{j} \cdot \mathbf{A}) + \mathbf{k}(\mathbf{k} \cdot \mathbf{A}) \tag{14.75}$$

Equation (14.75) is valid for any set of basis vectors $\{\mathbf{i}, \mathbf{j}, \mathbf{k}\}$ satisfying the orthonormality. The choice of such a set is often very important in solving a particular problem. For example, in treating the motion of a falling body we may choose the vertical direction to coincide with the direction of **k**.

In general, specification of an abstract quantity, such as a vector, by a suitable set of numbers is called a *representation* of that quantity. In the above example, we used the orthonormal basis vectors $\{\mathbf{i}, \mathbf{j}, \mathbf{k}\}$ to obtain the representation.

For later convenience, let us introduce new notations for the orthonormal vectors

$$\{\mathbf{i}, \mathbf{j}, \mathbf{k}\} \equiv \{\mathbf{e}_1, \mathbf{e}_2, \mathbf{e}_3\} \tag{14.76}$$

We can then express (14.76) as

$$\mathbf{A} = \mathbf{e}_1(\mathbf{e}_1 \cdot \mathbf{A}) + \mathbf{e}_2(\mathbf{e}_2 \cdot \mathbf{A}) + \mathbf{e}_3(\mathbf{e}_3 \cdot \mathbf{A})$$

$$= \sum_{j=1}^{3} \mathbf{e}_j(\mathbf{e}_j \cdot \mathbf{A}) \tag{14.77}$$

Let us now represent the set of the orthonormal vectors $(\mathbf{e}_1, \mathbf{e}_2, \mathbf{e}_3)$ by the normalized kets:

$$|e_1\rangle = \begin{pmatrix} 1 \\ 0 \\ 0 \end{pmatrix}, \quad |e_2\rangle = \begin{pmatrix} 0 \\ 1 \\ 0 \end{pmatrix}, \quad |e_3\rangle = \begin{pmatrix} 0 \\ 0 \\ 1 \end{pmatrix} \tag{14.78}$$

The orthonormality can, then, be represented by

$$\langle e_i | e_j \rangle = \delta_{ij} \tag{14.79}$$

If we represent an ordinary vector **A** by the column matrix $|A\rangle$, we can express the vector decomposition property (14.77) in the form:

$$|A\rangle = |e_1\rangle \langle e_1|A\rangle + |e_2\rangle \langle e_2|A\rangle + |e_3\rangle \langle e_3|A\rangle$$

or

$$\boxed{|A\rangle = \sum_{j=1}^{3} |e_j\rangle \langle e_j|A\rangle} \qquad (14.80)$$

This relation means that we *can expand* an arbitrary ket vector $|A\rangle$ in terms of the orthonormal ket vectors $|e_1\rangle, |e_2\rangle, |e_3\rangle$. By re-expressing (14.80) in the form

$$|A\rangle = \left(\sum_j |e_j\rangle \langle e_j|\right) |A\rangle \qquad (14.81)$$

we may formally represent the "expandability" by

$$\boxed{1 = \sum_j |e_j\rangle \langle e_j|} \qquad (14.82)$$

This relations is often called the *completeness relation*. We will also say that $|e_1\rangle, |e_2\rangle$, and $|e_3\rangle$ form a *complete* set of orthonormal vectors.

We now wish to discuss an orthogonal representation of the generalized vector in Hilbert space. We can, in principle, use any set of n orthonormal vectors for this purpose. For a particular physical problem, however, we will do better if we choose a special set such that the problem can be handled in the simplest possible manner.

Let us take a certain Hermitean operator ξ. Assume that n orthonormal eigenvectors of ξ,

$$|\xi_1\rangle, |\xi_2\rangle, \ldots |\xi_n\rangle \qquad (14.83)$$

are found. We will in general use the same letter for both operator and its eigenvalues with the distinction being given by the subscripts or primes. The eigenvectors are denoted by the corresponding eigenvalues with ket or bra symbols according to whether they are column or row vectors.

Using the n vectors in (14.83), we can expand an arbitrary vector $|\psi\rangle$ as the sum of orthogonal vectors in the form

$$|\psi\rangle = |\xi_1\rangle \langle \xi_1|\psi\rangle + |\xi_2\rangle \langle \xi_2|\psi\rangle + \ldots + |\xi_n\rangle \langle \xi_n|\psi\rangle$$

or

$$\boxed{|\psi\rangle \equiv \sum_{j=1}^{n} |\xi_j\rangle \langle \xi_j|\psi\rangle} \qquad (14.84)$$

This is a straightforward generalization of (14.80). The orthonormal eigenvectors can be represented by column vectors having one nonvanishing element whose value is unity. If we choose

$$|\xi_1\rangle \equiv \begin{pmatrix} 1 \\ 0 \\ 0 \\ \vdots \end{pmatrix}, \quad |\xi_2\rangle \equiv \begin{pmatrix} 0 \\ 1 \\ 0 \\ \vdots \end{pmatrix}, \ldots, \quad |\xi_n\rangle \equiv \begin{pmatrix} 0 \\ 0 \\ 0 \\ \vdots \\ 1 \end{pmatrix} \qquad (14.85)$$

then we can express the ket-vector $|\psi\rangle$ as follows:

$$|\psi\rangle = \sum_j |\xi_j\rangle\langle\xi_j|\psi\rangle = \begin{pmatrix} \langle\xi_1|\psi\rangle \\ \langle\xi_2|\psi\rangle \\ \vdots \\ \langle\xi_n|\psi\rangle \end{pmatrix} \qquad (14.86)$$

Here, we see that the vector $|\psi\rangle$ is represented by the set of n complex numbers $\langle\xi_1|\psi\rangle, \langle\xi_2|\psi\rangle, \ldots, \langle\xi_n|\psi\rangle$ just as an ordinary vector $\mathbf{A}$ can be represented by the set of 3 real numbers

$$A_x \equiv \mathbf{i}\cdot\mathbf{A} \equiv \langle e_1|A\rangle, \quad A_y \equiv \langle e_2|A\rangle, \quad A_z \equiv \langle e_3|A\rangle \qquad (14.87)$$

From (14.86) we can write down the *completeness relation* in the form:

$$\boxed{\sum_{j=1}^{n} |\xi_j\rangle\langle\xi_j| = 1} \qquad (14.88)$$

We now wish to represent an arbitrary linear operator α in terms of orthonormal bases, $\{|\xi_j\rangle\}$.

Let us apply the linear Hermitean operator α to a ket-vector $|\psi\rangle$ and denote the resulting ket-vector by $|\phi\rangle$:

$$\alpha|\psi\rangle = |\phi\rangle \qquad (14.89)$$

We expand the rhs in the form

$$|\phi\rangle = \sum_{j=1}^{n} |\xi_j\rangle\langle\xi_j|\phi\rangle \qquad (14.90)$$

We note that this expansion can be obtained by simply multiplying by the completeness relation (14.88) from the left. In fact, this simple operation is the principal convenience of having the completeness relation.

We expand the lhs of (14.89) by using the completeness relation twice:

$$\alpha |\psi\rangle = \left(\sum_{j=1}^{n} |\xi_j\rangle\langle\xi_j|\right) \alpha \left(\sum_{k=1}^{n} |\xi_k\rangle\langle\xi_k|\right) |\psi\rangle$$

$$= \sum_{j=1}^{n} |\xi_j\rangle \sum_{k=1}^{n} \langle\xi_j| \alpha |\xi_k\rangle \langle\xi_k |\psi\rangle \tag{14.91}$$

Comparing the coefficients of the vector $|\xi_j\rangle$ in (14.90) and (14.91), we obtain

$$\langle\xi_j |\phi\rangle = \sum_{k=1}^{n} \langle\xi_j| \alpha |\xi_k\rangle \langle\xi_k |\psi\rangle , \quad j = 1, 2, \ldots, n \tag{14.92}$$

This means that if we represent the operator α by the $n \times n$ matrix

$$\begin{pmatrix} \langle\xi_1| \alpha |\xi_1\rangle & \langle\xi_1| \alpha |\xi_2\rangle & \cdots & \langle\xi_1| \alpha |\xi_n\rangle \\ \langle\xi_2| \alpha |\xi_1\rangle & \langle\xi_2| \alpha |\xi_2\rangle & \cdots & \langle\xi_2| \alpha |\xi_n\rangle \\ \vdots & \vdots & \ddots & \vdots \\ \langle\xi_n| \alpha |\xi_1\rangle & \langle\xi_n| \alpha |\xi_2\rangle & \cdots & \langle\xi_n| \alpha |\xi_n\rangle \end{pmatrix} \tag{14.93}$$

then the elements of $|\phi\rangle$ and $|\psi\rangle$ are connected by the known matrix multiplication rule as exhibited in (14.92).

In summary, an arbitrary vector in the Hilbert space can be represented by a column matrix of the form (14.86). An arbitrary linear operator α can be represented by a square matrix as shown in (14.93). The orthogonal bases in which vectors and linear operators are represented, can be chosen so that the particular problem in question may be treated in the simplest manner.

15
Quantum Mechanics for a Particle

Following Dirac, we construct a quantum theory in analogy with classical Hamiltonian mechanics. For a particle moving in one dimension the observable dynamical variables are: the position x, the momentum p, and the Hamiltonian H. These observables, by postulate, are represented by Hermitian operators: $H^\dagger = H$. The quantum state of a particle is represented by a ket vector $|\psi\rangle$ or equivalently by a bra vector $\langle\psi|$. The state $|\psi, t\rangle$ changes, following the Schrödinger equation of motion, (15.19). The dynamics is quantized through the fundamental commutation rules: $[x, p] \equiv xp - px = i\hbar$, where $\hbar$ is the Planck constant h divided by 2π: $\hbar \equiv h/(2\pi)$. The expectation value of an observable ξ is given by $\langle \xi \rangle = \langle \psi | \xi | \psi \rangle$.

15.1
Quantum Description of a Linear Motion

The quantum description of a dynamical system is quite different from, but also much related to, the classical description. We may develop a quantum theory in close analogy with the classical Hamiltonian mechanics. This was done by Dirac in his classic book [1], *Principles of Quantum Mechanics*. Following this development, we will present a quantum theory of a one-dimensional system in the present section.

Let us consider a particle moving along a straight line of length L in the range $(0, L)$. In classical mechanics, the system is characterized by the Hamiltonian

$$H = \frac{1}{2m} p^2 + V(x) \tag{15.1}$$

In Section 3.2 we saw that the dynamical state of the particle can be represented by the set of canonical variables x and p, which change with time in accordance with Hamilton's equations of motion:

$$\dot{x} = \frac{\partial H(x, p)}{\partial p} = \frac{p}{m}$$

$$\dot{p} = -\frac{\partial H(x, p)}{\partial x} = -\frac{dV}{dx} \tag{15.2}$$

Mathematical Physics. Shigeji Fujita and Salvador V. Godoy
Copyright © 2010 WILEY-VCH Verlag GmbH & Co. KGaA, Weinheim
ISBN: 978-3-527-40808-5

In quantum mechanics, *observable dynamical quantities* such as position, momentum, and energy, by assumption, are represented by *Hermitean operators*. In particular, let x be the Hermitean operator representing the position. *The eigenvalue equation for the position x is*

$$x|x'\rangle = x'|x'\rangle \tag{15.3}$$

where x' is the eigenvalue and may take any value between 0 and L.

If we take any two eigenstates associated with two different eigenvalues x' and x'' and form their scalar product, then the product vanishes according to the orthogonality:

$$\langle x''|x'\rangle = 0 \quad \text{if} \quad x' \neq x'' \tag{15.4}$$

When the eigenvalues form a continuum, as in the present case, we cannot normalize the ket-vectors $|x'\rangle$ simply. This is so because the scalar product $\langle x'|x'\rangle$ becomes infinitely large, as we will see presently. It is then found convenient to express the *orthonormality relation* by

$$\langle x''|x'\rangle = \delta(x' - x'') \tag{15.5}$$

where $\delta(y)$ is the Dirac delta function, which was defined earlier. The principal definition properties of this function are [see (4.2.5)]

$$\delta(y) = 0 \quad \text{if } y \neq 0$$

$$\int_{-\infty}^{\infty} f(y)\delta(y)\,dy = f(0) \tag{15.6}$$

where $f(y)$ is an arbitrary function that is continuous at $y = 0$.

The convenience of (15.5) can be demonstrated as follows. Let us assume that the vectors $\{|x'\rangle\}$, for $0 \leq x' \leq L$, form a complete set such that an arbitrary ket $|\psi\rangle$ can be expanded as

$$|\psi\rangle = \int_0^L dx' |x'\rangle\langle x'|\psi\rangle \tag{15.7}$$

The absolute square of $|\psi\rangle$ is

$$\langle\psi|\psi\rangle = \left(\int_0^L dx'' \langle\psi|x''\rangle\langle x''|\right) \cdot \left(\int_0^L dx' |x'\rangle\langle x'|\psi\rangle\right)$$

$$= \int_0^L dx'' \langle\psi|x''\rangle \left[\int_0^L dx' \langle x''|x'\rangle\langle x'|\psi\rangle\right]$$

Consider now the quantity in the square brackets:

$$\int_0^L dx' \langle x'' | x' \rangle \langle x' | \psi \rangle$$

According to (15.4) the factor $\langle x'' | x' \rangle$ in the integrand vanishes unless $x' = x''$. The integral, however, must have the value $\langle x'' | \psi \rangle$ to be consistent with the relation (15.7). This is possible only if $\langle x'' | x' \rangle$ is infinitely large at $x' = x''$. Furthermore, the desired result can be obtained if we assume (15.5) and apply (15.6), that is,

$$\int_0^L dx' \langle x'' | x' \rangle \langle x' | \psi \rangle = \int_0^L dx' \delta(x'' - x') \langle x' | \psi \rangle = \langle x'' | \psi \rangle$$

We now consider the matrix representation in terms of the eigenstates of the position operator x. This representation is called the *position representation*. Using (15.3) and (15.5), we obtain

$$\langle x'' | x | x' \rangle = \langle x'' | x' | x' \rangle = x' \langle x'' | x' \rangle = x' \delta(x'' - x') \tag{15.8}$$

We can easily see that

$$\langle x'' | x^n | x' \rangle = \langle x'' | (x^{n-1}) x | x' \rangle = x' \langle x'' | x^{n-1} | x' \rangle = x'^n \langle x'' | x' \rangle$$
$$= x'^n \delta(x'' - x') \quad n = 1, 2, \ldots \tag{15.9}$$

If $f(x)$ represents a polynomial or series in x:

$$f(x) = a_0 + a_1 x + a_2 x^2 + \ldots \tag{15.10}$$

then we have

$$\langle x'' | f(x) | x' \rangle = f(x') \delta(x'' - x') \tag{15.11}$$

In particular, the potential energy operator $V(x)$, which depends on x only, has nonvanishing values only along the diagonal ($x' = x''$) in the position representation.

In sharp contrast to classical mechanics, the quantum operators x and p representing the position and momentum of the particle, do not commute but, by postulate, satisfy the following *commutation relation*:

$$\boxed{xp - px = i\hbar} \tag{15.12}$$

where $\hbar$ is Planck's constant divided by 2π: $\hbar \equiv h/2\pi$. It can be shown from (15.12) [Problem 6.1.1.] that the operator p is equivalent to the following operator except for an unimportant phase factor:

$$\boxed{p = -i\hbar \frac{d}{dx}} \tag{15.13}$$

15 Quantum Mechanics for a Particle

where the operator d/dx is defined by

$$\langle x'| \frac{d}{dx} |\psi\rangle \equiv \frac{d}{dx'} \langle x'| \psi\rangle \quad \text{for any} \quad |\psi\rangle \tag{15.14}$$

Note that the d/dx is a quantum operator and not a differential operator. After a series of steps

$$\begin{aligned}
\langle \psi| \frac{d}{dx} |x'\rangle &= \int \langle \psi| x''\rangle\, dx''\, \langle x''| \frac{d}{dx} |x'\rangle && \text{[use of (15.7)]} \\
&= \int \langle \psi| x''\rangle\, dx''\, \frac{d}{dx''} \langle x''| x'\rangle && \text{[use of (15.14)]} \\
&= \int \langle \psi| x''\rangle\, dx''\, \frac{d}{dx''} \delta(x'' - x') && \text{[use of (15.5)]} \\
&= \int \langle \psi| x''\rangle\, dx'' \left[-\frac{d}{dx'} \delta(x'' - x') \right] \\
&= -\frac{d}{dx'} \langle \psi| x'\rangle
\end{aligned}$$

we obtain

$$\langle \psi| \frac{d}{dx} |x'\rangle = -\frac{d}{dx'} \langle \psi| x'\rangle \tag{15.15}$$

Using the last two equations, we obtain

$$\begin{aligned}
\left(\langle x'| \frac{d}{dx} |\psi\rangle \right)^* &= \left(\frac{d}{dx'} \langle x'| \psi\rangle \right)^* && \text{[use of (15.14)]} \\
&= \frac{d}{dx'} \langle \psi| x'\rangle \\
&= \langle \psi| \left(-\frac{d}{dx} \right) |x'\rangle
\end{aligned}$$

where we used (15.15) in the last step. This means that the operator d/dx is not Hermitean but anti-Hermitean:

$$\left(\frac{d}{dx} \right)^\dagger = -\frac{d}{dx} \tag{15.16}$$

Since

$$\left(-i\hbar \frac{d}{dx} \right)^\dagger = (-i\hbar)^* \left(\frac{d}{dx} \right)^\dagger = (i\hbar)\left(-\frac{d}{dx} \right) = -i\hbar \frac{d}{dx}$$

The operator $-i\hbar\, d/dx$, which is equivalent to p, is Hermitean.
Using (15.13), we obtain

$$\langle x'| f(x, p) |\psi\rangle = f\left(x', -i\hbar \frac{d}{dx'} \right) \langle x'| \psi\rangle \tag{15.17}$$

15.1 Quantum Description of a Linear Motion

For example, by choosing $f = p^2/2m + V(x)$, we have

$$\langle x'| p^2/2m + V(x) |\psi\rangle = \left[\frac{1}{2m}\left(-i\hbar \frac{d}{dx'}\right)^2 + V(x')\right]\langle x'| \psi\rangle \tag{15.18}$$

In classical mechanics the dynamical state is represented by a point in the phase space. In quantum mechanics, the *dynamical state*, by postulate, will be represented by a *ket-vector*, say $|\psi\rangle$, which may be viewed as a multidimensional column vector.

The dynamical state in general changes with time. In classical mechanics, this change is governed by Hamilton's equations of motion (15.2). In quantum mechanics, the change is ruled, by postulate, by Schrödinger's equation of motion :

$$i\hbar \frac{d}{dt}|\psi, t\rangle = H(x, p)|\psi, t\rangle \tag{15.19}$$

where $H(x, p)$ is the *Hamiltonian operator*, the Hermitean operator that has the same functional form as the classical Hamiltonian. The d/dt, unlike d/dx, is a differential operator. Multiplying (15.19) from the left by $\langle x'|$, we obtain

$$i\hbar \langle x'| \frac{d}{dt}|\psi, t\rangle = \langle x'| H(x, p)|\psi, t\rangle \tag{15.20}$$

We may write the term on the lhs as

$$i\hbar \frac{\partial}{\partial t}\psi(x', t) \tag{15.21}$$

where

$$\boxed{\psi(x', t) \equiv \langle x'| \psi, t\rangle} \tag{15.22}$$

is a function of both x' and t, and because of this the partial time derivative $\partial/\partial t$ was indicated. Using (15.18) we can write the rhs of (15.20) as

$$\left[-\frac{\hbar^2}{2m}\frac{\partial^2}{\partial x'^2} + V(x')\right]\psi(x', t)$$

Therefore, (15.20) can be re-expressed as follows:

$$\boxed{i\hbar \frac{\partial}{\partial t}\psi(x, t) = \left[-\frac{\hbar^2}{2m}\frac{\partial^2}{\partial x^2} + V(x)\right]\psi(x, t)} \tag{15.23}$$

where we dropped the primes indicating the position eigenvalues.

Equation (15.23) is called Schrödinger's equation of motion. The function $\psi(x, t)$ is called the *wave function*. Its absolute square

$$|\psi(x, t)|^2 = |\langle x| \psi, t\rangle|^2 \equiv P(x, t) \tag{15.24}$$

by postulate, represents the *probability distribution function* (in position.) That is,

$$|\psi(x, t)|^2 \, dx = \text{the probability of finding the particle in } (x, x + dx) \tag{15.25}$$

normalized such that

$$\int_0^L dx\, \psi^*(x,t)\psi(x,t) = \int_0^L dx\, P(x,t) = 1 \qquad (15.26)$$

The *average position* of the particle is then given by

$$\langle x \rangle \equiv \int_0^L dx\, x\, P(x,t) = \int_0^L dx\, x\, \psi^*(x,t)\psi(x,t) \qquad (15.27)$$

More generally if the dynamical quantity ξ is given as a function of x and p, then its average is given by

$$\langle \xi \rangle_t = \int_0^L dx\, \psi^*(x,t)\, \xi\!\left(x, -i\hbar\frac{\partial}{\partial x}\right) \psi(x,t) \qquad (15.28)$$

For example, if we take the Hamiltonian H for ξ, we then obtain

$$\langle H \rangle_t = \int_0^L dx\, \psi^*(x,t)\left[-\frac{\hbar^2}{2m}\frac{\partial^2}{\partial x^2} + V(x)\right]\psi(x,t) \qquad (15.29)$$

Since

$$\psi^*(x,t) = \langle \psi, t | x \rangle, \qquad \langle x | \xi | \psi, t \rangle = \xi\!\left(x, -i\hbar\frac{\partial}{\partial x}\right) \langle x | \psi, t \rangle$$

we can rewrite (15.28) as follows:

$$\langle \xi \rangle_t = \int_0^L dx\, \langle \psi, t | x \rangle \langle x | \xi | \psi, t \rangle$$

$$= \langle \psi, t | \xi | \psi, t \rangle \quad \text{use of (15.7)}$$

or

$$\boxed{\langle \xi \rangle_t = \langle \psi, t | \xi | \psi, t \rangle} \qquad (15.30)$$

The average of the dynamical function ξ, $\langle \xi \rangle_t$, as given by (15.28) or by (15.30) is called the *expectation value* of the dynamical function ξ. It gives the average value of the dynamical function ξ after repeated experiments when the particle is in the state $|\psi\rangle$.

Problem 15.1.1

Show that

$$\langle x' | [x(-i\hbar d/dx) - (-i\hbar d/dx)x] | \psi \rangle = i\hbar \langle x' | \psi \rangle$$

for any ket $|\psi\rangle$.

Problem 15.1.2

The operators x and p for the position and momentum, are, by definition, Hermitean.

1. Is the product xp Hermitean?
2. Find the Hermitean conjugates of $xp - px$ and $xp + px$. Are they Hermitean?

Problem 15.1.3

Show that the normalization condition for a wavefunction $\psi(x, t)$:

$$f(t) \equiv \int_{-\infty}^{\infty} dx\, \psi^*(x, t) \psi(x, t) = 1$$

is maintained for all time. Hint: By using Schrödinger's wave equation and its Hermitean conjugate, show that

$$\frac{df}{dt} = 0$$

15.2 The Momentum Eigenvalue Problem

The momentum p is a basic dynamical variable. In quantum mechanics it is just as important as the position x. This can be expected from the fact that they both appear in the commutation rule (a fundamental postulate)

$$[x, p] \equiv xp - px = i\hbar \tag{15.31}$$

The *eigenvalue equation for the momentum* (operator) p is

$$p|p'\rangle = p'|p'\rangle \tag{15.32}$$

Multiplying this equation from the left by $\langle x'|$, we obtain

$$\langle x'|p|p'\rangle = p'\langle x'|p'\rangle$$

using the equivalence relation: $p = -i\hbar\, d/dx$ we obtain

$$-i\hbar \frac{\partial}{\partial x'} \langle x'|p'\rangle = p'\langle x'|p'\rangle \tag{15.33}$$

This is simply a differential equation for the ordinary, c-number (complex number) function $\langle x'|p'\rangle$. We can write its solution in the form:

$$\langle x'|p'\rangle = c \exp(i p' x'/\hbar) \tag{15.34}$$

Hereafter, we drop the prime from the position eigenvalue with the convention that the symbols within a ket or a bra vector mean eigenvalues and not operators. The factor c does not depend on x but may possibly depend on p'. This possibility, however, can be excluded since the fundamental commutation rule (15.31) is symmetric in x and p. Thus, c must be a numerical constant.

In many branches of physics including statistical physics, we deal with properties that do not depend on specific boundary conditions. In such cases it is advantageous to choose a boundary condition that makes the theory as simple as possible. The *periodic boundary condition*

$$\langle x + L | p' \rangle = \langle x | p' \rangle, \quad -\infty < x < \infty \tag{15.35}$$

is such a boundary condition. Substitution of (15.34) in (15.35) yields

$$\exp(i p'(x + L)/\hbar) = \exp(i p' x/\hbar) \quad \text{for all } x$$

from which we obtain

$$p' = \frac{2\pi\hbar}{L} n \equiv p_n, \quad n = 0, \pm 1, \pm 1, \ldots \tag{15.36}$$

The eigenvalues p_n for the momentum p are discrete, and are given by integral multiples of $2\pi\hbar/L$.

The constant c in (15.34) can be determined from the orthonormality relation:

$$\langle p_n | p_m \rangle = \delta_{n,m} \tag{15.37}$$

Since

$$\langle p_n | p_n \rangle = \int_0^L \langle p_n | x \rangle \, dx \, \langle x | p_n \rangle$$

$$= \int_0^L \left[c^* \exp(-i p_n x/\hbar) \right] dx \left[c \exp(i p_n x/\hbar) \right]$$

$$= |c|^2 L = 1$$

we may choose the normalization constant $c = L^{-1/2}$. Thus, we obtain

$$\langle x | p_n \rangle = L^{-1/2} \exp(i p_n x/\hbar)$$

$$\langle p_n | x \rangle = L^{-1/2} \exp(-i p_n x/\hbar) \tag{15.38}$$

These are called *transformation functions*, which are useful when we change from the position to the momentum representation or vice versa.

The eigenvalues of the momentum, $\{p_n\}$, form a complete set. That is, an arbitrary ket can be expanded in the form:

$$|\psi\rangle = \sum_n |p_n\rangle \langle p_n | \psi\rangle \tag{15.39}$$

15.2 The Momentum Eigenvalue Problem

Or equivalently we may represent this property by the *completeness relation*:

$$\sum_n |p_n\rangle \langle p_n| = 1 \tag{15.40}$$

We note that the completeness of the momentum eigenstates $\{p_n\}$ is related to that of the position eigenstates characterized by the continuous eigenvalue x, $0 \leq x \leq L$, by the Fourier transformation [3]. In fact, we have

$$\langle p_n | \psi \rangle = \int_0^L \langle p_n | x \rangle \, dx \, \langle x | \psi \rangle$$

$$= \int_0^L dx \, L^{-1/2} \exp(-i p_n x / \hbar) \langle x | \psi \rangle \tag{15.41}$$

$$\langle x | \psi \rangle = \sum_n \langle x | p_n \rangle \langle p_n | \psi \rangle$$

$$= \sum_n L^{-1/2} \exp(i p_n x / \hbar) \langle p_n | \psi \rangle \tag{15.42}$$

When the length L is made greater, the spacing in momentum, $2\pi\hbar/L$, becomes smaller. In the limit, the momentum eigenvalues form a continuum. There, the situation becomes similar to the case of the continuous position eigenvalues. We must, then, reformulate the orthonormality condition since the length of the momentum ket becomes infinitely large.

In analogy with (15.40), let us assume the completeness relation of the form

$$\int_{-\infty}^{\infty} dp' \, |p'\rangle \langle p'| = 1 \tag{15.43}$$

Using this relation, we obtain

$$\langle p'' | \psi \rangle = \int_{-\infty}^{\infty} dx \int_{-\infty}^{\infty} dp' \, \langle p'' | x \rangle \langle x | p' \rangle \langle p' | \psi \rangle$$

$$= \int_{-\infty}^{\infty} dx \int_{-\infty}^{\infty} dp' [c^* \exp(-i p'' x / \hbar)][c \exp(i p' x / \hbar)] \langle p' | \psi \rangle$$

$$= |c|^2 \int_{-\infty}^{\infty} dx \int_{-\infty}^{\infty} dp' \exp[-i(p'' - p') x / \hbar] \langle p' | \psi \rangle \tag{15.44}$$

Let us compare this expression with the identity (*Fourier's integral theorem*)

$$f(k'') = \frac{1}{2\pi} \int_{-\infty}^{\infty} dx \int_{-\infty}^{\infty} dk' \exp[-i(k'' - k')x] f(k') \tag{15.45}$$

where f represents a continuous function of k. Assuming that

$$p' = \hbar k', \quad p'' = \hbar k'', \quad \langle p'' | \psi \rangle \equiv \langle \hbar k'' | \psi \rangle = f(k'')$$

$$\langle p' | \psi \rangle \equiv \langle \hbar k' | \psi \rangle = f(k'), \quad dp' = \hbar dk'$$

we obtain $|c|^2 = (2\pi\hbar)^{-1}$. We may, therefore, choose $c = (2\pi\hbar)^{-1/2}$, and obtain for continuous p:

$$\langle x | p \rangle = (2\pi\hbar)^{-1/2} \exp(i p x / \hbar)$$

$$\langle p | x \rangle = (2\pi\hbar)^{-1/2} \exp(-i p x / \hbar) \tag{15.46}$$

Here, we dropped the primes on p.

Let us now go back and look at (15.45) representing the Fourier integral theorem. A way of obtaining the lhs in a formal manner from the rhs is: (a) to allow the change of the order at integrations and (b) introduce the delta function $\delta(k'' - k')$ by

$$\delta(k'' - k') = \frac{1}{2\pi} \int_{-\infty}^{\infty} dx \exp[-i(k'' - k')x] \tag{15.47}$$

which makes (15.45) appear as

$$f(k'') = \int_{-\infty}^{\infty} dk' \delta(k'' - k') f(k') \tag{15.48}$$

This formal manipulation is often useful. It is stressed here that the validity of (15.47) depends on the validity of the Fourier integral theorem and not on the formal identity between the left- and right-hand sides.

It is convenient to express the orthonormality in the form

$$\langle p' | p'' \rangle = \delta(p' - p'') \quad \text{continuous } p' \text{ and } p'' \tag{15.49}$$

In fact, this can be verified formally as follows:

$$\langle p' | p'' \rangle = \int_{-\infty}^{\infty} dx \, \langle p' | x \rangle \langle x | p'' \rangle$$

$$= \int_{-\infty}^{\infty} dx \, [(2\pi\hbar)^{-1/2} \exp(i p' x / \hbar)] [(2\pi\hbar)^{-1/2} \exp(i p'' x / \hbar)]$$

$$= \frac{1}{2\pi\hbar} \int_{-\infty}^{\infty} dx \exp[-i(p' - p'') x / \hbar]$$

$$= \delta(p' - p'')$$

Note that the orthonormality relation (15.49) for the continuous momentum eigenvalues is quite analogous to the corresponding relation (15.5) for the position.

15.3
The Energy Eigenvalue Problem

The Hamiltonian H plays a central role in the description of the dynamics. Most often, the Hamiltonian H represents the total energy, that is, the sum of the kinetic and potential energies. In the present section, we will discuss the eigenvalue problem for the Hamiltonian operator H.

Let us start with the *energy eigenvalue equation*:

$$H | E \rangle = E | E \rangle \tag{15.50}$$

where H is the Hamiltonian and E its eigenvalue. Multiplying this from the left by $\langle x |$ we obtain

$$\langle x | H(x, p) | E \rangle = H\left(x, -i\hbar \frac{d}{dx}\right) \langle x | E \rangle = E \langle x | E \rangle \tag{15.51}$$

If we regard the c-number $\langle x | E \rangle$ as a function of x and write

$$\psi_E(x) \equiv \langle x | E \rangle \tag{15.52}$$

we can rewrite (15.50) in the form

$$H\left(x, -i\hbar \frac{d}{dx}\right) \psi_E(x) = E \psi_E(x) \tag{15.53}$$

If

$$H = \frac{1}{2m}p^2 + V(x),$$

we, then, obtain

$$\left(-\frac{\hbar^2}{2m}\frac{d^2}{d^2x} + V(x)\right)\psi_E(x) = E\psi_E(x) \tag{15.54}$$

This equation is called *Schrödinger's equation for the energy-eigenvalue problem*. The function ψ_E defined in (15.52) is called the *probability amplitude* or the wavefunction. By postulate, its absolute square $|\psi_E|^2$ gives the (relative) position–space distribution function when the system is in the eigenstate E.

16
Fourier Series and Transforms

Fourier series and Fourier transforms are important in many areas of physics. We shall summarize the elementary theory in this chapter. The position and momentum eigenvalues in quantum theory are connected by the Fourier transformation. Heisenberg's uncertainty principle, which is the most important feature in quantum mechanics, will be discussed.

16.1
Fourier Series

The 1D classical wave equation is

$$\frac{\partial^2 \psi(x,t)}{\partial t^2} = c^2 \frac{\partial^2 \psi}{\partial x^2} \tag{16.1}$$

where c is the propagation speed. After using the separation-of-variables method, we obtain the 1D Helmholtz equation for the position variable x:

$$\frac{d^2 u(x)}{dx^2} = -k^2 u \tag{16.2}$$

From now on we assume a *periodic boundary* such that

$$u(x+L) = u(x), \quad 0 < x < L \tag{16.3}$$

Then, the solutions of (16.2) with the periodic boundary condition (16.3) are

$$u_n(x) \equiv \exp(ik_n x) \tag{16.4}$$

with

$$k_n = \frac{2\pi n}{L}, \quad n = 0, \pm 1, \pm 2, \ldots \tag{16.5}$$

We note that there are an infinite number of solutions represented by $\{k_n\}$. These states are *orthogonal* to each other:

$$\int_0^L dx\, u_n^*(x) u_m(x) = 0 \quad \text{if} \quad n \neq m \tag{16.6}$$

Mathematical Physics. Shigeji Fujita and Salvador V. Godoy
Copyright © 2010 WILEY-VCH Verlag GmbH & Co. KGaA, Weinheim
ISBN: 978-3-527-40808-5

For $n = m$, we have

$$\int_0^L dx\, u_n^*(x) u_m(x) = L \tag{16.7}$$

Using Kronecker's delta

$$\delta_{nm} = \begin{cases} 1 & \text{if } n = m \\ 0 & \text{otherwise} \end{cases} \tag{16.8}$$

we may write (16.6) and (16.7) together as

$$\int_0^L dx\, u_n^*(x) u_m(x) = L\delta_{nm} \tag{16.9}$$

We may expand an arbitrary function $f(x)$ in a Fourier series:

$$f(x) = \sum_{n=-\infty}^{\infty} c_n \exp(-i2\pi n x/L) \tag{16.10}$$

Using the orthogonality relation (16.9), we find the expansion coefficient

$$c_n = \frac{1}{L} \int_0^L dx\, \exp(i2\pi n x/L) f(x) \tag{16.11}$$

The expansion coefficients $\{c_n\}$ are in general complex for a real function $f(x)$. Alternatively, we can expand $f(x)$ in sine and cosine series:

$$f(x) = a_0 + \sum_{n=1}^{\infty} \left[a_n \cos(2\pi n x/L) + b_n \sin(2\pi n x/L) \right] \tag{16.12}$$

This form may be used for some problems.

16.2
Fourier Transforms

From now on we assume that the function f has no constant terms in the expansion (16.10):

$$c_0 = \frac{1}{L} \int_0^L dx\, f(x) = 0 \tag{16.13}$$

The Fourier integrals are obtained from the Fourier series in (16.10) by passing to the limit of an infinite integral.

We assume that the interval is

$$-L/2 < x < L/2 \tag{16.14}$$

This choice does not change (16.5)–(16.7). We then let $L \to \infty$. We write

$$k = \frac{2\pi n}{L} \tag{16.15}$$

and note that the unit interval is

$$\Delta k = \frac{2\pi(n+1)}{L} - \frac{2\pi n}{L} = \frac{2\pi}{L}\Delta n \tag{16.16}$$

between allowed k-values become smaller and smaller in the small-L limit. We may replace the series (16.10) by an integral:

$$\sum_n \to \frac{L}{2\pi} \int_{-\infty}^{\infty} dk \tag{16.17}$$

In the limit, we obtain from (16.10) and (16.11)

$$f(x) = \frac{L}{2\pi} \int_{-\infty}^{\infty} dk\, c(k) e^{-ikx} \tag{16.18}$$

$$c(k) = \frac{1}{L} \int_{-\infty}^{\infty} dx\, e^{ikx} f(x) \tag{16.19}$$

16.3
Bra and Ket Notations

Bra and ket notations invented by Dirac for quantum physics are also useful in classical physics.

We denote a function representing a property f

$$f(x) \equiv \langle x | f \rangle \tag{16.20}$$

1. For a finite interval L, the k-values are discrete $\{k_n\}$. The Fourier transform of $\langle x | f \rangle$ is written as

$$\langle k_n | f \rangle = \int_0^L dx\, \langle k_n | x \rangle \langle x | f \rangle = L^{-1/2} \int_0^L dx\, e^{ik_n x} \langle x | f \rangle \tag{16.21}$$

where

$$\langle k_n | x \rangle = L^{-1/2} e^{ik_n x} \tag{16.22}$$

The *orthonormalities* are given by

$$\langle k_n | k_m \rangle = \delta_{n,m}$$

$$\langle x | x' \rangle = \delta(x - x') \tag{16.23}$$

The completeness relations are

$$\sum_{k_n} |k_n\rangle \langle k_n| = 1 \tag{16.24}$$

$$\int dx \, |x\rangle \langle x| = 1 \tag{16.25}$$

The Fourier series expansion is written down as

$$\langle x | f \rangle = \sum_{k_n} \langle x | k_n \rangle \langle k_n | f \rangle \tag{16.26}$$

$$\langle x | k_n \rangle = (\langle k_n | x \rangle)^* = L^{-1/2} e^{-i k_n x} \tag{16.27}$$

The main advantage is that the property f is represented by $\langle x | f \rangle$ in the x-space and by $\langle k_n | f \rangle$ in the (Fourier) k-space. The only one symbol f is used. The transformation functions $\langle k_n | x \rangle$ and $\langle x | k_n \rangle$ carry symmetric normalization factors $L^{-1/2}$.

2. For an infinite interval $(-\infty, \infty)$, the k-values are continuous. We then have:
The Fourier transformation:

$$\langle k | f \rangle = \int_{-\infty}^{+\infty} dx \, \langle k | x \rangle \langle x | f \rangle \tag{16.28}$$

$$\langle k | x \rangle = (2\pi)^{-1/2} e^{ikx}, \qquad \langle x | k \rangle = (2\pi)^{-1/2} e^{-ikx} \tag{16.29}$$

The orthonormality:

$$\langle k | k' \rangle = \delta(k - k')$$

$$\langle x | x' \rangle = \delta(x - x') \tag{16.30}$$

The Fourier inverse transformation:

$$\langle x | f \rangle = \int_{-\infty}^{+\infty} dk \, \langle x | k \rangle \langle k | f \rangle \tag{16.31}$$

We note that the normalization is different for discrete and continuous k.

16.4
Heisenberg's Uncertainty Principle

The most significant distinction of quantum mechanics from classical mechanics lies in the commutation rule for position operator x and the momentum operator p:

$$[x, p] = i\hbar \tag{16.32}$$

This leads to a *fundamental uncertainty with regard to the simultaneous observation of the position and momentum*. This is known as Heisenberg's uncertainty principle, and will be discussed in the present section.

For demonstration, let us take a wavefunction $\psi(x)$ of the form

$$\psi(x) \equiv \langle x|\psi\rangle = \left(\frac{1}{\pi a^2}\right)^{1/4} \exp\left[-\frac{(x-x_0)^2}{2a^2}\right] \tag{16.33}$$

where x_0 and a are both real constants. This function $\psi(x)$ represents a ground-state wavefunction for an harmonic oscillator with the center at x_0. The position distribution function $P(x)$ corresponding to this wavefunction $\psi(x)$ is given by

$$P(x) = |\psi(x)|^2 = \left(\frac{1}{\pi a^2}\right)^{1/2} \exp\left[-\frac{(x-x_0)^2}{a^2}\right] \tag{16.34}$$

After simple calculations, we find that (Problem 16.4.1)

$$\langle x \rangle \equiv \int_{-\infty}^{\infty} dx\, x\, P(x) = x_0 \tag{16.35}$$

$$\langle (\Delta x)^2 \rangle \equiv \langle (x-x_0)^2 \rangle = \frac{1}{2}a^2 \tag{16.36}$$

Equations (16.34) and (16.36) mean that the wavefunction $\psi(x)$ represents a quantum state ψ of a particle in which the particle is localized around the position x_0 with the root-mean-square deviation:

$$(\Delta x)_{\text{rms}} \equiv \sqrt{\langle (\Delta x)^2 \rangle} = \frac{a}{\sqrt{2}} \tag{16.37}$$

We now wish to find the momentum distribution function for the quantum state ψ. To do this we first find the "wavefunction in the momentum space", $\langle p|\psi\rangle$, which is related to the wavefunction $\psi(x) \equiv \langle x|\psi\rangle$ by

$$\langle p|\psi\rangle = \int_{-\infty}^{\infty} dx\, \langle p|x\rangle \langle x|\psi\rangle \tag{16.38}$$

By assuming the infinite-length normalization, we can calculate this quantity as follows:

$$\langle p|\psi\rangle = \int_{-\infty}^{\infty} dx \frac{1}{(2\pi\hbar)^{1/2}} \exp\left[-\frac{ipx}{\hbar}\right] \left(\frac{1}{\pi a^2}\right)^{1/4} \exp\left[-\frac{(x-x_0)^2}{2a^2}\right]$$

$$= \frac{1}{(2\pi\hbar)^{1/2}} \left(\frac{1}{\pi a^2}\right)^{1/4} \exp\left[-\frac{a^2 p^2}{2\hbar^2}\right]$$

$$\times \int_{-\infty}^{\infty} dx \exp\left[-\frac{1}{2a^2}\left(x - x_0 + \frac{ia^2 p}{\hbar}\right)^2\right]$$

$$= \left(\frac{a}{\sqrt{\pi}\hbar}\right)^{1/2} \exp\left[-\frac{a^2 p^2}{2\hbar^2}\right] \tag{16.39}$$

Taking the absolute square of this expression, we obtain the momentum distribution function $P(p)$ as

$$P(p) \equiv |\langle p|\psi\rangle|^2 = \frac{a}{\pi^{1/2}\hbar} \exp\left[-\frac{a^2 p^2}{\hbar^2}\right] \tag{16.40}$$

which is normalized such that

$$\int_{-\infty}^{\infty} dp\, P(p) = 1 \tag{16.41}$$

Using (16.40), we obtain

$$\langle p \rangle \equiv \int_{-\infty}^{\infty} dp\, p\, P(p) = 0 \quad \text{(integrand is odd)} \tag{16.42}$$

and (Problem 16.4.1)

$$\langle (\Delta p)^2 \rangle \equiv \langle (p - \langle p \rangle)^2 \rangle = \langle p^2 \rangle = \frac{1}{2}\frac{\hbar^2}{a^2} \tag{16.43}$$

Equation (16.42) means that the average momentum for the quantum state vanishes. The root-mean-square deviation of the momentum, $(\Delta p)_{\text{rms}}$, calculated from (16.43), is given by

$$(\Delta p)_{\text{rms}} \equiv \sqrt{\langle (\Delta p)^2 \rangle} = \frac{1}{\sqrt{2}}\frac{\hbar}{a} \tag{16.44}$$

Multiplying this expression by (16.37) we obtain

$$(\Delta x)_{\text{rms}} (\Delta p)_{\text{rms}} = \frac{1}{2}\hbar \tag{16.45}$$

This equation implies that the uncertainties (or standard deviations) in position and momentum are correlated, and their product is finite. In fact, if we reduce the uncertainty in the position $(\Delta x)_{\text{rms}} = a/\sqrt{2}$ by halving the parameter a, the uncertainty in momentum $(\Delta p)_{\text{rms}} = \hbar/\sqrt{2}a$, doubles. This most remarkable feature of a quantum state is known as *Heisenberg's uncertainty principle*.

The uncertainty relation (16.45) is not restricted to the wavefunction (16.33). In fact, for any noncommuting observables q and p satisfying a commutation relation of the form (16.32) the following inequality can be established

$$\Delta q \Delta p \geq \frac{1}{2}\hbar \qquad (16.46)$$

where Δq and Δp are the uncertainties associated with the measurement of q and p for any quantum state.

Problem 16.4.1

Verify (16.35), (16.36) and (16.43).

Problem 16.4.2

Assume that $\langle x|\psi\rangle \equiv \psi(x) = c\dfrac{a}{x^2 + a^2}$.

1. Find the constant c from the normalization

$$\int_{-\infty}^{\infty} dx\, |\psi(x)|^2 = 1$$

2. Evaluate

$$\langle x \rangle \equiv x_0, \quad \text{and} \quad \langle (\Delta x)^2 \rangle \equiv \langle (x - x_0)^2 \rangle$$

3. Find the corresponding momentum distribution function

$$|\langle p|\psi\rangle|^2 \quad \text{where} \quad \langle p|\psi\rangle = \int_{-\infty}^{\infty} dx\, \langle p|x\rangle \langle x|\psi\rangle$$

4. Evaluate

$$\langle p \rangle \equiv p_0, \quad \text{and} \quad \langle (\Delta x)^2 \rangle \equiv \langle (x - x_0)^2 \rangle$$

5. Evaluate

$$(\Delta x)_{\text{rms}} (\Delta p)_{\text{rms}} \equiv \sqrt{\langle (\Delta x)^2 \rangle} \sqrt{\langle (\Delta p)^2 \rangle}$$

17
Quantum Angular Momentum

Quantum angular momentum components (j_x, j_y, j_z) do not commute with each other. The eigenvalues of the squared angular momentum $\mathbf{j}^2 \equiv \mathbf{j} \cdot \mathbf{j}$ and the z-component j_z are obtained in this chapter.

17.1
Quantum Angular Momentum

Consider a particle whose motion is described by the Cartesian coordinates and momenta, (x, y, z, p_x, p_y, p_z). Its angular momentum $\mathbf{j}$ about the origin is defined by

$$j_x \equiv y p_z - z p_y, \quad j_y \equiv z p_x - x p_z, \quad j_z \equiv x p_y - y p_x \qquad (17.1)$$

These components (operators) (j_x, j_y, j_z) are Hermitian if x, y, z, p_x, p_y and p_z satisfy the fundamental commutation relations: $[q_\mu, p_\nu] = i\hbar \delta_{\mu\nu}$, $[q_\mu, q_\nu] = [p_\mu, p_\nu] = 0$. [Problem 17.1.1].

Let us evaluate the *commutators* of angular momentum components and dynamical variables $x, p_x, \ldots$, and so on. We obtain

$$\begin{aligned}
[j_z, x] &\equiv [x p_y - y p_x, x] \\
&= x[p_y, x] + [x, x]p_y - y[p_x, x] - [y, x]p_x \\
&= 0 + 0 - y(-i\hbar) - 0 = i\hbar y \\
[j_z, y] &\equiv [x p_y - y p_x, y] = x[p_y, y] = -i\hbar x \\
[j_z, p_x] &\equiv [x p_y - y p_x, p_x] = [x, p_x]p_y = i\hbar p_y, \text{etc.}
\end{aligned}$$

In summary, we have

$$[j_z, x] = i\hbar y, \quad [j_z, y] = -i\hbar x, \quad [j_z, z] = 0 \qquad (17.2)$$

$$[j_z, p_x] = i\hbar p_y, \quad [j_z, p_y] = -i\hbar p_x, \quad [j_z, p_z] = 0 \qquad (17.3)$$

with corresponding relations for j_x and j_y, which can be written down by cyclic permutation $x \to y \to z \to x$.

Mathematical Physics. Shigeji Fujita and Salvador V. Godoy
Copyright © 2010 WILEY-VCH Verlag GmbH & Co. KGaA, Weinheim
ISBN: 978-3-527-40808-5

Using these results, we obtain

$$[j_y, j_z] \equiv [zp_x - xp_z, j_z] = z[p_x, j_z] - [x, j_z]p_z$$
$$= -i\hbar z p_y + i\hbar y p_z = i\hbar j_x$$

$$[j_y, j_z] = i\hbar j_x, \quad [j_z, j_x] = i\hbar j_y, \quad [j_x, j_y] = i\hbar j_z \qquad (17.4)$$

Here, we see that the angular momentum components (j_x, j_y, j_z) do not commute with each other, unlike the position (or momentum) components. The signs in (17.2)–(17.4) may be memorized with the rule that the + sign occurs when the three variables, consisting of the two in the brackets on the lhs and the one on the rhs, are in the cyclic order (xyz), and the − sign occurs otherwise. Equations (17.2)–(17.4) have analogs in the classical theory with the correspondence: $(i\hbar)^{-1}[A, B] \leftrightarrow \{A, B\}$, where $\{A, B\}$ are the Poisson's brackets.

Equation (17.4) may be put in the vector form:

$$\boxed{\mathbf{j} \times \mathbf{j} = i\hbar \mathbf{j}} \qquad (17.5)$$

This is truly a quantum expression since the classical vector product of any vector with itself should vanish.

We can express the three sets of equations, (17.2), (17.3) and (17.4), in a unified manner:

$$\boxed{[j_z, A_x] = i\hbar A_y, \quad [j_z, A_y] = -i\hbar A_x, \quad [j_z, A_z] = 0} \qquad (17.6)$$

where $\mathbf{A} = (A_x, A_y, A_z)$ represents $\mathbf{r}$, $\mathbf{p}$ or $\mathbf{j}$. In fact, we can show that (17.6) hold for any vector $\mathbf{A}$ that can be constructed from $\mathbf{r}$ and $\mathbf{p}$ (see Problem 17.1.2)

Let B_x, B_y, B_z be a second set of components of a vector $\mathbf{B}$ that satisfy (17.6). We obtain

$$[j_z, A_x B_x + A_y B_y + A_z A_z]$$
$$= [j_z, A_x]B_x + A_x[j_z, B_x] + [j_z, A_y]B_y + A_y[j_z, B_y]$$
$$+ [j_z, A_z]B_z + A_z[j_z, B_z]$$
$$= i\hbar A_y B_x + i\hbar A_x B_y - i\hbar A_x B_y - i\hbar A_y B_x = 0$$

or

$$\boxed{[j_z, \mathbf{A} \cdot \mathbf{B}] = 0} \qquad (17.7)$$

This means that any scalar product commutes with j_z.

The results obtained for a single particle can simply be generalized for a system composed of several particles. Let their angular momenta be

$$\mathbf{j}_1 \equiv \mathbf{r}_1 \times \mathbf{p}_1, \quad \mathbf{j}_2 \equiv \mathbf{r}_2 \times \mathbf{p}_2, \ldots \qquad (17.8)$$

In classical mechanics the Poisson brackets $\{A_1, B_2\}$, where the suffixes (1,2) indicate the particles, vanish:

$$\{A_1, B_2\} = 0 \qquad (17.9)$$

We postulate that the dynamical variables describing different particles commute. For example

$$[x_1, p_{2x}] = [x_2, p_{1x}] = [y_1, p_{2y}] = 0 \tag{17.10}$$

Using these, we obtain

$$\mathbf{j}_k \times \mathbf{j}_l = -\mathbf{j}_l \times \mathbf{j}_k, \quad l \neq k \tag{17.11}$$

We note that these relations resemble the properties of the classical vectors. On the other hand, for the same particle, we have, from (17.5),

$$\mathbf{j}_k \times \mathbf{j}_k = i\hbar \mathbf{j}_k, \quad k = 1, 2, \ldots \tag{17.12}$$

The *total angular momentum* **J** is defined by

$$\mathbf{J} \equiv \sum_k \mathbf{j}_k \tag{17.13}$$

We then obtain

$$\mathbf{J} \times \mathbf{J} = \left(\sum_k \mathbf{j}_k\right) \times \left(\sum_l \mathbf{j}_l\right)$$

$$= \sum_k \mathbf{j}_k \times \mathbf{j}_k + \sum_{k \neq l}\sum (\mathbf{j}_k \times \mathbf{j}_l + \mathbf{j}_l \times \mathbf{j}_k)$$

$$= i\hbar \sum_k \mathbf{j}_k + 0 = i\hbar \mathbf{J}$$

or

$$\boxed{\mathbf{J} \times \mathbf{J} = i\hbar \mathbf{J}} \tag{17.14}$$

This result has the same form as (17.5).

We can further show that relations similar to (17.6) and (17.7) hold for **J**. That is,

$$[J_z, A_x] = i\hbar A_y, \quad [J_z, A_y] = -i\hbar A_x, \quad [J_z, A_z] = 0 \tag{17.15}$$

$$\boxed{[J_z, \mathbf{A} \cdot \mathbf{B}] = 0} \tag{17.16}$$

where **A** and **B** are vectors constructed in terms of $\mathbf{r}_1, \mathbf{p}_1, \mathbf{r}_2, \mathbf{p}_2, \ldots$.

Problem 17.1.1

Show that j_x, j_y, j_z are Hermitean if x, y, z, p_x, p_y and p_z satisfy the fundamental commutation relations.

Problem 17.1.2

Show that if (A_x, A_y, A_z) and (B_x, B_y, B_z) satisfy (17.6), then the vector $\mathbf{C} = \mathbf{A} \times \mathbf{B}$ also satisfies the same equations.

17.2
Properties of Angular Momentum

We have seen in the preceding section that the three components j_x, j_y, j_z, of angular momentum do not commute but do satisfy the following relations:

$$[j_x, j_y] = i\hbar j_z, \quad [j_y, j_z] = i\hbar j_x, \quad [j_z, j_x] = i\hbar j_y \tag{17.17}$$

The square of angular momentum $\mathbf{j}$, that is,

$$j^2 \equiv j_x^2 + j_y^2 + j_z^2 \equiv \mathbf{j} \cdot \mathbf{j} \tag{17.18}$$

is a scalar. According to (17.7) it must, then, commute with j_x, j_y or j_z:

$$[j_x, j^2] = [j_y, j^2] = [j_z, j^2] = 0 \tag{17.19}$$

From (17.17) and (17.19), we can derive important properties of the quantum angular momentum, which will be discussed in the present section.

Let us suppose that we have a fictitious system for which j_x, j_y and j_z are the only dynamical variables. We know from Problem 17.1.1 that the angular momentum constructed by the rule $\mathbf{r} \times \mathbf{p}$ is Hermitean. So is the square of the angular momentum j^2. Since $[j_z, j^2] = 0$, there exist *simultaneous eigenstates for j^2 and j_z* such that

$$j^2 |\beta, m\rangle = \beta \hbar^2 |\beta, m\rangle \tag{17.20}$$

$$j_z |\beta, m\rangle = m\hbar |\beta, m\rangle \tag{17.21}$$

where $\beta\hbar^2$ and $m\hbar$ are the eigenvalues for j^2 and j_z, respectively. Note that β and m are dimensionless real numbers. Furthermore, β must be a nonnegative number since j_x, j_y and j_z all have real eigenvalues.

We now wish to show that: (a)

$$j'(j'+1) \tag{17.22}$$

where j' denotes a *nonnegative integer or a half-integer*, and (b) for a given j', the possible values of m are

$$-j', -j'+1, -j'+2, \ldots, j'-1, j' \tag{17.23}$$

For example, for $j' = 1/2$, $\beta = 1/2(1/2 + 1) = 3/4$, and the possible values of m are $-1/2$ and $1/2$; for $j' = 1$, $\beta = 1(1+1) = 2$, and the possible values of m are -1, 0, and 1.

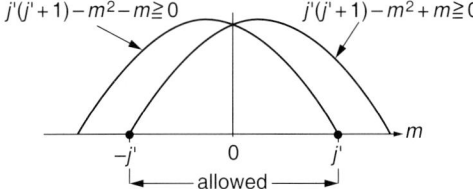

Fig. 17.1 The values of m are restricted to $-j', -j'+1, \ldots, j'-1, j'$. The limits are imposed by the inequalities $j'(j'+1) - m^2 + m \geq 0$ and $j'(j'+1) - m^2 - m \geq 0$.

and hence,

$$\beta - m^2 - m \geq 0 \tag{17.37}$$

or

$$j'(j'+1) - m^2 - m \geq 0 \tag{17.38}$$

Plotting $j'(j'+1) - m^2 - m$ against m in the same figure we see that the highest permissible value of m is j'. We know from the previous study that all possible values of m are separated by integers. The three restrictions on m are satisfied only if j' is either an integer or a half-integer. Q. E. D.

We have seen that the possible eigenvalues for the square of the angular momentum, j^2 are $\beta\hbar^2 = j'(j'+1)\hbar^2$, where j' are either an integer or a half-integer. It is customary to call $j'\hbar$, rather than $(\beta)^{1/2}\hbar = [j'(j'+1)]^{1/2}\hbar$, the *magnitude of the quantum angular momentum*. This is convenient because the possible values for j' are

$$0, \frac{1}{2}, 1, \frac{3}{2}, \ldots \tag{17.39}$$

while the possible values of $\beta^{1/2}$ are more complicated numbers.

Problem 17.2.1

Assume that $|j', m\rangle$, $|j', m-1\rangle$ and $|j', m+1\rangle$ are all normalized to unity. Calculate the following matrix elements:
1. $\langle j', m+1| j_+ |j', m\rangle$
2. $\langle j', m-1| j_- |j', m\rangle$

18
Spin Angular Momentum

The spin angular momentum of a quantum particle arises from an internal motion of the particle. The magnitude of the electron spin is $\hbar/2$. The magnetic moment $\boldsymbol{\mu}$ is proportional to the angular momentum. The electron spin can point in two directions, the up or down direction along an applied magnetic field.

18.1
The Spin Angular Momentum

The angular momentum, arising from the motion of a particle and defined by the rule $\mathbf{r} \times \mathbf{p}$, will be called *orbital angular momentum*. In quantum mechanics there exists, by postulate, another kind of angular momentum called *spin angular momentum*. The spin angular momentum of a particle arises from a certain *internal motion* of the particle, but has no analog in classical mechanics. We will hereafter denote the orbital angular momentum by $\mathbf{l}$ and the spin angular momentum by $\mathbf{s}$. The theory dealing with the spin will be developed in parallel to that dealing with the orbital angular momentum.

We postulate that the components of the spin accompanying the particle are represented by Hermitean operators (s_x, s_y, s_z) that satisfy the following commutation relations:

$$[s_x, s_y] = i\hbar s_z, \quad [s_y, s_z] = i\hbar s_x, \quad [s_z, s_x] = i\hbar s_y \tag{18.1}$$

These relations, which are analogous to (17.4), can also be written by the vector equation:

$$\mathbf{s} \times \mathbf{s} = i\hbar \mathbf{s} \tag{18.2}$$

which are analogs of (17.5), the corresponding equations for orbital angular momentum. Each of the spin components s_x, s_y, s_z by postulate, commute with all dynamical variables x, y, z, p_x, p_y, p_z, describing the motion of the particle:

$$[s_x, x] = [s_x, y] = [s_x, z] = 0$$

$$[s_x, p_x] = [s_x, p_y] = [s_x, p_z] = 0 \tag{18.3}$$

and similar relations for s_y and s_z. From these, we obtain

$$\mathbf{s} \times \mathbf{l} = -\mathbf{l} \times \mathbf{s} \tag{18.4}$$

It is further postulated that the spin and orbital angular momentum, $\mathbf{s}$ and $\mathbf{l}$, can be added as vectors:

$$\mathbf{j} \equiv \mathbf{s} + \mathbf{l} \tag{18.5}$$

This sum defines the *total angular momentum* $\mathbf{j}$.

We now look at

$$\begin{aligned} \mathbf{j} \times \mathbf{j} &\equiv (\mathbf{s}+\mathbf{l}) \times (\mathbf{s}+\mathbf{l}) \\ &= \mathbf{s} \times \mathbf{s} + (\mathbf{s} \times \mathbf{l} + \mathbf{l} \times \mathbf{s}) + \mathbf{l} \times \mathbf{l} \\ &= i\hbar \mathbf{s} + 0 + i\hbar \mathbf{l} = i\hbar(\mathbf{s}+\mathbf{l}) = i\hbar \mathbf{j} \end{aligned}$$

or

$$\mathbf{j} \times \mathbf{j} = i\hbar \mathbf{j} \tag{18.6}$$

This has the same form as (17.6). We can further show that

$$[j_z, A_x] = i\hbar A_y, \quad [j_z, A_y] = i\hbar A_x, \quad [j_z, A_z] = 0 \tag{18.7}$$

where A_x, A_y, A_z are the three components of any vector $\mathbf{A}$ that is constructed from $\mathbf{r}$, $\mathbf{p}$, and $\mathbf{s}$. (See Problem 18.1.1.)

We can also show that

$$[j_z, \mathbf{A} \cdot \mathbf{B}] = 0 \tag{18.8}$$

where $\mathbf{A}$ and $\mathbf{B}$ are any two vectors that are constructed from $\mathbf{r}$, $\mathbf{p}$, and $\mathbf{s}$.

In summary, the properties observed for the orbital angular momentum hold quite generally for the total angular momentum $\mathbf{j}$. This means that the properties of angular momentum concerning its eigenvalues are also valid for $\mathbf{j}$.

By direct calculation, we can show that

$$[s_i, \mathbf{s} \cdot \mathbf{s}] \equiv \left[s_i, s_x^2 + s_y^2 + s_z^2 \right] = 0, \quad i = x, y, \text{ or } z \tag{18.9}$$

[This can also be regarded as a special case of relation (18.8).] Because of this and (18.1), all results for the eigenvalues in Section 18.2 can be applied to the spin angular momentum.

It has been found that elementary particles have a spin whose magnitude is equal to either half-integral or integral multiples of Planck's constant $\hbar$. For example, electrons, positrons (antielectrons with positive charge), protons, neutrons, μ-mesons, all have a spin whose magnitude is $\hbar/2$. Photons have a spin of magnitude $\hbar$. Gravitons, massless quanta corresponding to the gravitational wave, are believed to have spin of magnitude $2\hbar$. Those particles with half-integer spins obey the Fermi–Dirac statistics and are called *fermions*, and those with integer spins obey the Bose–Einstein statistics and are called *bosons*. If particles carry electric charge, they have intrinsic magnetic moments that are closely associated with their spins. This topic will be discussed in the following section.

In contrast, the orbital angular momentum arises from the motion of the particle about a certain origin. Its magnitude may vary, but must have an *integral* multiple of $\hbar$. The fact that its magnitude cannot be half-integers in units of $\hbar$, which is not excluded according to our study in Section 18.2, can be argued for on general physical grounds, but will not be discussed here. The interested reader should study this point in a standard graduate-level textbook on quantum mechanics [1] or read Dirac's book [2], pp. 144–148.

Problem 18.1.1

Prove (18.7) by choosing $\mathbf{A} = \mathbf{r}, \mathbf{p}$ and $\mathbf{s}$.

Problem 18.1.2

Using the results of Problem 18.1.1, prove (18.8).

18.2
The Spin of the Electron

Electrons have a spin of the magnitude $\hbar/2$. This is found from various experimental evidence.

In dealing with an angular momentum whose magnitude is $\hbar/2$, it is convenient to introduce the *Pauli spin operator* $\boldsymbol{\sigma}$ defined by

$$\mathbf{s} = \frac{\hbar}{2}\boldsymbol{\sigma} \tag{18.10}$$

Note that this $\boldsymbol{\sigma}$ is a dimensionless vector operator. After substituting its components into (18.1), we can obtain

$$[\sigma_x, \sigma_y] = 2i\sigma_z, \quad [\sigma_y, \sigma_z] = 2i\sigma_x, \quad [\sigma_z, \sigma_x] = 2i\sigma_y \tag{18.11}$$

The eigenvalues of s_z are $\hbar/2$ and $-\hbar/2$ according to our study in Section 18.2. From (18.10), the eigenvalues of σ_z are then equal to 1 and -1. From this, we can conclude that σ_z^2 has only one eigenvalue 1. This means that σ_z^2 is the unit operator I, which, by definition, has the unique eigenvalue 1. By symmetry, σ_x^2 and σ_y^2 must also equal I. Therefore, we have

$$\sigma_x^2 = \sigma_y^2 = \sigma_z^2 = I \tag{18.12}$$

The product of the unit operator I and any quantity generates the very same quantity, and can be looked upon as multiplication by one. [See Section 3.1.] From now on, we will denote the unit operator by 1.

Let us take $\sigma_y^2 \sigma_z - \sigma_z \sigma_y^2$, which vanishes since $\sigma_y^2 = 1$:

$$\sigma_y^2 \sigma_z - \sigma_z \sigma_y^2 = 0$$

Rewriting this, we obtain

$$\sigma_y \sigma_y \sigma_z - \sigma_y \sigma_z \sigma_y + \sigma_y \sigma_z \sigma_y - \sigma_z \sigma_y \sigma_y$$
$$= \sigma_y [\sigma_y, \sigma_z] + [\sigma_y, \sigma_z]\sigma_y = 2i\sigma_y \sigma_x + 2i\sigma_x \sigma_y$$
$$= 2i(\sigma_y \sigma_x + \sigma_x \sigma_y) = 0$$

or

$$\sigma_y \sigma_x + \sigma_x \sigma_y = 0$$

Similar equations can be obtained by permuting the indices (x, y, z). Thus, we have

$$\sigma_x \sigma_y = -\sigma_y \sigma_x, \quad \sigma_z \sigma_x = -\sigma_x \sigma_z, \quad \sigma_y \sigma_z = -\sigma_z \sigma_y \quad (18.13)$$

Two operators that commute except for the minus sign are said to *anticommute*. Thus, σ_y anticommutes with σ_z, and with σ_x. From (18.11)–(18.13) we also deduce that

$$\sigma_y \sigma_z = i\sigma_x, \quad \sigma_z \sigma_x = i\sigma_y, \quad \sigma_x \sigma_y = i\sigma_z \quad (18.14)$$

$$\sigma_x \sigma_y \sigma_z = i \quad (18.15)$$

Equations (18.12)–(18.15) are the fundamental properties satisfied by the operator σ, which describes a spin of magnitude $\hbar/2$.

Let us now find a matrix representation for σ_x, σ_y and σ_z. Since these three operators do not commute, we cannot diagonalize them simultaneously. By convention, let us choose that the z-component operator σ_z is diagonal. Since the eigenvalues of σ_z are +1 and –1, the matrix for σ_z can be represented by

$$\sigma_z = \begin{pmatrix} 1 & 0 \\ 0 & -1 \end{pmatrix} \quad (18.16)$$

Using this and (18.12)–(18.15) and the fact that all σ_x, σ_y and σ_z are Hermitean operators, we can obtain matrix representations for σ_x, σ_y and σ_z. [See Problem 18.2.1.] The representations are not unique, but the following representation,

$$\sigma_x = \begin{pmatrix} 0 & 1 \\ 1 & 0 \end{pmatrix}, \quad \sigma_y = \begin{pmatrix} 0 & -i \\ i & 0 \end{pmatrix}, \quad \sigma_z = \begin{pmatrix} 1 & 0 \\ 0 & -1 \end{pmatrix} \quad (18.17)$$

is one of the simplest and most commonly used. These matrices are called *Pauli's spin matrices*.

For a complete description of the motion of an electron, we need the spin variables as well as Cartesian coordinates x, y, z and momenta p_x, p_y, p_z. The spin

18.2 The Spin of the Electron

variables commute with these coordinates and momenta. Therefore, we can choose (x, y, z, σ_z) as a complete set of commuting observables. *The corresponding eigenvalues (x', y', z', σ'_z) can be used to characterize the quantum state for the electron.* Then, any quantum-state vector $|\psi\rangle$ can be expanded in terms of the eigenkets $|x', y', z', \sigma'_z\rangle$ as follows:

$$|\psi\rangle = \iiint dx'dy'dz' \sum_{\sigma'_z = \pm 1} |x', y', z', \sigma'_z\rangle\langle x', y', z', \sigma'_z | \psi\rangle$$

$$= \int d^3r' \sum_{\sigma'_z} |\mathbf{r}', \sigma'_z\rangle\langle \mathbf{r}', \sigma'_z | \psi\rangle \tag{18.18}$$

This is a generalization of the relation (8.6.7):

$$|\psi\rangle = \int dx' |x'\rangle\langle x' | \psi\rangle$$

We note that the spin quantum number σ'_z can take on the values ± 1. In the wave-mechanical description, the quantum states are usually characterized by the two sets of wavefunctions:

$$\psi_+(x, y, z) \equiv \langle x, y, z, \sigma'_z = +1 | \psi\rangle$$

$$\psi_-(x, y, z) \equiv \langle x, y, z, \sigma'_z = -1 | \psi\rangle \tag{18.19}$$

Alternatively, we may chose $(p_x, p_y, p_z, \sigma_z)$ as a complete set of commuting observables. In terms of their eigenvalues, an arbitrary state vector $|\psi\rangle$ can be expanded as follows:

$$|\psi\rangle = \sum_{p'_x}\sum_{p'_y}\sum_{p'_z}\sum_{\sigma'_z} |p'_x, p'_y, p'_z, \sigma'_z\rangle\langle p'_x, p'_y, p'_z, \sigma'_z | \psi\rangle$$

$$= \sum_{\mathbf{p}'}\sum_{\sigma'_z} |\mathbf{p}', \sigma'_z\rangle\langle \mathbf{p}', \sigma'_z | \psi\rangle \tag{18.20}$$

Problem 18.2.1

(a) Assume that σ_x is of the form

$$\sigma_x = \begin{pmatrix} a_1 & a_2 \\ a_3 & a_4 \end{pmatrix}$$

Using the fact that σ_x is Hermitean, and the relation $\sigma_z \sigma_x = -\sigma_x \sigma_z$, show that:

$$a_1 = a_4 = 0, \quad a_2 = a_3^*, \quad |a_2| = 1$$

Therefore, σ_x is of the form

$$\sigma_x = \begin{pmatrix} 0 & e^{i\alpha} \\ e^{-i\alpha} & 0 \end{pmatrix}$$

where a is a real number. (b) By using $\sigma_y = i\sigma_x\sigma_z$, and choosing

$$\sigma_x = \begin{pmatrix} 0 & 1 \\ 1 & 0 \end{pmatrix}$$

show that σ_y is of this form

$$\sigma_y = \begin{pmatrix} 0 & -i \\ i & 0 \end{pmatrix}$$

18.3
The Magnetogyric Ratio

Let us consider a classical electron describing a circle in the *xy*-plane as shown in Figure 18.1. The angular momentum $\mathbf{j} \equiv \mathbf{r} \times \mathbf{p}$ points in the positive *z*-axis and its magnitude is given by

$$mrv \tag{18.21}$$

According to the electromagnetic theory a current loop generates a magnetic moment $\boldsymbol{\mu}$ (vector) whose magnitude equals the current times the area of the loop and whose direction is specified by the right-hand screw rule. The magnitude of the moment generated by the electron motion, therefore, is given by

$$(\text{current}) \times (\text{area}) = \left(\frac{ev}{2\pi r}\right)(\pi r^2) = \frac{1}{2}evr \tag{18.22}$$

and the direction is along the negative *z*-axis. We observe here that the *magnetic moment $\boldsymbol{\mu}$ is proportional to the angular momentum $\mathbf{j}$*. We may express this relation by

$$\boxed{\boldsymbol{\mu} = a\mathbf{j}} \tag{18.23}$$

This relation, in fact, holds not only for this circular motion but in general. The proportionality factor

$$a = \frac{-e}{2m} \tag{18.24}$$

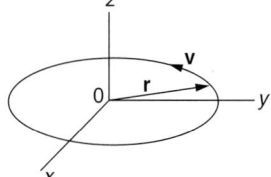

Fig. 18.1 An electron in a circular motion generates a magnetic moment $\boldsymbol{\mu}$ proportional to its angular momentum $\mathbf{j}$.

is called the *magnetogyric* or *magnetomechanical* ratio. We note that the ratio is inversely proportional to the mass m and proportional to the charge $-e$. Let us assume that a magnetic field **B** is applied along the positive z-axis. The potential energy V of a magnetic dipole with moment **μ** is given by

$$V = -\mu B \cos\theta = -\mu_z B \tag{18.25}$$

where θ is the angle between the vectors **μ** and **B**.

We may expect that a general relation such as (18.23) holds also in quantum mechanics. In the preceding sections we saw that the angular momentum (eigenvalues) is quantized in units of $\hbar$. Let us now take simple cases.

18.3.1
A. Free Electron

The electron has a spin **s** whose z-component can assume either $\hbar/2$ or $-\hbar/2$. Let us write

$$s'_z = \frac{\hbar}{2}\sigma'_z \equiv \frac{\hbar}{2}\sigma \tag{18.26}$$

where

$$\sigma \equiv \sigma'_z = \pm 1 \tag{18.27}$$

In analogy with (18.23) we may assume that

$$\mu_z \propto s'_z \propto \sigma \tag{18.28}$$

We will write this relation in the form:

$$\mu_z = \frac{1}{2} g \mu_B \sigma \tag{18.29}$$

where

$$\mu_B \equiv \frac{e\hbar}{2mc} = 0.927 \times 10^{-20} \text{ erg gauss}^{-1} \tag{18.30}$$

called the *Bohr magneton*, has the dimensions of magnetic moment. The constant g in (18.29) is a numerical factor of order 1, and is called the *g factor*. If the magnetic moment of the electron is accounted for by the "spinning" of the charge around a certain axis, the g-factor should be exactly one.

Comparison with the experiments, however, shows that this factor is 2. This so-called *spin anomaly* is an important indication of the quantum nature of the spin.

In the presence of a magnetic field **B**, the electron whose spin is directed along **B**, that is, the electron with the "up-spin" will have a lower energy than the electron whose spin is directed against **B**, that is, the electron with the "down-spin". The difference is, according to (18.25) and (18.29),

$$\Delta\varepsilon = \frac{1}{2}g\mu_B(+1) - \frac{1}{2}g\mu_B(-1) = g\mu_B B \tag{18.31}$$

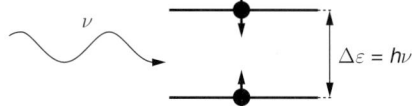

Fig. 18.2 An electron with the up-spin may absorb a photon of energy $h\nu$ and jump to the upper energy level by flipping its spin, provided that $\Delta \varepsilon = h\nu$ (electron spin resonance).

For $B = 7000$ G and $g = 2$, we obtain the numerical estimate: $\Delta \varepsilon \simeq 1°k_B$.

If an electromagnetic wave with the frequency ν satisfying $h\nu = \Delta \varepsilon$ is applied, then the electron will absorb a photon of the energy $h\nu$, and jump up to the upper energy level. See Figure 18.2. This phenomenon is known as electron spin resonance. The frequency corresponding to $\Delta \varepsilon = 1°k_B$ is

$$\nu = 2.02 \times 10^{10} \text{ cycles s}^{-1} \qquad \lambda = \frac{c}{\nu}$$

This frequency falls in the microwave region of the electromagnetic radiation spectrum.

18.3.2
B. Free Proton

The proton carries a positive charge e, whose magnitude is equal to that of the electron charge, and a spin of magnitude $\hbar/2$. We can therefore expect that it has a magnetic moment proportional to the spin quantum number s'_z. According to (18.24), the classical magnetogyric ratio is inversely proportional to the mass. Since the proton has about 1840 times heavier mass than the electron, the magnetic moment will be smaller by the same factor.

The natural unit of magnetic moment for the proton is defined by

$$\mu'_B \equiv \frac{e\hbar}{2m_p c} = \frac{e\hbar}{2m_e c}\left(\frac{m_e}{m_p}\right)$$

$$= \mu_B (m_e/m_p) = 5.05 \times 10^{-24} \text{ erg gauss}^{-1} \qquad (18.32)$$

where m_p is the mass of a proton. This unit is sometimes called the *nuclear magneton*. In analogy with the case of electron, we may assume that

$$\mu_z = \frac{1}{2}g_p \mu'_B \sigma \qquad (18.33)$$

The *proton g-factor*, g_p, defined here and compared with experiments, is

$$g_p = 2.78 \qquad (18.34)$$

This value, different from the classical value 2, indicates the quantum nature of the proton spin.

Under a magnetic field, the energy levels of a proton will be split into two, allowing a possibility of *nuclear magnetic resonance* (NMR). The energy difference is smaller by three orders of magnitude compared with a similar condition for an electron.

18.3.3
C. Free Neutron

The neutron carries a spin of magnitude $\hbar/2$ but no charge. In the classical picture, no neutral particle should carry a magnetic moment. In reality, the neutron is a quantum particle and has a nuclear magnetic moment characterized by (18.33) with the *neutron g-factor*

$$g_n = -1.913 \qquad (18.35)$$

How can a neutral particle like a neutron have a magnetic moment? This is a good question. Study of this question, however, must involve difficult quantum-mechanical calculations [4]. It is still an unsolved problem.

18.3.4
D. Atomic Nuclei

An atomic nucleus is composed of a number of nucleons. For example, the α-particle, the nucleus of He^4, is composed of two protons and two neutrons. When a nucleus contains even numbers of protons and neutrons, the stable nucleus (in the lowest-energy state) is most likely to have zero net spin and no magnetic moment. On the other hand, the He^3 nucleus has two protons and one neutron; it has a net spin of magnitude $\hbar/2$ and consequently a finite magnetic moment. As in this example, if a nucleus consists of an odd number of nucleons, it must have a spin of nonzero magnitude and a finite magnetic moment.

18.3.5
E. Atoms and Ions

An atom contains a nucleus and a certain number of electrons orbiting around the nucleus. For example, the He atom has two electrons in the 1s states, which will be denoted $(1s)^2$ [see [1] for atomic orbital notations such as $(ls)^2$]; the Ne atom has ten electrons in the state, $(1s)^2(2s)^2(2p)^6$. If the atomic orbitals are filled to the fullest, as in the cases of He or Ne in the ground state, the net spin of the electrons is zero; in more detail, the 1s orbitals are filled by a pair of electrons with up and down spins; so are the 2s orbitals. Such atoms do not possess orbital angular momentum.

Ions formed from elements of the so-called *transition groups* in the Periodic Table usually have large magnetic moments. These groups are located at regions of the Periodic Table where *inner* electron shells grow from one inert configuration to a larger one. Examples are rare-earth ions (Ce^{3+}, Gd^{3+}, Yb^{3+}, ...) and iron-group ions (Fe^{3+}, Co^{++}, Ni^{++}, ...). Ions outside the transition groups normally have zero resultant angular momentum (whether spin or orbital in nature) and zero magnetic moment. Readers interested in learning about these features are encouraged to look at the introductory section of the excellent book, *Paramagnetic Resonance* by Pake [5].

19
Time-Dependent Perturbation Theory

The quantum evolution equations are linear. If the Hamiltonian H of a system can be split into two, one major part H_0 and a small perturbation λV: $H = H_0 + \lambda V$, where λ is a small number parameter, the evolution equation can be solved in the form of a power series in λ. This is shown in this chapter.

19.1
Perturbation Theory 1; The Dirac Picture

Most quantum problems cannot be solved exactly with the present resources of mathematics. For such problems we can often use a perturbation method [1]. This consists of splitting the Hamiltonian into two parts, one of which must be simple and the other small. The first part may then be considered as the Hamiltonian of a simple or unperturbed system, and the addition of the second then give rise to small corrections. If the perturbing Hamiltonian contains a small number parameter λ, we can solve the equation of motion for the perturbed system in a power series in λ.

Suppose we have an unperturbed system characterized by a time-independent Hamiltonian H_0 *and* a perturbing Hamiltonian V, which can be an arbitrary function of the time t. The Hamiltonian H for the system is

$$H = H_0 + \lambda V(t) \tag{19.1}$$

where λ is the coupling constant (number). We assume that at the initial time 0 the system is distributed over various unperturbed states $\{\alpha\}$ with the probabilities $\{Q_\alpha\}$. The density operator corresponding to this distribution is given by

$$\rho(0) = \sum_\alpha |\alpha\rangle Q_\alpha \langle\alpha| , \quad \sum_\alpha Q_\alpha = 1 \tag{19.2}$$

If there were no perturbation, this distribution would be stationary. The perturbation causes it to change. At a time t each Schrödinger ket $|\alpha\rangle$ will change, following the Schrödinger equation of motion:

$$i\hbar \frac{\partial}{\partial t} |\alpha, t\rangle = H |\alpha, t\rangle \tag{19.3}$$

Mathematical Physics. Shigeji Fujita and Salvador V. Godoy
Copyright © 2010 WILEY-VCH Verlag GmbH & Co. KGaA, Weinheim
ISBN: 978-3-527-40808-5

19 Time-Dependent Perturbation Theory

It is convenient to introduce an *evolution operator* $U(t)$ such that

$$|a, t\rangle = U(t)|a, 0\rangle = U(t)|a\rangle, \quad U(0) = 1 \tag{19.4}$$

where U satisfies

$$i\hbar \frac{\partial}{\partial t} U(t) = [H_0 + \lambda V(t)] U(t) \tag{19.5}$$

Each ket $|a\rangle$ at $t = 0$ changes to $U(t)|a\rangle$ and each bra $\langle a|$ changes to $\langle a| U^\dagger(t)$. Then, the density operator $\rho(t)$ at the time t will be

$$\rho(t) = \sum_a U(t)|a\rangle Q_a \langle a| U^\dagger(t) \tag{19.6}$$

The probability of the system being in a state $|a'\rangle$, is given by

$$\langle a'| \rho(t) |a'\rangle = \sum_a P(a', a, t) Q_a \tag{19.7}$$

$$P(a', a, t) = |\langle a'| U(t) |a\rangle|^2 \tag{19.8}$$

Thus, the problem of calculating the *transition probability* $P(a', a, t)$ is reduced to determining the probability amplitude $\langle a'| U(t) |a\rangle$.

We can simplify our calculation by working in the *Dirac picture* (DP). Define

$$U_D(t) \equiv \exp\left(\frac{it H_0}{\hbar}\right) U(t) \tag{19.9}$$

After simple calculations we obtain (Problem 19.1.1).

$$i\hbar \frac{d}{dt} U_D(t) = \lambda V_D(t) U_D(t) \tag{19.10}$$

$$V_D(t) \equiv \exp\left(\frac{it H_0}{\hbar}\right) V(t) \exp\left(\frac{-it H_0}{\hbar}\right) \tag{19.11}$$

Equation (19.10) is more convenient to handle than (19.5) because (19.10) makes the change in U_D depend entirely on the perturbation V_D. From (19.9) we obtain

$$\langle a'| U_D(t) |a\rangle = \exp\left(\frac{it E_{a'}}{\hbar}\right) \langle a'| U(t) |a\rangle \tag{19.12}$$

where E_a is the eigenvalue of H_0. Using (19.8) and (19.12), we obtain

$$P(a', a, t) \equiv |\langle a'| U_D(t) |a\rangle|^2 \tag{19.13}$$

showing that U_D and U are equally good for determining the transition probability P.

From (19.4) and (19.9) we obtain

$$U_D(0) = 1 \tag{19.14}$$

We can easily show (Problem 19.1.2) that the solution uniquely determined from (19.10) and (19.14) is the solution of the following integral equation:

$$U_D(t) = 1 - \frac{i\lambda}{\hbar} \int_0^t d\tau\, V_D(\tau)\, U_D(\tau) \tag{19.15}$$

This equation allows a solution by iteration (Problem 19.1.3):

$$U_D(t) = 1 - \frac{i\lambda}{\hbar} \int_0^t d\tau\, V_D(\tau) + \left(\frac{-i\lambda}{\hbar}\right)^2 \int_0^t d\tau \int_0^\tau d\tau'\, V_D(\tau) V_D(\tau') + \cdots \tag{19.16}$$

In some applications it is useful to retain only the first-order term in λ. The transition probability P is then given by

$$P(a', a, t) \cong \left(\frac{\lambda}{\hbar}\right)^2 \left| \int_0^t d\tau\, \langle a | V_D(t) | a' \rangle \right|^2 \tag{19.17}$$

The DP is very important in quantum theory. We briefly discuss its connection with the Schrödinger picture (SP). In the SP we have observable ξ_S and the time-dependent ket $|\psi, t\rangle_S$, which moves following the Schrödinger equation of motion. The expectation value of ξ at the time t is given by

$$\langle \xi \rangle_t \equiv {}_S\langle \psi, t | \xi_S | \psi, t \rangle_S \tag{19.18}$$

In the DP we define the Dirac ket $|\psi, t\rangle_D$ and the Dirac observable $\xi_D(t)$ by

$$|\psi, t\rangle_D \equiv \exp\left(\frac{it H_0}{\hbar}\right) |\psi, t\rangle_S \tag{19.19}$$

$$\xi_D(t) \equiv \exp\left(\frac{it H_0}{\hbar}\right) \xi_S \exp\left(\frac{-it H_0}{\hbar}\right) \tag{19.20}$$

both of which change in time as follows: (Problem 19.1.4)

$$i\hbar \frac{d}{dt} |\psi, t\rangle_D = \lambda V_D(t) |\psi, t\rangle_D \tag{19.21}$$

$$i\hbar \frac{d}{dt} \xi_D(t) = [\xi_D, H_0] \tag{19.22}$$

The expectation value of ξ can be expressed in the standard form:

$$\boxed{\langle \xi \rangle_t \equiv {}_S\langle \psi, t | \xi_S | \psi, t \rangle_S = {}_D\langle \psi, t | \xi_D | \psi, t \rangle_D} \tag{19.23}$$

Problem 19.1.1

Derive (19.10) by using (19.3) and (19.9).

Problem 19.1.2

(a) Assume (19.15) and verify (19.10). (b) Integrate (19.10) from 0 to t' and derive (19.15).

Problem 19.1.3

(a) Obtain the first-order and second-order solutions from (19.15). (b) Verify that the expansion solution (19.16) satisfies (19.15).

Problem 19.1.4

(a) Verify (19.21). (b) Verify (19.22).

19.2
Scattering Problem; Fermi's Golden Rule

In this section we treat a scattering problem, using the time-dependent perturbation method. (This method has a wide applicability because it can be applied to nonstationary problems.)

Suppose at the initial time $t = 0$, particles of momenta all nearly equal to a momentum $\mathbf{p}_0$ are distributed uniformly in the whole space, except possibly in the neighborhood of the origin, where a fixed scatterer with a short-range potential $v(\mathbf{r})$ is located. Because of this perturbation v, the distribution of particles will change with time. After a long time and at a point far from the origin we may then observe a steady flux of particles with momentum $\mathbf{p}$ deflected by the potential. What will then be the intensity of this flux as a function of $\mathbf{p}$ and $\mathbf{p}_0$? We answer this question by the time-dependent perturbation method. Here, we are interested in calculating the following transition probability [see (19.8)]:

$$P(\mathbf{p}, \mathbf{p}_0, t) \equiv |\langle \mathbf{p}| U_D(t) |\mathbf{p}_0\rangle|^2 \qquad (19.24)$$

Let us first consider the lowest- (second-) order approximation. From (19.17)

$$P(\mathbf{p}, \mathbf{p}_0, t) \cong \left(\frac{\lambda}{\hbar}\right)^2 \left|\int_0^t d\tau \langle \mathbf{p}| V_D(t) |\mathbf{p}_0\rangle\right|^2 \qquad (19.25)$$

In our problem the potential v does not depend on the time t. Thus, from (19.11)

$$\langle \mathbf{p}| V_D(t) |\mathbf{p}_0\rangle = \exp\left[\frac{it(\epsilon - \epsilon_0)}{\hbar}\right] \langle \mathbf{p}| v |\mathbf{p}_0\rangle, \quad \epsilon = \frac{p^2}{2m}, \quad \epsilon_0 = \frac{p_0^2}{2m} \qquad (19.26)$$

We then obtain

$$\int_0^t d\tau \langle \mathbf{p}| V_D(t) |\mathbf{p}_0\rangle = \langle \mathbf{p}| v |\mathbf{p}_0\rangle \frac{\exp[it(\epsilon - \epsilon_0)/\hbar] - 1}{i(\epsilon - \epsilon_0)} \tag{19.27}$$

provided that $\epsilon \neq \epsilon_0$. Therefore, the transition probability P becomes

$$\begin{aligned} P(\mathbf{p}, \mathbf{p}_0, t) &= \left(\frac{\lambda}{\hbar}\right)^2 |\langle \mathbf{p}| v |\mathbf{p}_0\rangle|^2 \\ &\quad \times \frac{\{\exp[i(\epsilon - \epsilon_0)t/\hbar] - 1\}\{\exp[-i(\epsilon - \epsilon_0)t/\hbar] - 1\}}{(\epsilon - \epsilon_0)^2/\hbar} \\ &= 2\left(\frac{\lambda}{\hbar}\right)^2 |\langle \mathbf{p}| v |\mathbf{p}_0\rangle|^2 \frac{1 - \cos[(\epsilon - \epsilon_0)t/\hbar]}{(\epsilon - \epsilon_0)^2/\hbar} \end{aligned} \tag{19.28}$$

If ϵ differs appreciably from ϵ_0 this probability P is small and remains so for all t, which is required by the law of energy conservation. In fact, the small perturbation λv is disregarded, the energy h_0 is approximately equal to the total energy $h = h_0 + \lambda v$. Since this h is constant, the energy h_0 must be approximately constant. This means that if h_0 initially has the numerical value ϵ_0, the probability of its having a value different from ϵ_0 at any later time must be small. On the other hand, if there exists a group of states $\{\mathbf{p}\}$ having nearly the same energy as ϵ_0, the probability of a transition to such a group can be finite. In our scattering problem, there is a continuous range of states $\mathbf{p}$ having a continuous range of energy ϵ surrounding the value ϵ_0.

Let us now introduce $P_c(\mathbf{p}, \mathbf{p}_0, t)$ a coarse-grained probability distribution function and the density of states $\mathcal{D}(\mathbf{p})$ such that the (relative) probability of a transition to a final group of states within the small range $\Delta^3 p$ be given by

$$P_c(\mathbf{p}, \mathbf{p}_0, t)\mathcal{D}(\hat{\mathbf{p}})\Delta^3 p \equiv \int_{\Delta^3 p} P_c(\mathbf{p}', \mathbf{p}_0, t)\mathcal{D}(\hat{\mathbf{p}}) d^3 p' \tag{19.29}$$

where $\Delta^3 p$ is chosen such that the phase-space volume $V\Delta^3 p$ contains a large number of quantum states, and $\hat{\mathbf{p}}$ is a typical value in $\Delta^3 p$.

The momentum state $|\mathbf{p}\rangle$ may alternatively be characterized by energy ϵ and values of polar and azimuthal angles $(\theta, \phi) \equiv \omega$: $|\mathbf{p}\rangle \equiv |\epsilon, \omega\rangle$. Now, the total probability of a transition to the final group of states $\mathbf{p}$ for which the ω have the value ω and the ϵ has any value (there will be a strong probability of its having a value near ϵ_0) will be given by

$$\begin{aligned} \int P(\mathbf{p}, \mathbf{p}_0, t) d\epsilon &= 2\int_0^\infty d\epsilon |\langle \epsilon, \omega| v |\mathbf{p}_0\rangle|^2 \frac{1 - \cos[(\epsilon - \epsilon_0)t/\hbar]}{(\epsilon - \epsilon_0)^2} \\ &= \frac{2t}{\hbar} \int_{-\epsilon_0 t/\hbar}^\infty dx |\langle \epsilon_0 + (\hbar x/t), \omega| v |\mathbf{p}_0\rangle|^2 \frac{1 - \cos x}{x^2} \end{aligned} \tag{19.30}$$

where we set $\lambda = 1$. For large values of t this expression reduces to

$$\int P(\mathbf{p}, \mathbf{p}_0, t) d\epsilon = \frac{2\pi t}{\hbar} |\langle \epsilon, \omega| v |\mathbf{p}_0\rangle|^2 \left[\int_{-\infty}^{\infty} dx \frac{1-\cos x}{x^2} = \pi \right] \quad (19.31)$$

Thus, the total probability up to the time t of the transition to a final group of states $\mathbf{p}$ for which the angles ω have the value ω is proportional to t. There is therefore a finite transition rate (transition probability per unit time) equal to

$$\frac{2\pi t}{\hbar} |\langle \epsilon, \omega| v |\mathbf{p}_0\rangle|^2 = \frac{2\pi t}{\hbar} |\langle \mathbf{p}| v |\mathbf{p}_0\rangle|^2 \quad (19.32)$$

Recalling that the great contribution to this rate comes from the states having energy values near ϵ_0, we may write the transition rate as

$$\lim_{t \to \infty} \frac{P(\mathbf{p}, \mathbf{p}_0, t)}{t} = \frac{2\pi t}{\hbar} |\langle \mathbf{p}| v |\mathbf{p}_0\rangle|^2 \delta(\epsilon - \epsilon_0) \quad (19.33)$$

We substitute this into (19.29) and obtain for the *transition rate* R:

$$R \equiv P_c(\mathbf{p}, \mathbf{p}_0, t) \mathcal{D}(\hat{\mathbf{p}}) \Delta^3 p = \frac{2\pi t}{\hbar} \int_{\Delta^3 p} |\langle \mathbf{p}'| v |\mathbf{p}_0\rangle|^2 \delta(\epsilon - \epsilon_0) \mathcal{D}(\hat{\mathbf{p}}) d^3 p' \quad (19.34)$$

Now, the number of states in $\Delta^3 p$ can be expressed by

$$\mathcal{D}(\hat{\mathbf{p}}) \Delta^3 p = \int_{\Delta^3 p} \mathcal{D}(\hat{\mathbf{p}}) d^3 p' = \int_{\Delta^3 p} d\epsilon \mathcal{N}(\epsilon) \quad (19.35)$$

where $\mathcal{N}(\epsilon)$ is the density of states in energy and $\Delta \epsilon$ the range of energy associated with the momenta in $\Delta^3 p$. Using this identity we obtain from (19.34) (Problem 19.2.1).

$$\boxed{R = \frac{2\pi}{\hbar} |\langle \mathbf{p}| v |\mathbf{p}_0\rangle|^2 \mathcal{N}_f(\epsilon_0)} \quad (19.36)$$

where the subscript f denotes the final state $\mathbf{p}$. For the approximations used in deriving (19.31) and (19.32) be valid, the time t must be neither too small nor too large. It must be large compared with the duration of collision for the approximation evaluation of the integral leading to (19.30) to be valid. It must not be excessively large or else probability (19.25) will lose the meaning of a transition probability, since we could make the probability (19.30) greater than 1 by making t large enough. However, there is no difficulty in t satisfying these restrictions simultaneously if the potential is weak and of short range. Equation (19.36) is called *Fermi's golden rule*.

Equation (19.36) is valid to the second order in λ. If the potential v is singular in a small region as for a hard-core potential, the Fourier transform of $v(r)$ does not exist. It is, however, possible to generalize the theory to obtain the following formula:

$$R = \frac{2\pi}{\hbar} |\langle \mathbf{p}| \, |\mathbf{p}_0\rangle|^2 \mathcal{N}_f(\epsilon_0) \qquad (19.37)$$

where T is the transition operator, (see [2]). The T-matrix elements $\langle \mathbf{p}| \, T \, |\mathbf{p}_0\rangle$ are finite for a hard-core potential, and (19.37) can be used.

Problem 19.2.1

Verify (19.36).

19.3
Perturbation Theory 2. Second Intermediate Picture

We stated earlier that quantum many-body problems can best be treated in the Heisenberg picture (HP). This necessitates solving the equation of motion:

$$-i\hbar \frac{d}{dt} \xi(t) = [H, \xi] \qquad (19.38)$$

A perturbation theory similar to that used to solve the Schrödinger evolution equation will be developed and discussed in the present section.

Let us introduce a *second intermediate* (I) *picture*. We define new ket $|\psi, t\rangle_I$ and observables $\xi_I(t)$ by

$$|\psi, t\rangle_I \equiv e^{-it H_0/\hbar} |\psi\rangle_H \qquad (19.39)$$

$$\xi_I(t) \equiv e^{-it H_0/\hbar} \xi_H e^{it H_0/\hbar} \qquad (19.40)$$

which change in time, following (Problem 19.3.1):

$$i\hbar \frac{d}{dt} |\psi, t\rangle_I = H_0 |\psi, t\rangle_I \qquad (19.41)$$

$$i\hbar \frac{d}{dt} \xi_I(t) = -\lambda [V_I(t), \xi_I(t)] \qquad (19.42)$$

$$V_I(t) \equiv e^{-it H_0/\hbar} V(t) e^{it H_0/\hbar} \qquad (19.43)$$

The expectation value $\langle \xi \rangle_t$ can be expressed in the standard form:

$$\langle \xi \rangle_t \equiv {}_S\langle \psi, t| \xi_S |\psi, t\rangle_S = {}_I\langle \psi, t| \xi_I |\psi, t\rangle_I = {}_H\langle \psi, t| \xi_H |\psi, t\rangle_H \qquad (19.44)$$

To solve (19.42), we introduce "script" operators $\mathcal{V}_I$, which generate commutators acting on a quantum operator ξ such that [2]

$$\mathcal{V}_I(t) \xi \equiv [V_I(t), \xi] = V_I \xi - \xi V_I \qquad (19.45)$$

The commutator-generating operator is sometimes called the *quantum Liouville operator*, see (19.51) below. We can then rewrite (19.42) as

$$i\hbar \frac{d}{dt}\xi_I(t) = -\lambda V_I(t)\xi, \quad \xi_I(0) = \xi_H(0) = \xi_S \tag{19.46}$$

which is similar to (19.10). We convert it into the following integral equation:

$$\xi_I(t) - \xi_I(0) = \xi_I(t) - \xi_S = \frac{i\lambda}{\hbar} \int_0^t d\tau V_I(\tau)\xi_I(\tau) \tag{19.47}$$

By iteration, we obtain

$$\xi_I(t) - \xi_S = \frac{i\lambda}{\hbar} \int_0^t d\tau V_I(\tau)\xi_S + \left(\frac{i\lambda}{\hbar}\right)^2 \int_0^t d\tau_1 \int_0^{\tau_1} d\tau_2 V_I(\tau_1)V_I(\tau_2)\xi_S + \cdots \tag{19.48}$$

Consider a physical property of the form:

$$X = \sum_j \xi^{(j)} \tag{19.49}$$

In the HP the variable $X(t)$ changes with the time t, following the Heisenberg equation of motion and the system density operator ρ is stationary. Thus, we may solve (19.38) by the perturbation method and compute the average $\langle X \rangle_t$ through

$$\langle X \rangle_t \equiv Tr\{X(t)\rho\} = tr\{\xi(t)n\} \tag{19.50}$$

In nonequilibrium quantum statistical mechanics, we conventionally carry out the computations in the SP, where the system density operator $\rho(t)$ changes in time following the *quantum Liouville equation*:

$$i\hbar \frac{d}{dt}\rho(t) = [H, \rho] \equiv \mathcal{H}\rho \tag{19.51}$$

where $\mathcal{H}$ is the quantum Liouville operator generating a commutator. Then, the time-dependent average $\langle X \rangle_t$ can be computed from

$$\langle X \rangle_t \equiv Tr\{X\rho(t)\} = tr\{\xi n(t)\} \tag{19.52}$$

If the Hamiltonian H is the sum of single-particle Hamiltonians $h^{(j)}$:

$$H = \sum_j h^{(j)} \tag{19.53}$$

then the one-body density operator $n(t)$ defined through

$$n_{ba} \equiv \langle \alpha_b | n(t) | \alpha_a \rangle \tag{19.54}$$

changes in time, following the one-body *quantum Liouville equation*:

$$i\hbar \frac{\partial}{\partial t} n(t) = [h, n] \tag{19.55}$$

Equation (19.55) is the same, except for the sign, as the Heisenberg equation of motion, implying that a similar perturbation method can be used for its solution.

The statistical-mechanical method of computing the average $\langle X \rangle_t$ with (19.51) rather than (19.50) has an advantage, since we solve an evolution equation for $n(t)$ only once instead of solving the equation of motion for $\xi(t)$ each time. This method has disadvantages, too. If the Hamiltonian H for the system is not of the form in (19.53), then the evolution equation for $n(t)$ is more complicated than (19.55), and may be hard to solve. Thus, choosing the computation method depends on the physical problem at hand.

Problem 19.3.1

Verify (19.41) and (19.42).

References

1 Dirac, P.A.M. (1958) *Principles of Quantum Mechanics*, Oxford University Press, London, UK, 4th edn, pp. 172–178.

2 Fujita, S. (1983) *Introduction to Non-Equilibrium Quantum Statistical Mechanics*, reprint edn, Krieger, Malabar, Fl. USA, pp. 143–144, 148–150.

20
Laplace Transformation

Laplace transformation and operator algebras are discussed in this chapter.

20.1
Laplace Transformation

Many time-dependent problems can be handled simply by use of the Laplace transformation techniques.

Let $f(t)$ be a function of time t specified for $t \geq 0$. Its *Laplace transform* $F(s)$, denoted symbolically by $F(s) \equiv \mathcal{L}\{f(t)\}$, is defined by

$$\mathcal{L}\{f(t)\} = F(s) \equiv \int_0^\infty e^{-st} f(t) \, dt \tag{20.1}$$

For derivatives, we obtain after simple calculations (Problem 20.1.1)

$$\mathcal{L}\{f'(t)\} = -f(0) + s\mathcal{L}\{f(t)\} \tag{20.2}$$

$$\mathcal{L}\{f''(t)\} = -f'(0+) - s f(0+) + s^2 \mathcal{L}\{f(t)\} \tag{20.3}$$

and so on. The arguments 0+ mean that we approach 0 from the positive side; $f(t)$ is undefined for negative t. By using (20.1)–(20.3), we can reduce a linear time differential equation with constant coefficients to an algebraic equation in the Laplace space, a major advantage, see Section 20.2.

By direct calculation, we obtain (Problem 20.1.2)

$$\mathcal{L}\{1\} = \frac{1}{s}, \quad \mathcal{L}\left\{\frac{t^n}{n!}\right\} = \frac{1}{s^{n+1}} \tag{20.4}$$

where s^{-1} and s^{-n-1} are analytic (differentiable) in the right-half complex plane:

$$\text{Real part of } s \equiv \text{Re}(s) > 0 \tag{20.5}$$

Mathematical Physics. Shigeji Fujita and Salvador V. Godoy
Copyright © 2010 WILEY-VCH Verlag GmbH & Co. KGaA, Weinheim
ISBN: 978-3-527-40808-5

The original function $f(t)$ can be obtained from its transform $F(s)$ by the inverse Laplace transformation $\mathcal{L}^{-1}$:

$$f(t) = \frac{1}{2\pi i} \int_{c-i\infty}^{c+i\infty} ds\, e^{st} F(s) \equiv \mathcal{L}^{-1}\{F(s)\} \tag{20.6}$$

where c is chosen such that $F(s)$ is analytic on the right of the integration path.

For example, take

$$X \equiv \frac{1}{2\pi i} \int_{c-i\infty}^{c+i\infty} ds\, e^{st} \frac{1}{s} \tag{20.7}$$

We may choose c to be a small positive number, which is often denoted by 0+. The integrand e^{st}/s is vanishingly small if $s = Re^{i\theta}$ has an argument θ such that $\pi/2 < \theta < 3\pi/2$ and R is far away from the origin:

$$\frac{e^{st}}{s} = \frac{e^{Rt(\cos\theta + i\sin\theta)}}{R(\cos\theta + i\sin\theta)} \to 0 \quad \text{as} \quad R \to \infty \quad (\cos\theta < 0) \tag{20.8}$$

We may then choose a closed semicircle of an infinite radius, C, as shown in Figure 20.1.

There is a pole of order one at the origin inside the closed path. By using the *residue theorem*, we then obtain

$$X = \frac{1}{2\pi i} \oint_C ds\, \frac{e^{st}}{s} = \frac{1}{2\pi i} \times (2\pi i) \text{ Residue of } \frac{e^{st}}{s} = 1$$

Note: We obtained the desired result without carrying out the integration in (20.7).

As we see in this example, an elementary knowledge of the theory of functions of complex variables is very useful.

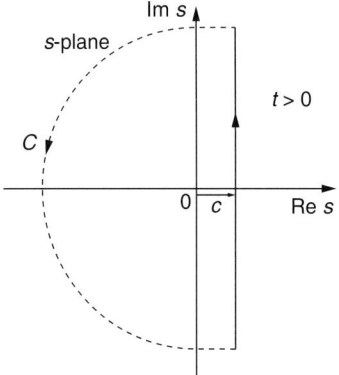

Fig. 20.1 A closed contour for evaluating (20.7).

Problem 20.1.1

Derive (20.2) and (20.3).

Problem 20.1.2

Verify (20.4) and (20.5).

Problem 20.1.3

Show that the Laplace inverse of $s^{-(n+1)}$ is equal to $t^n/n!$. Use (20.6).

20.2
The Electric Circuit Equation

Let us consider an electric circuit shown in Figure 20.2. The following differential equation with constant coefficients may be assumed:

$$L\frac{d^2q}{dt^2} + R\frac{dq}{dt} + \frac{1}{C}q = v(t) \tag{20.9}$$

where q is the charge stored in the capacitor, L the inductance, R the resistance, C the capacitance and $v(t)$ the applied voltage.

We apply the Laplace transform operator $\mathcal{L}$ to (20.9), use (20.1)–(20.3), and obtain

$$L\left[-q'(0) - sq(0) + s^2 Q(s)\right] + R\left[-q(0) + sQ(s)\right] + \frac{1}{C}Q(s) = V(s) \tag{20.10}$$

This is an algebraic equation for $Q(s) \equiv \mathcal{L}\{q(t)\}$, which is not difficult to solve. Then, the problem of solving the original differential equation is reduced to the problem of inverting the transform $Q(s)$. Some applications are given as the reader's exercises.

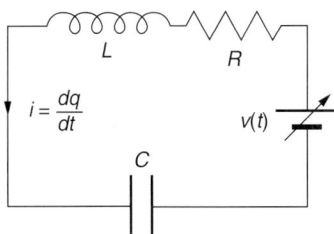

Fig. 20.2 An electric circuit.

20.3
Convolution Theorem

Let us consider an integral:

$$h(t) \equiv \int_0^\infty d\tau\, f(t-\tau)g(\tau) \tag{20.11}$$

The Laplace transform of $h(t)$ can be written simply in terms of the Laplace transforms of $f(t)$ and $g(t)$ as follows:

$$\mathcal{L}\{h(t)\} = \mathcal{L}\{f(t)\}\mathcal{L}\{g(t)\} \quad \text{or} \quad H(s) = F(s)G(s) \tag{20.12}$$

This is called the *convolution theorem*. It is very important in quantum theory. See Section 20.4.

Equation (20.12) may be proved as follows. We take the rhs, use the definition, introduce a new variable $u + v = t$, and obtain

$$F(s)G(s) = \int_0^\infty du \int_0^\infty dv\, e^{-su-sv} f(u)g(v)$$

$$= \int_0^\infty du \int_0^\infty dt\, f(u)g(t-u)e^{-st} \tag{20.13}$$

The integration domain is shown in Figure 20.3.

We now change the order of integration and obtain

$$F(s)G(s) = \int_0^\infty dt\, e^{-st} \int_0^t du\, f(u)g(t-u)$$

$$= \mathcal{L}\left\{\int_0^t du\, f(u)g(t-u)\right\}$$

$$= \mathcal{L}\left\{\int_0^t du'\, f(t-u')g(u')\right\} = \mathcal{L}\{h(t)\} \tag{20.14}$$

We give a more direct proof, using the contour integration techniques, in Appendix B.

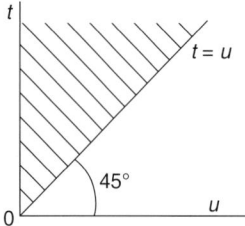

Fig. 20.3 The integration domain in (t, u) for the double integral in (20.13) is shown as the shaded area.

20.4
Linear Operator Algebras

Quantum linear operators do not in general commute. Furthermore, they may not have inverses; we must therefore treat these operators with care. The perturbation theory developed in Section 19.1 can be applied to a general case in which the perturbation V depends on time; There is considerable simplifications if V is time independent. We treat here a special case in which the total Hamiltonian H of a system is time independent.

The evolution operator $U(t)$ for the system satisfies the Schrödinger equation

$$i\hbar \frac{d}{dt} U(t) = H U(t) \qquad (20.15)$$

The formal solution of this equation with the initial condition that $U(0) = 1$ is

$$U(t) = \exp(-itH/\hbar) \qquad (20.16)$$

This solution holds as though H were an ordinary algebraic variable (instead of a linear operator) because there is no quantity that does not commute with H. If the system had a time-dependent Hamiltonian, the solution of (20.15) is more complicated.

To proceed in a general manner, we assume that a function of a linear operator ξ is defined by either a polynomial or an infinite series in powers of some parameters (numbers). For example we may regard $\exp(-itH/\hbar)$ as a series in powers of t:

$$\exp(-itH/\hbar) = 1 - itH/\hbar + (1/2)(itH/\hbar)^2 + \cdots \qquad (20.17)$$

The evolution operators U_D and V_D in the Dirac picture (DP) are defined as, see (19.9) and (19.11),

$$U_D \equiv e^{itH_0/\hbar} U(t) = e^{itH_0/\hbar} e^{-itH/\hbar} \qquad (20.18)$$

$$V_D = e^{itH_0/\hbar} V e^{-itH_0/\hbar} \qquad (20.19)$$

Multiplying (19.15) from the left by $\exp(-itH_0/\hbar)$ and using (20.18)–(20.19), we obtain

$$e^{-itH/\hbar} = e^{-itH_0/\hbar} - i\lambda\hbar^{-1} \int_0^t d\tau \, e^{-i(t-\tau)H_0/\hbar} V e^{i\tau H/\hbar} \qquad (20.20)$$

Laplace transforming this equation, we obtain (Problem 20.4.1):

$$\frac{1}{H_0 + \lambda V - z} = \frac{1}{H_0 - z} - \lambda \frac{1}{H_0 - z} \frac{1}{H_0 + \lambda V - z} \qquad (20.21)$$

Equation (20.21) is a special case of the general identity (Problem 20.4.2):

$$\frac{1}{a+b} = \frac{1}{a} - \frac{1}{a} b \frac{1}{a+b} \tag{20.22}$$

where a and b are arbitrary operators that may not commute; in addition, the existence of a^{-1} is assumed. This identity may be proved by multiplying from the right by $a+b$.

We solve (20.21) by iteration and obtain:

$$\frac{1}{H_0 + \lambda V - z} = \frac{1}{H_0 - z} - \lambda \frac{1}{H_0 - z} V \frac{1}{H_0 - z}$$

$$+ \lambda^2 \frac{1}{H_0 - z} V \frac{1}{H_0 - z} V \frac{1}{H_0 - z} + \ldots \tag{20.23}$$

Thus, the perturbation expansion in the Laplace space can be handled concisely.

Problem 20.4.1

Verify (20.21). Use the convolution theorem (20.12).

Problem 20.4.2

1. Prove (20.22).
2. Prove

$$\frac{1}{a+b} = \frac{1}{a} - \frac{1}{a+b} b \frac{1}{a}$$

where we assumed that a^{-1} and $(a+b)^{-1}$ exist.

21
Quantum Harmonic Oscillator

The quantum harmonic oscillator is treated first in the Schrödinger picture, and then by using creation and annihilation operators. The eigenvalues and eigenstates are obtained explicitly in the latter case.

21.1
Energy Eigenvalues

We can obtain the energies of the harmonic oscillator by solving the Schrödinger equation. We start with the Hamiltonian

$$H(x, p) = \frac{1}{2m}p^2 + \frac{1}{2}kx^2 = \frac{1}{2m}p^2 + \frac{1}{2}m\omega^2 x^2 \tag{21.1}$$

where $\omega \equiv \sqrt{k/m}$ is the angular frequency. Schrödinger's equation for the energy eigenvalue can be written down as

$$\left[-\frac{\hbar^2}{2m}\frac{d^2}{dx^2} + \frac{1}{2}m\omega^2 x^2\right]\varphi(x) = E\varphi(x) \tag{21.2}$$

where E is the energy eigenvalue.

In order to solve (21.2), let us try the function

$$\varphi_0(x) = e^{-\alpha x^2} \tag{21.3}$$

After differentiation, we obtain

$$\frac{d\varphi_0}{dx} = -2\alpha x \, e^{-\alpha x^2}$$

$$\frac{d^2\varphi_0}{dx^2} = -2\alpha e^{-\alpha x^2} + 4\alpha^2 x^2 e^{-\alpha x^2} \tag{21.4}$$

By choosing

$$\alpha = m\omega^2/2\hbar$$

Mathematical Physics. Shigeji Fujita and Salvador V. Godoy
Copyright © 2010 WILEY-VCH Verlag GmbH & Co. KGaA, Weinheim
ISBN: 978-3-527-40808-5

Equation (21.4) can be rewritten in the form:

$$-\frac{\hbar^2}{2m}\frac{d^2\varphi_0}{dx^2} = \left(\frac{1}{2}\hbar\omega - \frac{1}{2}m\omega^2 x^2\right)\varphi_0$$

or

$$-\frac{\hbar^2}{2m}\frac{d^2\varphi_0}{dx^2} + \frac{1}{2}m\omega^2 x^2 \varphi_0 = \frac{1}{2}\hbar\omega\varphi_0 \tag{21.5}$$

This equation is the same as (21.2) if we take

$$E_0 = \frac{1}{2}\hbar\omega \tag{21.6}$$

That is, our trial function φ_0 is an eigenfunction with the eigenvalue $\hbar\omega_0/2$.

Now consider another function

$$\varphi_1 = xe^{-\alpha x^2} \tag{21.7}$$

The reader may verify that this function φ_1 is also an eigenfunction with the eigenvalue (Problem 21.1.1)

$$E_1 = \frac{3}{2}\hbar\omega \tag{21.8}$$

The eigenvalues and eigenfunctions can, of course, be found by a more systematic procedure. This is somewhat lengthy, however, and may be referred to in a standard textbook of quantum mechanics [1]. Systematic calculations yield the set of energy eigenvalues given by

$$\boxed{E_n = \left(n + \frac{1}{2}\right)\hbar\omega, \quad n = 0, 1, 2, \ldots} \tag{21.9}$$

Here, we see that the energy eigenvalues are *discrete*. The eigenvalues are equally spaced with the distance-in-energy $\hbar\omega_0$, where ω_0 is the classical angular frequency, see Figure 21.1.

The eigenfunctions φ_0 and φ_1 in (21.3) and (21.7) are orthogonal to each other:

$$\int_{-\infty}^{\infty} dx\, \varphi_n^*(x)\varphi_m(x) = 0, \quad n \neq m \tag{21.10}$$

However, they are not normalized. Using the normalization condition,

$$\int_{-\infty}^{\infty} dx\, \varphi_n^*(x)\varphi_n(x) = 1, \quad n = 1, 2 \tag{21.11}$$

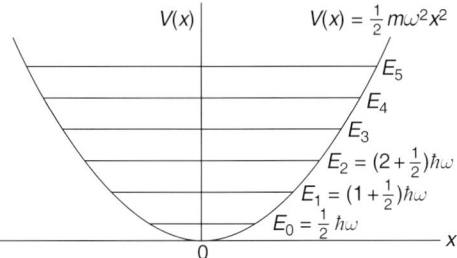

Fig. 21.1 The harmonic potential $V = \frac{1}{2}kx^2 = \frac{1}{2}m\omega^2 x^2$ and the energy eigenvalues $E_n = (n + \frac{1}{2})\hbar\omega$ for the harmonic oscillator.

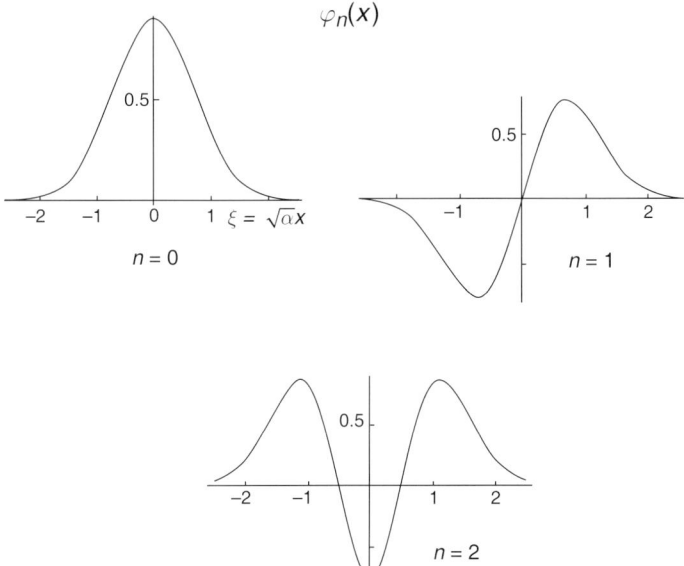

Fig. 21.2 The wavefunction of the lowest three energies for the harmonic oscillator.

we can obtain the normalized wave amplitude (or wave) function as follows: (Problem 21.1.2)

$$\varphi_0(x) = \left(\frac{2\alpha}{\pi}\right)^{1/4} e^{-\alpha x^2} \tag{21.12}$$

$$\varphi_0(x) = 2\left(\frac{2\alpha^3}{\pi}\right)^{1/4} x e^{-\alpha x^2} \tag{21.13}$$

These wavefunctions are real. The wavefunctions of the lowest three energies are shown in Figure 21.2.

It is interesting to observe that the *ground-state wavefunction* $\varphi_0(x)$ with the lowest energy $\frac{1}{2}\hbar\omega$ is positive and therefore has a single sign, while the first excited-state wavefunction $\varphi_1(x)$ changes sign once (at the origin). In general, the wave func-

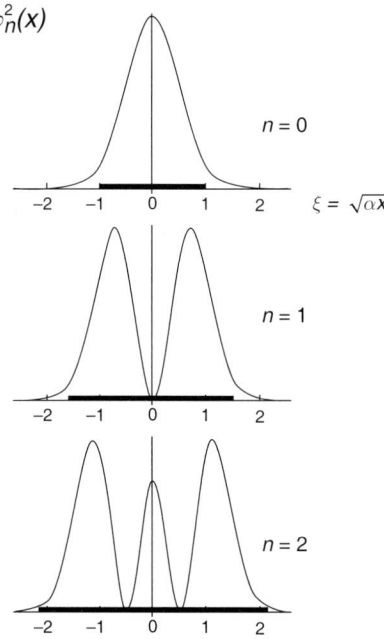

Fig. 21.3 Probability distribution functions $P_n(x) = |\varphi_n(x)|^2$ plotted against the dimensionless displacement $\xi = \sqrt{\alpha}x$. The horizontal bars indicate the classical ranges of motion.

tions with higher energies have more zeros and change sign more times. This feature is connected with the orthogonality relation for the wavefunctions.

The probability distribution functions

$$P_n(x) = |\varphi_n(x)|^2 \tag{21.14}$$

corresponding to these wavefunctions are shown in Figure 21.3.

It is interesting to observe that the quantum probability distributions extend further than the permissible ranges of the classical motion of the same total energies; these ranges are indicated by horizontal line segments in the figure.

Another point worth noting is as follows: according to classical mechanics, the oscillator could have zero energy in which case the particle would be at rest at the origin. In quantum mechanics, however, this is impossible. The lowest energy that the oscillator can have is $\frac{1}{2}\hbar\omega_0$, which is called the *zero-point energy* of the oscillator. This fact has important practical consequences. For example, the atoms (or ions) in a crystal lattice vibrate like a collection of harmonic oscillators. At the absolute zero of temperature, when classical theory would have predicted a state in which all the atoms are at rest, they must still move about with their zero-point energies. Such zero-point motion leads to the striking existence of quantum liquids, liquid He^3 and He^4, which do not freeze even at 0 K (at atmospheric pressure).

Problem 21.1.1

Show that

$$\varphi_1(x) = x\, e^{-\alpha x^2}, \quad \alpha \equiv m\omega/2\hbar$$

satisfies the eigenvalue equation (21.2) with the energy $E_1 = 3\hbar\omega/2$.

Problem 21.1.2

Verify the normalization factors for φ_0 and φ_1 in (21.12) and (21.13).

21.2 Quantum Harmonic Oscillator

Let us consider a classical simple harmonic oscillator characterized by

$$H = H(x, p) = \frac{1}{2m} p^2 + \frac{1}{2} k x^2 \tag{21.15}$$

The system oscillates with the *angular frequency* $\omega = \sqrt{k/m}$. We introduce a new set of canonical variables

$$q_1 = \sqrt{m}\, x, \quad p_1 = p/\sqrt{m} \tag{21.16}$$

and write the Hamiltonian H as

$$H = H(q_1, p_1) = \frac{1}{2}\left(p_1^2 + \omega^2 q_1^2\right) \tag{21.17}$$

Using Hamilton's equations of motion,

$$\dot{q}_1 = \frac{\partial H}{\partial p_1} = p_1, \quad \dot{p}_1 = -\frac{\partial H}{\partial q_1} = -\omega^2 q_1 \tag{21.18}$$

we obtain

$$\ddot{q}_1 = -\omega^2 q_1 \tag{21.19}$$

showing that the coordinate q_1 oscillates with the angular frequency ω (Problem 21.2.1). We drop the suffix 1 hereafter.

For the quantum problem, if we assume the same Hamiltonian H,

$$H = \frac{1}{2}(p^2 + \omega^2 q^2) \tag{21.20}$$

where (q, p) are now operators that satisfy the basic commutation relations:

$$[q, p] = i\hbar, \quad [q, q] = [p, p] = 0 \tag{21.21}$$

the quantum description of the harmonic oscillator is complete.

The equations of motion are

$$\dot{q} = (1/i\hbar)[q, H] = p, \quad \dot{p} = (1/i\hbar)[p, H] = -\omega^2 q \qquad (21.22)$$

(Problem 21.2.2). We introduce the dimensionless complex dynamic variables:

$$a^\dagger \equiv (2\hbar\omega)^{-1/2}(p + i\omega q), \quad a \equiv (2\hbar\omega)^{-1/2}(p - i\omega q) \qquad (21.23)$$

Using (21.22) and (21.23) we obtain

$$\dot{a}^\dagger = (2\hbar\omega)^{-1/2}(-\omega^2 q + i\omega p) = i\omega a^\dagger, \quad \dot{a} = -i\omega a \qquad (21.24)$$

We can express (q, p) in terms of $(a, a^\dagger)$:

$$q = -i\left(\frac{\hbar}{2\omega}\right)^{1/2}(a^\dagger - a), \quad p = \left(\frac{\hbar\omega}{2}\right)^{1/2}(a^\dagger + a) \qquad (21.25)$$

Thus, we may work entirely in terms of $(a, a^\dagger)$. After straightforward calculations, we obtain (Problem 21.2.3):

$$\hbar\omega a^\dagger a = \frac{1}{2}(p + i\omega q)(p - i\omega q) = \frac{1}{2}[p^2 + \omega^2 q^2 + i\omega(qp - pq)]$$

$$= H - \frac{1}{2}\hbar\omega \qquad (21.26)$$

$$\hbar\omega a a^\dagger = H + \frac{1}{2}\hbar\omega \qquad (21.27)$$

Taking the difference between (21.26) and (21.27), we obtain

$$a a^\dagger - a^\dagger a \equiv [a, a^\dagger] = 1 \qquad (21.28)$$

Adding (21.26) and (21.27) together and dividing by two, we obtain

$$H = \frac{1}{2}\hbar\omega(a^\dagger a + a a^\dagger) = \hbar\omega\left(a^\dagger a + \frac{1}{2}\right)$$

$$\equiv \hbar\omega\left(n + \frac{1}{2}\right) \qquad (21.29)$$

where $n \equiv a^\dagger a = n^\dagger$ is called the number operator.

Using the commutation relation (21.28), we shall show that the eigenvalues of n are

$$n' = 0, 1, 2, \ldots \qquad (21.30)$$

We start with the eigenvalue equation:

$$n|n'\rangle = n'|n'\rangle, \quad \langle n'|n'\rangle = 1 \qquad (21.31)$$

Multiplying this equation by the bra $\langle n'|$ from the left, we obtain

$$\langle n'|a^\dagger a|n'\rangle = n'\langle n'|n'\rangle \qquad (21.32)$$

Now $\langle n'| a^\dagger a |n'\rangle$ is the squared length of the ket $a |n'\rangle$ and hence it must be non-negative:

$$\langle n'| a^\dagger a |n'\rangle = n' \langle n' |n'\rangle \geq 0 \tag{21.33}$$

Also, by construction $\langle n' |n'\rangle > 0$. Hence, we obtain

$$n' \geq 0 \tag{21.34}$$

the case of equality occurs only if

$$a |n'\rangle = 0 \tag{21.35}$$

Consider now $[a, n] \equiv [a, a^\dagger a]$. We may use the following formulas (Problem 21.2.4)

$$[A, BC] = B[A, C] + [A, B]C, \quad [AB, C] = A[B, C] + [A, C]B \tag{21.36}$$

and obtain

$$[a, a^\dagger a] = a^\dagger [a, a] + a[a, a^\dagger] = a$$

or

$$an - na = a \tag{21.37}$$

Hence, we obtain

$$na |n'\rangle = (an - a) |n'\rangle = (n' - 1)a |n'\rangle \tag{21.38}$$

Now, if $a |n'\rangle \neq 0$, then $a |n'\rangle$ is according to (21.38), an eigenket of n belonging to the eigenvalue $n' - 1$. Hence, for nonzero n', $n' - 1$ is another eigenvalue. We can repeat the argument and deduce that, if $n' - 1 \neq 0$, $n' - 2$ is another eigenvalue of. Continuing in this way, we obtain a series of eigenvalues $n', n' - 1, n' - 2, \ldots$ that can terminate *only* with the value 0 because of inequality (21.34).

By a similar process, we can show from the Hermitian conjugate of (21.37):

$$na^\dagger - a^\dagger n = a^\dagger \tag{21.39}$$

that the eigenvalue of n has no upper limit (Problem 21.2.5). Hence, the eigenvalues of n are non-negative integers: $0, 1, 2, \ldots$ (Q. E. D.).

Let $|0\rangle$ be a normalized eigenket of n belonging to the eigenvalue 0 so that

$$n |0\rangle = a^\dagger a |0\rangle = 0 \quad \text{or} \quad a |0\rangle = 0 \tag{21.40}$$

Using the commutation rules $[a, a^\dagger] = 1$ we obtain a relation

$$a(a^\dagger)^{n'} - (a^\dagger)^{n'} a = n'(a^\dagger)^{n'-1} \tag{21.41}$$

that can be proved by induction (Problem 21.2.6). Multiply (21.41) by a from the left and operate the result to $|0\rangle$. Using (21.40) we obtain

$$n(a^\dagger)^{n'} |0\rangle = n'(a^\dagger)^{n'} |0\rangle \tag{21.42}$$

indicating that $(a^\dagger)^{n'} |0\rangle$ is an eigenket belonging to the eigenvalue n'. The square length of $(a^\dagger)^{n'} |0\rangle$ is

$$\langle 0| a^{n'} (a^\dagger)^{n'} |0\rangle = n' \langle 0| a^{n'-1} (a^\dagger)^{n'-1} |0\rangle = \cdots = n'! \tag{21.43}$$

Hence, we may normalize the eigenket by

$$(n'!)^{-1/2} a^{n'} |0\rangle \equiv |n'\rangle \tag{21.44}$$

such that

$$\langle n' | n' \rangle = 1 \quad \text{for} \quad n' = 0, 1, 2, \ldots \tag{21.45}$$

Problem 21.2.1

Verify (21.18) and (21.19).

Problem 21.2.2

Verify (21.22).

Problem 21.2.3

Verify (21.26) and (21.27).

Problem 21.2.4

Prove the two formulas in (21.36). Assume that A, B and C are noncommutative.

Problem 21.2.5

Verify that the eigenvalue of n has no upper limit.

Problem 21.2.6

Prove (21.41) by mathematical induction.

Reference

1 Messiah, A. (1962) *Quantum Mechanics*, Interscience Publishers, New York.

22
Permutation Group

Permutation group is discussed in this chapter. A permutation is said to be odd or even according to whether it can be built-up by odd or even number of exchanges.

22.1
Permutation Group

Let us consider the set of ordered numerals [1, 2, 3]. By interchanging 1 and 2, we obtain [2, 1, 3]. Let us denote this operation by

$$(2,1)[1,2,3] \equiv (1,2)[1,2,3] = [2,1,3] . \tag{22.1}$$

The operator $(2, 1) \equiv (1, 2)$ is called the *exchange* or *transposition* between 1 and 2. We express the same operation in another way:

$$\begin{pmatrix} 1 & 2 & 3 \\ 2 & 1 & 3 \end{pmatrix}[1,2,3] = [2,1,3] \tag{22.2}$$

where

$$\begin{pmatrix} 1 & 2 & 3 \\ 2 & 1 & 3 \end{pmatrix}$$

indicates the change in order from [1, 2, 3] to [2, 1, 3] and will be called a *permutation*. The order of writing the numerals in columns is immaterial, so that

$$\begin{pmatrix} 1 & 2 & 3 \\ 2 & 1 & 3 \end{pmatrix} \equiv \begin{pmatrix} 1 & 3 & 2 \\ 2 & 3 & 1 \end{pmatrix} \equiv \begin{pmatrix} 2 & 1 & 3 \\ 1 & 2 & 3 \end{pmatrix} \equiv \begin{pmatrix} 2 & 3 & 1 \\ 1 & 3 & 2 \end{pmatrix}, \text{etc.} \tag{22.3}$$

We can define other permutations in a similar manner. Distinct permutations of three numerals are given by

$$\begin{pmatrix} 1 & 2 & 3 \\ 1 & 2 & 3 \end{pmatrix}, \begin{pmatrix} 1 & 2 & 3 \\ 2 & 3 & 1 \end{pmatrix}, \begin{pmatrix} 1 & 2 & 3 \\ 3 & 1 & 2 \end{pmatrix},$$

$$\begin{pmatrix} 1 & 2 & 3 \\ 2 & 1 & 3 \end{pmatrix}, \begin{pmatrix} 1 & 2 & 3 \\ 1 & 3 & 2 \end{pmatrix}, \begin{pmatrix} 1 & 2 & 3 \\ 3 & 2 & 1 \end{pmatrix} \tag{22.4}$$

which will be symbolically denoted by $\{P_1, P_2, \ldots, P_6\}$.

Mathematical Physics. Shigeji Fujita and Salvador V. Godoy
Copyright © 2010 WILEY-VCH Verlag GmbH & Co. KGaA, Weinheim
ISBN: 978-3-527-40808-5

The product of any two permutations P_j and P_k is defined by

$$(P_j P_k)[1, 2, 3] \equiv P_j(P_k[1, 2, 3]) \tag{22.5}$$

The set of six elements $\{P_1, P_2, \ldots, P_6\}$ in (22.4) has the group properties (see below) and is called a *permutation* or *symmetry group* of degree 3. The said group properties are as follows:

(i) Composition

Any two members of the set, P_j and P_k have the *composition* property when the product $P_j P_k$ is also a member of the set, say, P_i:

$$P_j P_k = P_i \tag{22.6}$$

We show this property by examples. Let us choose

$$P_j = \begin{pmatrix} 1 & 2 & 3 \\ 2 & 1 & 3 \end{pmatrix} \quad \text{and} \quad P_k = \begin{pmatrix} 1 & 2 & 3 \\ 1 & 3 & 2 \end{pmatrix} \tag{22.7}$$

By direct calculations, we obtain

$$\left\{ \begin{pmatrix} 1 & 2 & 3 \\ 2 & 1 & 3 \end{pmatrix} \begin{pmatrix} 1 & 2 & 3 \\ 1 & 3 & 2 \end{pmatrix} \right\} [1, 2, 3]$$

$$= \begin{pmatrix} 1 & 2 & 3 \\ 2 & 1 & 3 \end{pmatrix} \left\{ \begin{pmatrix} 1 & 2 & 3 \\ 1 & 3 & 2 \end{pmatrix} [1, 2, 3] \right\}$$

$$= \begin{pmatrix} 1 & 2 & 3 \\ 2 & 1 & 3 \end{pmatrix} [1, 3, 2] = [2, 3, 1] \tag{22.8}$$

Since

$$\begin{pmatrix} 1 & 2 & 3 \\ 2 & 1 & 3 \end{pmatrix} [1, 3, 2] = [2, 3, 1] \tag{22.9}$$

we can express the product

$$\begin{pmatrix} 1 & 2 & 3 \\ 2 & 1 & 3 \end{pmatrix} \begin{pmatrix} 1 & 2 & 3 \\ 1 & 3 & 2 \end{pmatrix}$$

by the single permutation

$$\begin{pmatrix} 1 & 2 & 3 \\ 2 & 3 & 1 \end{pmatrix}$$

which is a member of the set.

We may obtain the same result by examining the permutations directly as follows. Start with the permutation

$$P_k = \begin{pmatrix} 1 & 2 & 3 \\ 1 & 3 & 2 \end{pmatrix}$$

This moves 1 to 1. Then, the permutation

$$P_j = \begin{pmatrix} 1 & 2 & 3 \\ 2 & 1 & 3 \end{pmatrix}$$

moves 1 to 2. The net move is therefore $1 \rightarrow 1 \rightarrow 2$. We repeat the same reasoning to observe that $2 \rightarrow 3 \rightarrow 3$ and $3 \rightarrow 2 \rightarrow 1$. We can then represent the net move by the permutation

$$\begin{pmatrix} 1 & 2 & 3 \\ 2 & 3 & 1 \end{pmatrix}$$

(ii) Association

If P_i, P_j and P_k are any three elements of the set, then (see Problem 22.2.1):

$$(P_i P_j) P_k = P_i (P_j P_k) = P_i P_j P_k \tag{22.10}$$

For example, by applying the second method of calculation, we obtain

$$\left\{ \begin{pmatrix} 1 & 2 & 3 \\ 2 & 1 & 3 \end{pmatrix} \begin{pmatrix} 1 & 2 & 3 \\ 1 & 3 & 2 \end{pmatrix} \right\} \begin{pmatrix} 1 & 2 & 3 \\ 2 & 1 & 3 \end{pmatrix}$$

$$= \begin{pmatrix} 1 & 2 & 3 \\ 2 & 1 & 3 \end{pmatrix} \begin{pmatrix} 1 & 2 & 3 \\ 2 & 1 & 3 \end{pmatrix} = \begin{pmatrix} 1 & 2 & 3 \\ 3 & 2 & 1 \end{pmatrix}$$

$$\begin{pmatrix} 1 & 2 & 3 \\ 2 & 1 & 3 \end{pmatrix} \left\{ \begin{pmatrix} 1 & 2 & 3 \\ 1 & 3 & 2 \end{pmatrix} \begin{pmatrix} 1 & 2 & 3 \\ 2 & 1 & 3 \end{pmatrix} \right\}$$

$$= \begin{pmatrix} 1 & 2 & 3 \\ 2 & 1 & 3 \end{pmatrix} \begin{pmatrix} 1 & 2 & 3 \\ 3 & 1 & 2 \end{pmatrix} = \begin{pmatrix} 1 & 2 & 3 \\ 3 & 2 & 1 \end{pmatrix}$$

Thus, we obtain

$$\left\{ \begin{pmatrix} 1 & 2 & 3 \\ 2 & 1 & 3 \end{pmatrix} \begin{pmatrix} 1 & 2 & 3 \\ 1 & 3 & 2 \end{pmatrix} \right\} \begin{pmatrix} 1 & 2 & 3 \\ 2 & 1 & 3 \end{pmatrix}$$

$$= \begin{pmatrix} 1 & 2 & 3 \\ 2 & 1 & 3 \end{pmatrix} \left\{ \begin{pmatrix} 1 & 2 & 3 \\ 1 & 3 & 2 \end{pmatrix} \begin{pmatrix} 1 & 2 & 3 \\ 2 & 1 & 3 \end{pmatrix} \right\}$$

(iii) Identity

There exists a unique element, called the *identity*, and denoted by E, which has the property

$$P_j E = P_j = E P_j \tag{22.11}$$

in the present case,

$$E = \begin{pmatrix} 1 & 2 & 3 \\ 1 & 2 & 3 \end{pmatrix} \tag{22.12}$$

(iv) Inverse

For every element P_j, there exists an element P_j^{-1} called the *inverse* of P_j, in the set, which satisfies

$$P_j P_j^{-1} = E = P_j^{-1} P_j \qquad (22.13)$$

For example, if

$$P = \begin{pmatrix} 1 & 2 & 3 \\ 2 & 3 & 1 \end{pmatrix}, \quad P^{-1} = \begin{pmatrix} 1 & 2 & 3 \\ 3 & 1 & 2 \end{pmatrix}$$

It is stressed that the group properties (i)–(iv) all refer to the principal operation defined by (22.5).

Let us consider the set of all integers $0, \pm 1, \pm 2, \ldots$. With respect to the summation $j + k$, the four group properties are satisfied; the identity is represented by 0 and the inverse of j is $-j$. Therefore, the set of all integers form a group with respect to the summation. The same set, however, does not form a group with respect to multiplication since the inverse of j, say 2, cannot be found within the set.

It is important to note that the product $P_j P_k$ is not in general equal to the product of the reversed order:

$$P_j P_k \neq P_k P_j. \quad \text{(in general)} \qquad (22.14)$$

For example,

$$\begin{pmatrix} 1 & 2 & 3 \\ 1 & 3 & 2 \end{pmatrix} \begin{pmatrix} 1 & 2 & 3 \\ 2 & 1 & 3 \end{pmatrix} = \begin{pmatrix} 1 & 2 & 3 \\ 3 & 1 & 2 \end{pmatrix}$$

$$\begin{pmatrix} 1 & 2 & 3 \\ 2 & 1 & 3 \end{pmatrix} \begin{pmatrix} 1 & 2 & 3 \\ 1 & 3 & 2 \end{pmatrix} = \begin{pmatrix} 1 & 2 & 3 \\ 2 & 3 & 1 \end{pmatrix}$$

Therefore,

$$\begin{pmatrix} 1 & 2 & 3 \\ 1 & 3 & 2 \end{pmatrix} \begin{pmatrix} 1 & 2 & 3 \\ 2 & 1 & 3 \end{pmatrix} \neq \begin{pmatrix} 1 & 2 & 3 \\ 2 & 1 & 3 \end{pmatrix} \begin{pmatrix} 1 & 2 & 3 \\ 1 & 3 & 2 \end{pmatrix}$$

The number of elements in a group is called the *order* of the group. The group of permutations of degree N has an order of $N!$. In fact, let us write a permutation in the form

$$\begin{pmatrix} 1 & 2 & \ldots & N \\ j_1 & j_2 & \ldots & j_N \end{pmatrix}$$

where $\{j_k\}$ are different numerals taken from $(1, 2, \ldots, N)$. Clearly there are $N!$ ways of ordering these js. Such ordering generates $N!$ distinct permutations.

Problem 22.1.1

Reduce each of the following into a single permutation.

a) $\begin{pmatrix} 1 & 2 & 3 \\ 2 & 3 & 1 \end{pmatrix} \begin{pmatrix} 1 & 2 & 3 \\ 2 & 1 & 3 \end{pmatrix}$, b) $\begin{pmatrix} 1 & 2 & 3 \\ 1 & 3 & 2 \end{pmatrix} \begin{pmatrix} 1 & 2 & 3 \\ 3 & 2 & 1 \end{pmatrix}$

c) $\begin{pmatrix} 1 & 2 & 3 \\ 2 & 1 & 3 \end{pmatrix} \begin{pmatrix} 1 & 2 & 3 \\ 2 & 1 & 3 \end{pmatrix}$

Problem 22.1.2

Refer to Problem 22.1.1. Construct the product of the reserved order for each case and reduce it to a single permutation. For each case state if the two permutations are commutative or not.

Problem 22.1.3

Find the inverses of the following permutations:

a) $\begin{pmatrix} 1 & 2 & 3 \\ 3 & 1 & 2 \end{pmatrix}$, b) $\begin{pmatrix} 1 & 2 & 3 & 4 \\ 2 & 3 & 4 & 1 \end{pmatrix}$, c) $\begin{pmatrix} A & B & C & D \\ B & C & D & A \end{pmatrix}$

Problem 22.1.4

The set of permutations of degree 3 is given in (22.4).
1. Multiply each permutation by (22.2) from the left, and reduce it to a single permutation. Confirm that the set of all permutations so obtained is the same as the original set.
2. The property (a) can be verified no matter what permutation is used to construct the new set. Explain the reason for this.

22.2
Odd and Even Permutations

As we can see from the two alternative expressions (22.1) and (22.2) for the same operation, permutations and interchanges are related. The relation may be summarized by the following two theorems.

☐ Theorem

Any permutation can be equivalently expressed as a product of interchanges.

For example,

$$\begin{pmatrix} 1 & 2 & 3 \\ 2 & 1 & 3 \end{pmatrix} = (2, 1),$$

$$\begin{pmatrix} 1 & 2 & 3 \\ 2 & 3 & 1 \end{pmatrix} = (2, 3)(3, 1) = (1, 3)(1, 2) = (1, 2)(2, 3)$$

This theorem is intuitively obvious. As we see in the second example, there generally exist a number of equivalent expressions for a given permutation. However, these expressions are not arbitrary, but are subject to the following important restriction.

If a permutation P is expressed in terms of an odd (even) number of interchanges, all of the equivalent decompositions of P have odd (even) numbers of interchanges.

We may rephrase this property as follows:

Theorem

Any permutation may be classified as an odd or even permutation according to whether it can be built up from odd or even numbers of interchanges.

We will demonstrate this property as follows. Let us call a permutation of the form

$$\begin{pmatrix} i_1 & i_2 & \cdots & i_n \\ i_2 & i_3 & \cdots & i_1 \end{pmatrix}$$

that transforms $i_1 \to i_2, i_2 \to i_3, \ldots, i_n \to i_1$, a *cycle*, and denote it by $(i_1, i_2, \ldots, i_n)$, that is,

$$(i_1, i_2, \ldots, i_n) \equiv \begin{pmatrix} i_1 & i_2 & \cdots & i_n \\ i_2 & i_3 & \cdots & i_1 \end{pmatrix} \tag{22.15}$$

The number n is called the *length* of the cycle. Such a cycle can be decomposed into a product of interchanges (cycles of length 2). For example,

$$(i_1, i_2, \ldots, i_n) \equiv (i_1, i_2,)(i_2, i_3) \ldots (i_{n-1}, i_n) \tag{22.16}$$

which shows that a cycle of length n can be decomposed into a product of $n-1$ interchanges. Therefore, a cycle is even or odd according to whether its length is odd or even. Any permutation can be expressed as a product of cycles whose arguments are mutually exclusive. For example,

$$\begin{pmatrix} 1 & 2 & 3 & 4 & 5 \\ 5 & 3 & 2 & 1 & 4 \end{pmatrix} = \begin{pmatrix} 1 & 5 & 4 \\ 5 & 4 & 1 \end{pmatrix} \begin{pmatrix} 2 & 3 \\ 3 & 2 \end{pmatrix} = (1, 5, 4)(2, 3)$$

The parity of the permutation can now be determined by finding the parity of each cycle and applying the rules:

(even)(even) = (odd)(odd) = even

(odd)(even) = odd

which follows simply from the definitions of parity. In the above example, the parity of (1,5,4) is even and the parity of (2,3) is odd. Therefore the parity of

$$\begin{pmatrix} 1 & 2 & 3 & 4 & 5 \\ 5 & 3 & 2 & 1 & 4 \end{pmatrix}$$

is odd according to the rule: (even)(odd) = odd. Clearly, the parity of any permutation determined in this manner is unique. Q. E. D.

The *sign* δ_P *of the parity* of a permutation P is defined by

$$\delta_P = \begin{cases} 1 & \text{if } P \text{ is even} \\ -1 & \text{if } P \text{ is odd} \end{cases} \tag{22.17}$$

From this definition, we obtain

$$\delta_P^2 = (\pm 1)^2 = 1 \tag{22.18}$$

We can further establish that

$$\delta_{P^{-1}} = \delta_P \tag{22.19}$$

$$\delta_{PQ} = \delta_P \delta_Q \tag{22.20}$$

The last property, which is known as the *group representation property*, may be proved as follows. If P and Q are both odd, P and Q can be expressed in terms of odd numbers of interchanges, then $\delta_P \delta_Q = (-1)(-1) = 1$. We can express the product PQ in terms of interchanges. The number of interchanges here is even since the sum of two odd numbers is even. Hence $\delta_{PQ} = 1$. We can repeat similar arguments for each of the other possibilities.

Let us now consider a function of n variables, $f(x_1, x_2, \ldots, x_n)$. If this function satisfies

$$P f(x_1, x_2, \ldots, x_n) = f(x_1, x_2, \ldots, x_n) \quad \text{for all } P \tag{22.21}$$

where the Ps are permutations of n indices, then it is called a *symmetric* function. For example, $x_1 + x_2 + x_3$, $x_1^2 + x_2^2 + x_3^2$, and $x_1 x_2 x_3$ are all symmetric functions. If a function $f(x_1, x_2, \ldots, x_n)$ satisfies

$$P f(x_1, x_2, \ldots, x_n) = \delta_P f(x_1, x_2, \ldots, x_n) \quad \text{for all } P \tag{22.22}$$

then f is called an *antisymmetric* function. For example, $(x_1 - x_2)(x_1 - x_3)(x_2 - x_3)$ is antisymmetric. Note that some functions may satisfy neither (22.21) nor (22.22) (see Problem 22.2.4).

Consider now two functions $f(x_1, x_2)$ and $g(x_1, x_2)$. Applying the permutation

$$\begin{pmatrix} 1 & 2 \\ 2 & 1 \end{pmatrix}$$

on the product $f(x_1, x_2)g(x_1, x_2)$, we obtain

$$\begin{pmatrix} 1 & 2 \\ 2 & 1 \end{pmatrix} \{f(x_1, x_2)g(x_1, x_2)\} = f(x_2, x_1)g(x_2, x_1) \tag{22.23}$$

We may write the last member as

$$\left[\begin{pmatrix} 1 & 2 \\ 2 & 1 \end{pmatrix} f(x_1, x_2)\right]\left[\begin{pmatrix} 1 & 2 \\ 2 & 1 \end{pmatrix} g(x_1, x_2)\right]$$

where the angular brackets mean that the permutation should be completed within the brackets. We then obtain from (22.23)

$$\left[\begin{pmatrix} 1 & 2 \\ 2 & 1 \end{pmatrix} f(x_1, x_2)g(x_1, x_2)\right] = \left[\begin{pmatrix} 1 & 2 \\ 2 & 1 \end{pmatrix} f(x_1, x_2)\right]\left[\begin{pmatrix} 1 & 2 \\ 2 & 1 \end{pmatrix} g(x_1, x_2)\right]$$

This example indicates that, for a general P,

$$P(fg) = [Pf][Pg] \tag{22.24}$$

holds. The restriction implied by the angular brackets may be removed by the following device.

With the understanding that a permutation always acts toward the right, we may write (22.24) as

$$P(fg) = [Pf]Pg \tag{22.25}$$

but

$$fg = fP^{-1}(Pg) \equiv fP^{-1}Pg \tag{22.26}$$

Multiplying this equation by P from the left, we obtain

$$PfP^{-1}Pg = P(fg) = [Pf]Pg$$

Comparison between the first and last lines shows that

$$PfP^{-1} = [Pf] \tag{22.27}$$

Using this result, we may re-express (22.24) as

$$P(fg)P^{-1} = (PfP^{-1})(PgP^{-1}) \tag{22.28}$$

We note that the term on the rhs can be obtained from $P(fg)P^{-1}$ by inserting the identity PP^{-1} between f and g, and reassociating the resulting factors:

$$P(fg)P^{-1} = Pf(P^{-1}P)gP^{-1} = (PfP^{-1})(PgP^{-1})$$

Equation (22.28) suggests that the symmetry property of the product fg can be determined by studying each factor separately, that is, by calculating PfP^{-1} and PgP^{-1}. Thus, if a quantity to be studied appears as a factor then it is convenient to

22.2 Odd and Even Permutations

define the permutation symmetry in the following form:

$$P f(x_1, \ldots, x_n) P^{-1} = f(x_1, \ldots, x_n) \quad \text{(symmetric)} \tag{22.29}$$

$$P f(x_1, \ldots, x_n) P^{-1} = \delta_P f(x_1, \ldots, x_n) \quad \text{(antisymmetric)} \tag{22.30}$$

Multiplying (22.29) by P from the right, we obtain

$$P f = f P \quad \text{or} \quad [P, f] \equiv P f - f P = 0 \quad \text{(symmetric)} \tag{22.31}$$

This form can also be used for the definition of a symmetric function.

Problem 22.2.1

Decompose the following permutations into products of interchanges:

a) $\begin{pmatrix} 1 & 2 & 3 & 4 \\ 2 & 3 & 1 & 4 \end{pmatrix}$ b) $\begin{pmatrix} 1 & 2 & 3 & 4 & 5 \\ 3 & 1 & 2 & 5 & 4 \end{pmatrix}$

Problem 22.2.2

Find the parities of the following permutations:

a) $\begin{pmatrix} 1 & 2 & 3 & 4 \\ 2 & 3 & 1 & 4 \end{pmatrix}$ b) $\begin{pmatrix} 1 & 2 & 3 & 4 & 5 \\ 3 & 1 & 2 & 5 & 4 \end{pmatrix}$

c) $\begin{pmatrix} 1 & 2 & 3 & 4 & 5 \\ 5 & 4 & 3 & 2 & 1 \end{pmatrix}$ d) $\begin{pmatrix} a & b & c & d & e & f \\ b & c & a & d & f & e \end{pmatrix}$

Problem 22.2.3

For any permutations group, the number of even permutations is equal to the number of odd permutations.
1. Confirm this statement for $N = 2$ and $N = 3$.
2. Prove it for a general N.

Problem 22.2.4

Let $f(x_1, x_2)$ be a function of x_1 and x_2.
1. Show that $f(x_1, x_2) + f(x_2, x_1)$ and $f(x_1, x_2) - f(x_2, x_1)$ are, respectively, symmetric and antisymmetric.
2. Using the results from (a), express f as the sum of symmetric and antisymmetric functions.

23
Quantum Statistics

Indistinguishability and quantum statistics are defined using the permutation group in this chapter. A quantum particle is either a fermion or a boson. The fermions, such as electrons and protons, are subject to Pauli's exclusion principle, and their occupation numbers for a particle state are limited to 1 or 0. Bosons occupation numbers are unlimited: $0, 1, 2, \ldots$ The many-boson (-fermion) state is symmetric (antisymmetric).

23.1
Classical Indistinguishable Particles

When a system is composed of a number of particles of the same kind, these particles cannot be distinguished from each other. If numbered particle-coordinates are used in a theoretical formulation, a prediction from the theory must be independent of the particles. For example, the N-body distribution function for the system must be invariant under permutations of the particle indices. Such a concept of indistinguishability can be applied to both classical and quantum-mechanical systems. In this section we will consider classical systems only.

Let us take a system of three identical particles moving in one dimension. The dynamic state can be represented by three points in the μ-space as indicated in Figure 23.1. Let us denote the location of these points by $(x_1, p_1, x_2, p_2, x_3, p_3)$.

Consider an infinitesimal phase area $dx_j\, dp_j$ surrounding each point $(x_j\, p_j)$. The probability of finding the system in a state in which three particle points are within the element $dx_1\, dp_1\, dx_2\, dp_2\, dx_3\, dp_3$ can be written in the form

$$\rho(x_1, p_1, x_2, p_2, x_3, p_3) dx_1 dp_1 dx_2 dp_2 dx_3 dp_3 \frac{1}{(2\pi\hbar)^3 (3!)}$$
$$\equiv \rho(1, 2, 3) dx_1 dp_1 dx_2 dp_2 dx_3 dp_3 \frac{1}{(2\pi\hbar)^3 (3!)} \tag{23.1}$$

where ρ represents the distribution function for the system. For identical particles, we require that the distribution function $\rho(1, 2, 3)$ be symmetric, that is,

$$P\rho(1, 2, 3) = \rho(1, 2, 3) \quad \text{for all permutations } P \tag{23.2}$$

Mathematical Physics. Shigeji Fujita and Salvador V. Godoy
Copyright © 2010 WILEY-VCH Verlag GmbH & Co. KGaA, Weinheim
ISBN: 978-3-527-40808-5

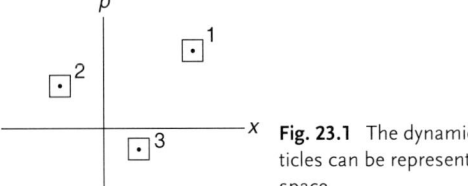

Fig. 23.1 The dynamic state for three identical classical particles can be represented by three particle points in the phase space.

Explicitly, this means that

$$\rho(1,2,3) = \rho(1,3,2) = \rho(3,2,1) = \rho(2,1,3) = \rho(3,1,2) = \rho(2,3,1)$$

In general, the distribution function ρ changes with time. However, *the symmetry of the distribution function is maintained if the Hamiltonian H of the system is symmetric*:

$$PH = HP \quad \text{for all } P \tag{23.3}$$

This may be shown as follows. The function ρ obeys the Liouville equation

$$\frac{\partial \rho}{\partial t} = \{H, P\} \equiv \sum_{k=1}^{3N} \left(\frac{\partial H}{\partial x_k} \frac{\partial \rho}{\partial p_k} - \frac{\partial \rho}{\partial x_k} \frac{\partial H}{\partial p_k} \right) \tag{23.4}$$

Let us write an arbitrary permutation P in the form:

$$P \equiv \begin{pmatrix} 1 & 2 & 3 \\ i_1 & i_2 & i_3 \end{pmatrix} \tag{23.5}$$

where (i_1, i_2, i_3) are numerals taken from $(1,2,3)$. By definition,

$$\frac{\partial H}{\partial x_1} \equiv \lim_{\Delta x \to 0} \frac{H(x_1 + \Delta x, p_1, x_2, p_2, x_3, p_3) - H(x_1, p_1, x_2, p_2, x_3, p_3)}{\Delta x}$$

Applying the permutation P from the left, we obtain

$$P \frac{\partial H}{\partial x_1} = \lim_{\Delta x \to 0} \frac{1}{\Delta x} \Big[H(x_{i_1} + \Delta x, p_{i_1}, x_{i_2}, p_{i_2}, x_{i_3}, p_{i_3})$$
$$\qquad - H(x_{i_1}, p_{i_1}, x_{i_2}, p_{i_2}, x_{i_3}, p_{i_3}) \Big] P$$

Since H is symmetric. we can reorder the arguments (i_1, i_2, i_3) to the natural order $(1,2,3)$. Subsequently, taking the limit $\Delta x \to 0$, we obtain

$$P \frac{\partial H}{\partial x_1} = \frac{\partial}{\partial x_{i_1}} H(x_1, p_1, x_2, p_2, x_3, p_3) P \equiv \frac{\partial H}{\partial x_{i_1}} P \tag{23.6}$$

or

$$P \frac{\partial H}{\partial x_1} P^{-1} = \frac{\partial H}{\partial x_{i_1}}$$

By similar calculations, we obtain

$$P\frac{\partial H}{\partial x_k}P^{-1} = \frac{\partial H}{\partial x_{i_k}}, \quad P\frac{\partial H}{\partial p_k}P^{-1} = \frac{\partial H}{\partial p_{i_k}} \tag{23.7}$$

In the same manner we can also show that

$$P\frac{\partial \rho}{\partial x_k}P^{-1} = \frac{\partial \rho}{\partial x_{i_k}}, \quad P\frac{\partial \rho}{\partial p_k}P^{-1} = \frac{\partial \rho}{\partial p_{i_k}} \tag{23.8}$$

Using (23.7) and (23.8), we obtain

$$P\sum_{\{k\}}\left(\frac{\partial H}{\partial x_k}\frac{\partial \rho}{\partial p_k} - \frac{\partial \rho}{\partial x_k}\frac{\partial H}{\partial p_k}\right)P^{-1} = \sum_{\{i_k\}}\left(\frac{\partial H}{\partial x_{i_k}}\frac{\partial \rho}{\partial p_{i_k}} - \frac{\partial \rho}{\partial x_{i_k}}\frac{\partial H}{\partial p_{i_k}}\right)$$

The summation over the set $\{i_k\}$ is equivalent to the summation over the set $\{k\}$. Therefore, it follows that

$$P\sum_{\{k\}}\left(\frac{\partial H}{\partial x_k}\frac{\partial \rho}{\partial p_k} - \frac{\partial \rho}{\partial x_k}\frac{\partial H}{\partial p_k}\right)P^{-1} = \sum_{\{k\}}\left(\frac{\partial H}{\partial x_k}\frac{\partial \rho}{\partial p_k} - \frac{\partial \rho}{\partial x_k}\frac{\partial H}{\partial p_k}\right)$$

or

$$P\{H,\rho\}P^{-1} = \{H,\rho\} \tag{23.9}$$

Using this result, we obtain from (23.4)

$$P\frac{\partial \rho}{\partial t}P^{-1} = \frac{\partial \rho}{\partial t} \tag{23.10}$$

By assumption, ρ is symmetric at the initial time, and we showed in (23.10) that $\partial \rho/\partial t$ is symmetric. Therefore, by mathematical induction for a continuous variable $\rho(t)$ is symmetric for all time.

A dynamical function ξ, such as the total momentum, depends on the particle variables:

$$\xi = \xi(x_1, p_1, x_2, p_2, x_3, p_3) \equiv \xi(1,2,3) \tag{23.11}$$

For identical particles, we require that ξ be symmetric,

$$P\xi(1,2,3) = \xi(1,2,3)P \quad \text{for all } P \tag{23.12}$$

This equation and (23.2) will insure that the average of ξ,

$$\langle \xi \rangle = Tr\{\xi\}$$

is independent of the numbering of the particles.

In summary, a system of identical particles can be properly treated if (a) the distribution function is symmetric, (23.2), and (b) all dynamical functions including the Hamiltonian are symmetric, (23.12) and (23.3). It is clear that we can extend our discussions to any number of particles.

The significance of the symmetry requirements may be clearly understood by considering distinguishable particles. Let us take a hydrogen atom, which consists of an electron with mass m_e, and a proton with mass m_p. The Hamiltonian H for the system is

$$H(\mathbf{r}_1, \mathbf{p}_1, \mathbf{r}_2, \mathbf{p}_2) \equiv H(1,2) \equiv \frac{1}{2m_e}\mathbf{p}_1^2 + \frac{1}{2m_p}\mathbf{p}_2^2 + V(|\mathbf{r}_1 - \mathbf{r}_2|) \tag{23.13}$$

Let us interchange (1,2) and obtain

$$(1,2)H(1,2)[(1,2)]^{-1} = H(2,1) = \frac{1}{2m_e}\mathbf{p}_2^2 + \frac{1}{2m_p}\mathbf{p}_1^2 + V(|\mathbf{r}_2 - \mathbf{r}_1|)$$
$$\neq H(1,2) \tag{23.14}$$

which shows that this Hamiltonian H is not symmetric.

This example shows that if the particles have different masses, and are therefore distinguishable, then the Hamiltonian H is not symmetric. There is no surprise here, the important point is that the Hamiltonian H can, and must, contain the built-in permutation symmetry appropriate for the system.

23.2
Quantum-Statistical Postulate. Symmetric States for Bosons

The symmetry requirements for a classical system of identical particles, given in the preceding section, can simply be extended to a quantum-mechanical many-particle system.

Let us consider a quantum system of N identical particles. We require that the Hamiltonian of the system be a symmetric function of the particle variables, $\zeta_j \equiv (\mathbf{r}_j, \mathbf{p}_j)$:

$$PH(\zeta_1, \zeta_2, \ldots, \zeta_N) = H(\zeta_1, \zeta_2, \ldots, \zeta_N)P \tag{23.15}$$

or

$$PH = HP \quad \text{for all } P \tag{23.16}$$

Since the quantum Hamiltonian is a linear operator, and appears as a factor in the definition equation, the symmetry requirement must necessarily be expressed in the form (23.16) instead of $H = H$.

We further require that all dynamical functions ξ be symmetric:

$$P\xi = \xi P. \tag{23.17}$$

Before discussing the symmetry requirement for the quantum system, we introduce a new postulate concerning quantum may-body systems.

23.2 Quantum-Statistical Postulate. Symmetric States for Bosons

Every particle in quantum physics is either a boson or a fermion. This is known as the *quantum-statistical postulate*, in addition to the postulates underlying quantum mechanics for a particle discussed in chapter 15, and should be treated as an independent postulate. As we will see later, a system of identical bosons or fermions behaves quite differently from the corresponding classical system.

In the remainder of this section we will discuss the quantum states for identical bosons. The quantum states for fermions will be discussed in the next section. For a quantum particle moving in one dimension, the eigenvalues for the momentum are

$$p_k = \frac{2\pi\hbar}{L} k, \quad k = 0, \pm 1, \pm 2, \ldots \quad (23.18)$$

The corresponding normalized eigenkets will be denoted by $\{|p_k\rangle\}$, which satisfy the orthonormality condition:

$$\langle p_k | p_l \rangle = \delta_{k,l} \quad (23.19)$$

When a system contains two different particles, say A and B, we may construct a ket for the system from the *direct product* of the single-particle kets for A and B. Such a ket

$$\left|p_k^{(A)}\right\rangle \left|p_l^{(B)}\right\rangle \equiv \left|p_l^{(B)}\right\rangle \left|p_k^{(A)}\right\rangle \equiv \left|p_k^{(A)} p_l^{(B)}\right\rangle \quad (23.20)$$

will represent a state in which the particles A and B occupy the momentum states p_k and p_l, respectively. The ket $\left|p_k^{(A)} p_l^{(B)}\right\rangle$ is normalized to unity since

$$\left\langle p_k^{(A)} p_l^{(B)} \middle| p_k^{(A)} p_l^{(B)} \right\rangle \equiv \left(\left\langle p_l^{(B)} \middle| \left\langle p_k^{(A)} \middle| \right.\right)\right) \left(\left| p_k^{(A)} \right\rangle \left| p_l^{(B)} \right\rangle\right)$$

$$= \left\langle p_l^{(B)} \middle| \left\langle p_k^{(A)} \middle| p_k^{(A)} \right\rangle \middle| p_l^{(B)} \right\rangle$$

$$= \left\langle p_l^{(B)} \middle| p_l^{(B)} \right\rangle \quad \text{(normalization)}$$

$$= 1 \quad (23.21)$$

We now consider a system of two identical particles. Let us construct a new ket

$$\frac{1}{\sqrt{2}} \left\{ \left|p_k^{(A)}\right\rangle \left|p_l^{(B)}\right\rangle + \left|p_k^{(B)}\right\rangle \left|p_l^{(A)}\right\rangle \right\} \equiv |p_k p_l\rangle_S \quad (23.22)$$

which is obtained by symmetrizing the ket (23.20). The new symmetric ket, by postulate, represents the sate for the system in which two states p_k and p_l are occupied by two particles with no further specifications. The factor $(2)^{-1/2}$ was given to facilitate the normalization. See below.

Let us now extend our theory to the case of N bosons moving in three dimensions. Let $\left|\alpha_a^{(j)}\right\rangle, \left|\alpha_b^{(j)}\right\rangle, \ldots$ be the kets for the jth particle occupying single-particle states $\alpha_a, \alpha_b, \ldots$. We construct a ket for N particles by taking the product of kets for each particle:

$$\left|\alpha_a^{(1)}\right\rangle \left|\alpha_b^{(2)}\right\rangle \cdots \left|\alpha_g^{(N)}\right\rangle \quad (23.23)$$

Multiplying this by the *symmetrizing operator*

$$S \equiv \frac{1}{\sqrt{N!}} \sum_P P \qquad (23.24)$$

where the summation is over all permutations, we obtain a ket for the system,

$$S\left(\left|a_a^{(1)}\right\rangle\left|a_b^{(2)}\right\rangle \cdots \left|a_g^{(N)}\right\rangle\right) \equiv \left|a_a a_b \ldots a_g\right) \qquad (23.25)$$

where the labels of particles are omitted on the rhs since they are no longer relevant. The ket in (23.25) is obviously symmetric:

$$PS\left(\left|a_a^{(1)}\right\rangle\left|a_b^{(2)}\right\rangle \cdots \left|a_g^{(N)}\right\rangle\right) = S\left(\left|a_a^{(1)}\right\rangle\left|a_b^{(2)}\right\rangle \cdots \left|a_g^{(N)}\right\rangle\right) \qquad (23.26)$$

The symmetric ket (23.25), by postulate, represents the state of the system of N bosons in which the single-particle states $a_a, a_b, \ldots, a_g$ are occupied.

Problem 23.2.1

Construct a symmetric ket for three bosons by means of (23.9).
1. Verify explicitly that the obtained ket is symmetric under the permutation $\begin{pmatrix} 1 & 2 & 3 \\ 2 & 3 & 1 \end{pmatrix}$.
2. Assuming that the occupied states, say $a_a \equiv a, b, c$, are all different, find the magnitude of the ket.
3. Assuming that $a = b = c$, find the magnitude of the ket.

23.3
Antisymmetric States for Fermions. Pauli's Exclusion Principle

Let us define the *antisymmetrizing operator* A by

$$A \equiv S \equiv \frac{1}{\sqrt{N!}} \sum_P \delta_P P \qquad (23.27)$$

where δ_P is the *sign of parity* of the permutation P defined as

$$\delta_P = \begin{cases} 1 & \text{if } P \text{ is even} \\ -1 & \text{if } P \text{ is odd} \end{cases} \qquad (23.28)$$

Applying the operator A on the ket in (23.23), we obtain a new ket

$$A\left(\left|a_a^{(1)}\right\rangle\left|a_b^{(2)}\right\rangle \cdots \left|a_g^{(N)}\right\rangle\right) \equiv \left|a_a a_b \ldots a_g\right)_A \qquad (23.29)$$

where the subscript A denotes the *antisymmetric* ket.

The fact that this ket is antisymmetric can be shown as follows. Multiply (23.29) by an arbitrary permutation P and obtain

$$PA\left(\left|\alpha_a^{(1)}\right\rangle\left|\alpha_b^{(2)}\right\rangle\cdots\left|\alpha_g^{(N)}\right\rangle\right)$$
$$=\frac{1}{\sqrt{N!}}\sum_Q \delta_Q PQ\left(\left|\alpha_a^{(1)}\right\rangle\left|\alpha_b^{(2)}\right\rangle\cdots\left|\alpha_g^{(N)}\right\rangle\right) \quad (23.30)$$

where we denoted the summation variable by Q. By the composition property, the product PQ is another element, say R, of the permutation group:

$$PQ \equiv R \quad (23.31)$$

or

$$Q = P^{-1} R \quad (23.32)$$

We now express δ_Q as follows:

$$\delta_Q = \delta_{P^{-1}R} = \delta_{P^{-1}}\delta_R = \delta_P \delta_R \quad (23.33)$$

Using this result and (23.31), we can rewrite the rhs of (23.30) as

$$\frac{1}{\sqrt{N!}}\sum_Q \delta_Q PQ\left(\left|\alpha_a^{(1)}\right\rangle\left|\alpha_b^{(2)}\right\rangle\cdots\left|\alpha_g^{(N)}\right\rangle\right)$$
$$= \frac{1}{\sqrt{N!}}\sum_R \delta_P \delta_R R\left(\left|\alpha_a^{(1)}\right\rangle\left|\alpha_b^{(2)}\right\rangle\cdots\left|\alpha_g^{(N)}\right\rangle\right)$$
$$= \delta_P \frac{1}{\sqrt{N!}}\sum_R \delta_R R\left(\left|\alpha_a^{(1)}\right\rangle\left|\alpha_b^{(2)}\right\rangle\cdots\left|\alpha_g^{(N)}\right\rangle\right)$$
$$= \delta_P A\left(\left|\alpha_a^{(1)}\right\rangle\left|\alpha_b^{(2)}\right\rangle\cdots\left|\alpha_g^{(N)}\right\rangle\right) \quad (23.34)$$

which establishes the desired antisymmetry property:

$$P\left[A\left(\left|\alpha_a^{(1)}\right\rangle\left|\alpha_b^{(2)}\right\rangle\cdots\left|\alpha_g^{(N)}\right\rangle\right)\right]$$
$$= \delta_P A\left(\left|\alpha_a^{(1)}\right\rangle\left|\alpha_b^{(2)}\right\rangle\cdots\left|\alpha_g^{(N)}\right\rangle\right) \quad (23.35)$$

or

$$P|\alpha_a \alpha_b \ldots \alpha_g\rangle_A = \delta_P |\alpha_a \alpha_b \ldots \alpha_g\rangle_A \quad (23.36)$$

We postulate that the antisymmetric ket in (23.29) represents the quantum state of N fermions in which particle states $a, b, \ldots, g$ are occupied. For illustration, let us take the case of two fermions moving in one dimension. Corresponding to the state in which the momentum states p_k and p_l are occupied, we construct the ket

$$|p_k p_l\rangle_A \equiv \frac{1}{\sqrt{2}}\sum_P \delta_P P\left[\left|p_k^{(1)}\right\rangle\left|p_l^{(2)}\right\rangle\right]$$
$$= \frac{1}{\sqrt{2}}\left\{\left|p_k^{(1)}\right\rangle\left|p_l^{(2)}\right\rangle - \left|p_k^{(2)}\right\rangle\left|p_l^{(1)}\right\rangle\right\} \quad (23.37)$$

This ket is clearly antisymmetric:

$$|p_k p_l\rangle_A = -|p_l p_k\rangle_A \qquad (23.38)$$

If $p_k = p_l$, we then obtain

$$|p_k p_k\rangle_A = 0 \qquad (23.39)$$

By postulate, the null ket does not correspond to any quantum state. Thus, the state for the system in which two fermions occupy the same state p_k cannot qualify as a quantum state. In other words, *two fermions must not occupy the same state*. This is known as *Pauli's exclusion principle*.

Let us shed more light on this point by treating the general case of N fermions. The antisymmetric state in (23.29) can be written in the determinant form:

$$|\alpha_a \alpha_b \ldots \alpha_g\rangle_A = \frac{1}{\sqrt{N!}} \begin{Vmatrix} |\alpha_a^{(1)}\rangle & |\alpha_a^{(2)}\rangle & \cdots & |\alpha_a^{(N)}\rangle \\ |\alpha_b^{(1)}\rangle & |\alpha_b^{(2)}\rangle & \cdots & |\alpha_b^{(N)}\rangle \\ \vdots & \vdots & \ddots & \vdots \\ |\alpha_g^{(1)}\rangle & |\alpha_g^{(2)}\rangle & \cdots & |\alpha_g^{(N)}\rangle \end{Vmatrix} \qquad (23.40)$$

It represents the system state in which single-particle states $a \equiv \alpha_a, \alpha_b, \ldots, \alpha_g$ are occupied. If two of the particle states are identical, the ket in (23.40) vanishes since two rows in the determinant become identical.

Pauli's principle applies to any single-particle state. For example, except for the spin degeneracy, two fermions can occupy neither the same momentum state nor the same space point. Historically, Pauli's exclusion principle was discovered from the study of atomic spectra. The original statement referred to atomic orbitals. No two electrons can occupy the same orbital with identical quantum numbers.

Problem 23.3.1

Construct an antisymmetric ket for three fermions by means of (23.29).
1. Multiply the result from the left by (1,2) and verify that the ket is indeed antisymmetric.
2. Show that the ket is normalized to unity.

23.4
Occupation-Number Representation

The quantum state for a system of bosons (or fermions) can most conveniently be represented by a set of occupation numbers $\{n'_a\}$, where n'_a are the numbers of bosons (or fermions) occupying the quantum particle states a. This representation

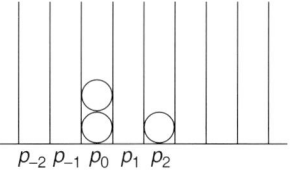

..., $n_{-2} = 0$, $n_{-1} = 0$, $n_0 = 2$, $n_1 = 0$, $n_2 = 1$, ...

Fig. 23.2 A many-boson state is represented by a set of boson numbers n_j occupying the states p_j.

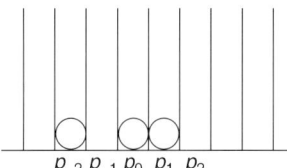

..., $n_{-2} = 1$, $n_{-1} = 0$, $n_0 = 1$, $n_1 = 1$, $n_2 = 0$, ...

Fig. 23.3 A many-fermion state can be represented by the set of the numbers $\{n_j\}$ occupying the states $\{p_j \equiv 2\pi\hbar L^{-1} j\}$. Each n_j is restricted to 0 or 1.

is called the *occupation-number representation* or simply the *number representation*. For bosons, the possible values for n'_a are zero, one, two, or any positive integers:

$$n'_a = 0, 1, 2, \ldots \quad \text{for bosons} \tag{23.41}$$

The many-boson state can best be represented by the distribution of particles (balls) in the states (boxes) as shown in Figure 23.2. The occupation numbers $\{n_a\}$ for fermions are limited to one or zero.

$$n_a = 0 \text{ or } 1 \quad \text{fermions} \tag{23.42}$$

The many-fermions state can be represented by the distribution of particles (balls) in the states (boxes) as shown in Figure 23.3.

24
The Free-Electron Model

The quantum theory of the free-electron model resolves the heat-capacity paradox, that is, the apparent absence of electron contribution to heat capacity. The linear-T low-temperature heat capacity is exactly calculated.

24.1
Free Electrons and the Fermi Energy

Let us consider a system of *free electrons* characterized by the Hamiltonian:

$$H = \sum \frac{p_j^2}{2m} \tag{24.1}$$

The *momentum eigenstates* for a quantum particle with a periodic cube-box boundary condition are characterized by three quantum numbers:

$$p_{x,j} = \left(\frac{2\pi\hbar}{L}\right) j, \quad p_{y,k} = \left(\frac{2\pi\hbar}{L}\right) k, \quad p_{z,l} = \left(\frac{2\pi\hbar}{L}\right) l \tag{24.2}$$

where L is the cube side length, and $\{j, k, l\}$ are integers. For simplicity, we indicate the momentum state by a single Greek letter κ:

$$\mathbf{p}_\kappa \equiv (p_{x,j}, p_{y,k}, p_{z,l}) \tag{24.3}$$

The quantum state of our many-electron system can be specified by the set of occupation numbers $\{n_\kappa\}$, with each $n_\kappa \equiv n_{\mathbf{p}_\kappa}$ taking on either one or zero. The ket vector representing such a state will be denoted by

$$|\{n\}\rangle \equiv |\{n_\kappa\}\rangle \tag{24.4}$$

The corresponding energy eigenvalue is given by

$$E(\{n\}) = \sum_\kappa \varepsilon_\kappa n_\kappa \tag{24.5}$$

where $\varepsilon_\kappa \equiv p_\kappa^2/2m$ is the kinetic energy of the electron with momentum $\mathbf{p}_\kappa$. The sum of the occupation numbers, n_κ equals the total number N of electrons:

$$\sum_\kappa n_\kappa = N \tag{24.6}$$

Mathematical Physics. Shigeji Fujita and Salvador V. Godoy
Copyright © 2010 WILEY-VCH Verlag GmbH & Co. KGaA, Weinheim
ISBN: 978-3-527-40808-5

24 The Free-Electron Model

We assume that the system is in thermodynamic equilibrium, which is characterized by temperature $T \equiv (k_B \beta)^{-1}$ and number density n. The thermodynamic properties of the system can then be computed in terms of the *grand canonical density operator*:

$$\rho_G = \frac{e^{\alpha N - \beta H}}{TR\{e^{\alpha N - \beta H}\}} \tag{24.7}$$

where TR means the sum of N-particle traces. The ensemble average of n_κ is represented by the *Fermi distribution function* f_F:

$$\boxed{\langle n_\kappa \rangle \equiv \frac{TR\{n_\kappa e^{\alpha N - \beta H}\}}{TR\{e^{\alpha N - \beta H}\}} = \frac{1}{\exp(\beta \varepsilon_\kappa - \alpha) + 1} \equiv f_F(\varepsilon_\kappa)} \tag{24.8}$$

[Problem 24.1.1]. The parameter α in this expression is determined from

$$n = \frac{\langle N \rangle}{V} = \frac{1}{V} \sum_\kappa \frac{1}{\exp(\beta \varepsilon_\kappa - \alpha) + 1} \equiv \frac{1}{V} \sum_\kappa f_F(\varepsilon_\kappa) \tag{24.9}$$

Hereafter, we drop the subscript F in f_F.

We now investigate the behavior of the Fermi distribution function $f(\varepsilon)$ at very low temperatures. Let us set

$$\alpha \equiv \beta \mu \equiv \frac{\mu}{k_B T} \tag{24.10}$$

Here, the quantity μ represents the *chemical potential*. In the low-temperature limit, the chemical potential μ approaches a positive constant μ_0: $\mu \to \mu_0 > 0$. We plot the Fermi distribution function $f(\varepsilon)$ at $T = 0$ against the energy ε by a solid curve in Figure 24.1. It is a step function with the step at $\varepsilon = \mu_0$. This means that every momentum state $\mathbf{p}_\kappa$ for which $\varepsilon_\kappa = p_\kappa^2/2m < \mu_0$ is occupied with probability 1, and all other states are unoccupied. This special energy μ_0 is called the *Fermi energy*; it is often denoted by ε_F, which can be calculated as follows:

From (24.9) we have

$$n = \frac{1}{V} \sum_\kappa [f(\varepsilon_\kappa)]_{T=0} = \frac{1}{V} \times \text{(the number of states } \kappa \text{ for which } \varepsilon_\kappa < \mu_0) \tag{24.11}$$

The momentum eigenstates in (24.2) can be represented by points in the three-dimensional momentum space, as shown in Figure 24.2. These points form a

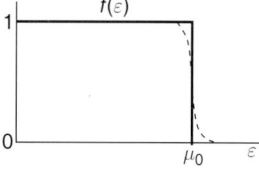

Fig. 24.1 The Fermi distribution function against energy ε. The solid line is for $T = 0$ and the broken line for a small T.

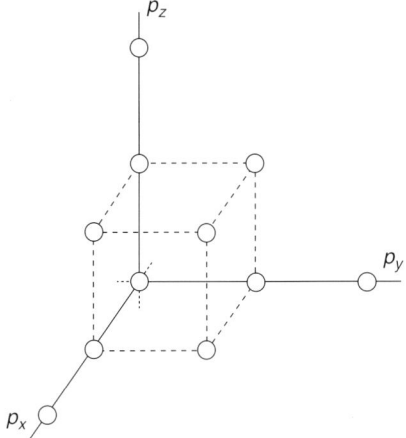

Fig. 24.2 The momentum states for a periodic boundary condition form a simple cubic lattice with the lattice constant $2\pi\hbar/L$.

simple cubic (sc) lattice with the lattice constant $2\pi\hbar/L$. Let us define the *Fermi momentum* p_F by

$$\mu_0 \equiv \varepsilon_F \equiv \frac{p_F^2}{2m} \tag{24.12}$$

The number of occupied states will be equal to the number of lattice points within the sphere of radius p_F. One lattice point corresponds to one unit cell for the sc lattice. Therefore, this number is equal to the volume of the sphere, $(4\pi/3)p_F^3$, divided by the volume of the unit cell, $(2\pi\hbar/L)^3$:

$$\text{Number of occupied states} = \left(\frac{4\pi}{3}\right) \frac{p_F^3}{(2\pi\hbar/L)^3} \tag{24.13}$$

Introducing this into (24.11), we obtain

$$n = \left(\frac{4\pi}{3}\right) \frac{p_F^3}{(2\pi\hbar/L)^3 L^3} = \left(\frac{4\pi}{3}\right) \frac{p_F^3}{(2\pi\hbar)^3} \tag{24.14}$$

This result was obtained under the assumption of a periodic cube-box boundary condition. The result obtained in the *bulk limit*, where

$$L^3 = V \to \infty, \quad N \to \infty \quad \text{such that } n \equiv \frac{N}{V} = \text{constant} \tag{24.15}$$

is valid independent of the type of the boundary (see Problem 24.1.2).

In our discussion so far, we have neglected the fact that an electron has a *spin angular momentum* (or simply spin) as an additional degree of freedom. It is known that any quantum state for an electron must be characterized not only by the quantum numbers $(p_{x,j}, p_{y,k}, p_{z,l})$ describing its motion in the ordinary space, but also by the quantum numbers describing its spin. It is further known that the electron has a permanent magnetic moment associated with its spin, and that the eigenvalues s_z of the z-component of the electronic spin are discrete and are restricted to the two values $\pm\hbar/2$. In the absence of a magnetic field, the magnetic potential energy is the same for both spin states. In the grand canonical ensemble, the states with the same energy are distributed with the same probability. In taking account of the spins, we must then multiply the rhs of (24.14) by the factor 2, called the *spin degeneracy factor*. We thus obtain

$$n = \frac{8\pi}{3} \frac{p_F^3}{(2\pi\hbar)^3} \qquad (24.16)$$

(including the spin degeneracy). After solving this equation for p_F, we obtain the Fermi energy as follows:

$$\boxed{\varepsilon_F = \frac{\hbar^2 (3\pi^2 n)^{2/3}}{2m}} \qquad (24.17)$$

Let us estimate the order of magnitude for ε_F by taking a typical metal Cu. This metal has a specific weight of $9\,\mathrm{g\,cm^{-3}}$ and a molecular weight of 63.5, yielding the number density $n = 8.4 \times 10^{22}$ electrons/cm^3. Using this value for n, we find that

$$\varepsilon_F \equiv k_B T_F, \qquad T_F \simeq 80\,000\,\mathrm{K} \qquad (24.18)$$

This T_F is called the *Fermi temperature*. The value found for the Fermi energy $\varepsilon_F \equiv k_B T_F$ is very high compared with the thermal excitation energy of the order $k_B T$, which we shall see later in Section 24.3. This makes the thermodynamic behavior of the conduction electrons at room temperature drastically different from that of a classical gas.

The Fermi energy by definition is the chemical potential at 0 K. We may look at this relation in the following manner. For a box of a finite volume V, the momentum states form a simple cubic lattice as shown in Figure 24.2. As the volume V is made greater, the unit-cell volume in the momentum space, $(2\pi\hbar/L)^3$, decreases as V^{-1}. However, we must increase the number of electrons N in proportion to V in the process of the bulk limit. Therefore, the radius of the *Fermi sphere* within which all momentum states are filled with electrons neither grows nor shrinks. Obviously, this configuration corresponds to the lowest-energy state for the system. The Fermi energy $\varepsilon_F \equiv p_F^2/2m$ represents the electron energy at the surface of the Fermi sphere. If we attempt to add an extra electron to the Fermi sphere, we must bring in an electron with an energy equal to, indicating that $\varepsilon_F = \mu_0$.

Problem 24.1.1

Verify (24.8).

Problem 24.1.2

The momentum eigenvalues for a particle in a periodic rectangular box with sides of unequal lengths (L_1, L_2, L_3) are given by $p_{x,j} = 2\pi\hbar j/L_1$, $p_{y,k} = 2\pi\hbar k/L_2$, $p_{z,l} = 2\pi\hbar l/L_3$. Show that the Fermi energy ε_F for free electrons is still given as (24.17) in the bulk limit.

24.2 Density of States

We must convert the *sum over quantum states* into an integral in many quantum-statistical calculations. This conversion becomes necessary when we first find discrete quantum states for a periodic box, and then seek the sum over states in the bulk limit. This conversion is a welcome procedure because the resulting integral is easier to handle than the sum. The conversion is purely mathematical in nature, but it is an important step in statistical-mechanical computations.

Let us first examine a sum over momentum states corresponding to a one-dimensional motion. We take

$$\sum_k A(p_k) \tag{24.19}$$

where $A(p)$ is an arbitrary function of p. The discrete momentum states are equally spaced, as shown by the short bars in Figure 24.3. As the normalization length L increases, the spacing (distance) between two successive states, $2p\hbar/L$, becomes smaller. This means that the number of states per unit momentum interval increases with L. We denote the number of states within a small momentum interval Δp by Δn. Consider the ratio $\Delta n/\Delta p$. Dividing both the numerator and denominator by Δp, we obtain

$$\frac{\Delta n}{\Delta p} = \frac{1}{\text{momentum spacing per state}} = \frac{L}{2\pi\hbar} \tag{24.20}$$

This ratio increases linearly with the *normalization length L*, Figure 24.3.

Let us now consider a sum:

$$\sum_l A(p_l) \frac{\Delta n}{\Delta_l p} \Delta_l p \tag{24.21}$$

where $\Delta_l p$ is the lth interval and p_l represents a typical value of p within the interval $\Delta_l p$, say the p-value at the midpoint of $\Delta_l p$. The two sums (24.19) and (24.21) have the same dimension, and they are close to each other if (i) the function $A(p)$ is a smooth function of p, and (ii) there exist many states in $\Delta_l p$ so that $\Delta n/\Delta_l p$ can be regarded as the *density of states*. The second condition is satisfied for the momentum states $\{p_k\}$ when the length L is made sufficiently large. In the bulk

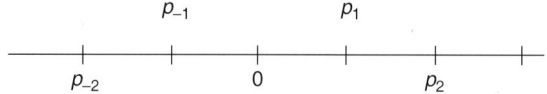
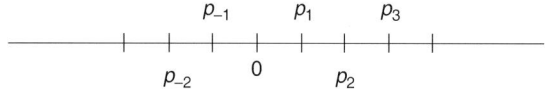
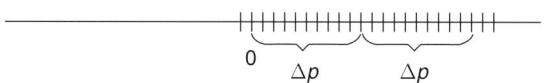

Fig. 24.3 The linear momentum states are represented by short bars forming a linear lattice with unit spacing equal to $2\pi\hbar/L$.

limit, (24.19) and (24.21) will be equal:

$$\lim_{L\to\infty} \sum_{k(\text{states})} A(p_k) = \sum_{\Delta_l p} A(p_l)\frac{\Delta n}{\Delta_l p}\Delta_l p \tag{24.22}$$

In the small-interval limit the sum on the rhs becomes the integral

$$\int dp\, A(p_l)\frac{dn}{dp}$$

where [using (24.20)]

$$\frac{dn}{dp} \equiv \lim_{\Delta p \to 0} \frac{\Delta n}{\Delta p} = \frac{L}{2\pi\hbar} \tag{24.23}$$

is the density of states in the momentum space (line). In summary, we therefore have

$$\boxed{\sum_k A(p_k) \to \int_{-\infty}^{\infty} dp\, A(p_l)\frac{dn}{dp}} \tag{24.24}$$

We stress that the condition (i) depends on the nature of the function A. Therefore, if $A(p)$ is singular at some point, this condition is not satisfied, which may invalidate the limit in (24.25). Such exceptional cases do occur. We further note that the density of states $dn/dp = L(2\pi\hbar)^{-1}$ does not depend on the momentum.

24.2 Density of States

The sum-to-integral conversion, which we have discussed, can easily be generalized for a multidimensional case. For example, we have

$$\sum_{p_k} A(p_k) \to \int_{-\infty}^{\infty} d^3 p \, A(\mathbf{p}) \mathcal{N}(\mathbf{p}) \quad \text{as } V \equiv L^3 \to \infty \tag{24.25}$$

The density of states $\mathcal{N}(\mathbf{p}) \equiv dn/d^3p$ can be calculated by extending the arguments leading to (24.20). We choose the periodic cubic box of side length L for the normalization, take the spin degeneracy into account and obtain

$$\mathcal{N}(\mathbf{p}) \equiv \frac{dn}{d^3 p} = \frac{2L^3}{(2\pi\hbar)^3} \quad \text{(with spin degeneracy)} \tag{24.26}$$

Let us use this result and simplify the *normalization condition* in (24.9). We then obtain

$$n = \lim_{V \to \infty} \frac{1}{V} \int d^3 p \, f(p^2/2m) \mathcal{N}(\mathbf{p}) = \frac{2}{(2\pi\hbar)^3} \int d^3 p \, f(p^2/2m) \quad (L^3 = V) \tag{24.27}$$

Next, consider the energy density of the system. Using (24.5) and (24.8) we obtain

$$e \equiv \lim_{V \to \infty} \frac{\langle H \rangle}{V} = \lim_{V \to \infty} \frac{1}{V} \sum_{\kappa} \varepsilon_\kappa f(\varepsilon_\kappa) = \frac{2}{(2\pi\hbar)^3} \int d^3 p \left(\frac{p^2}{2m} \right) f(p^2/2m) \tag{24.28}$$

Equations (24.27) and (24.28) were obtained, starting with the momentum eigenvalues corresponding to the *periodic cube-box boundary conditions*. The results in the bulk limit, however, do not depend on the type of boundary condition.

The concept of the density of states can also be applied to the energy domain. This is convenient when the sum over states has the form:

$$\sum_{\kappa} g(\varepsilon_\kappa) \tag{24.29}$$

where $g(\varepsilon_\kappa)$ is a function of the energy ε_κ associated with the state κ. The sums appearing in (24.5) and (24.9) are precisely in this form.

Let dn be the number of the states within the energy interval $d\varepsilon$. In the bulk limit this number dn will be proportional to the interval $d\varepsilon$, so that

$$dn = \mathcal{N}(\varepsilon) d\varepsilon \tag{24.30}$$

Here, the proportionality factor

$$\mathcal{N}(\varepsilon) \equiv \frac{dn}{d\varepsilon} \tag{24.31}$$

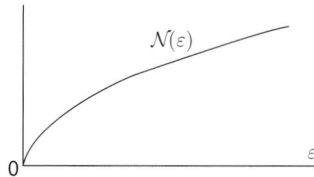

Fig. 24.4 The density of states in energy, $N(\varepsilon)$, for free electrons in three dimensions grows like $\varepsilon^{1/2}$.

is called the *density of states in the energy domain*. This quantity $\mathcal{N}(\varepsilon)$ generally depends on the location of the interval $d\varepsilon$. We may take the location to be the midpoint of the interval $d\varepsilon$. If the set of states $\{\kappa\}$ becomes densely populated in the bulk limit and the function g is smooth, then the sum may be converted into an integral of the form:

$$\sum_{\kappa(\text{states})} g(\varepsilon_\kappa) \to \int d\varepsilon g(\varepsilon) \mathcal{N}(\varepsilon) \tag{24.32}$$

Let us now calculate the *density of states* $\mathcal{N}(\varepsilon)$ for the system of free electrons. The number of states dn in the spherical shell in momentum space is obtained by dividing the volume of the shell, $4\pi p^2 dp$ by the unit-cell volume $(2\pi\hbar/L)^3$ and multiplying the result by the *spin-degeneracy factor* 2:

$$dn = \frac{2 \times 4\pi p^2 dp}{(2\pi\hbar/L)^3} = V\frac{8\pi p^2}{(2\pi\hbar)^3} dp \tag{24.33}$$

Since $p = (2m\varepsilon)^{1/2}$, we obtain

$$dp = \frac{dp}{d\varepsilon} d\varepsilon = \left(\frac{m}{2\varepsilon}\right)^{1/2} d\varepsilon \tag{24.34}$$

Using these equations we obtain

$$dn = V\frac{8\pi(2m\varepsilon)}{(2\pi\hbar)^3}\left(\frac{m}{2\varepsilon}\right)^{1/2} d\varepsilon = V\frac{8\pi 2^{1/2} m^{3/2}}{(2\pi\hbar)^3} \varepsilon^{1/2} d\varepsilon \tag{24.35}$$

or

$$\mathcal{N}(\varepsilon) \equiv \frac{dn}{d\varepsilon} = V\frac{2^{1/2} m^{3/2}}{\pi^2 \hbar^3} \varepsilon^{1/2} \tag{24.36}$$

The density of states $\mathcal{N}(\varepsilon)$ grows like $\varepsilon^{1/2}$ and is shown in Figure 24.4.
We may now restate the normalization condition (24.9) as follows:

$$n = \lim_{V\to\infty} \frac{1}{V} \sum_\kappa f(\varepsilon_\kappa) = \lim_{V\to\infty} \frac{1}{V} \int_0^\infty d\varepsilon\, f(\varepsilon) \mathcal{N}(\varepsilon)$$

$$= \frac{2^{1/2} m^{3/2}}{\pi^2 \hbar^3} \int_0^\infty d\varepsilon\, \varepsilon^{1/2} f(\varepsilon) \tag{24.37}$$

24.3 Qualitative Discussion

The density of states in the energy domain, defined by (24.31), is valid even when we have states other than the momentum states. We shall see such cases in later applications.

Problem 24.2.1

Obtain (24.37) directly from (24.27) by using the spherical polar coordinates (p, θ, ϕ) in the momentum space, integrating over the angles (θ, ϕ) and rewriting the p-integral in terms of the ε-integral.

24.3
Qualitative Discussion

At room temperature most metals have molar heat capacities of about $3R$ (where R is the gas constant) like nonmetallic solids. This experimental fact cannot be explained based on classical statistical mechanics. By applying the *Fermi–Dirac statistics* to conduction electrons, we can demonstrate the near absence of the electronic contribution to the heat capacity. In this section we treat this topic in a qualitative manner.

Let us consider highly degenerate electrons with a high Fermi temperature T_F ($\approx 80\,000$ K). At 0 K, the Fermi distribution function:

$$f(\varepsilon; \mu, T) \equiv \frac{1}{\exp[(\varepsilon - \mu)/k_B T] + 1} \tag{24.38}$$

is a step function, as indicated by the dotted line in the lower diagram in Figure 24.5. At a finite temperature T, the abrupt drop at $\varepsilon = \mu_0$ becomes a smooth drop, as indicated by a solid line in the same diagram. In fact, the change in the distribution function $f(\varepsilon)$ is appreciable only near $\varepsilon = \mu_0$. The function $f(\varepsilon; \mu, T)$ will drop from 1/2 at $\varepsilon = \mu$ to 1/101 at $\varepsilon = \mu + k_B T \ln(100)$ [which can be directly verified from (24.38)]. This value of $k_B T \ln(100) \simeq 4.6 k_B T$ is much less than the Fermi energy $\mu_0 = k_B T_F$ ($T_F \simeq 80\,000$ K). This means that only those electrons with energies close to the Fermi energy μ_0 are excited by the rise in temperature. In other words, the electrons with energies ε far below μ_0 are not affected. There are many such electrons, and in fact this group of unaffected electrons forms the great majority.

The number N_X of electrons that are thermally excited can be found in the following manner. The density of states $\mathcal{N}(\varepsilon)$ is shown in the upper diagram in Figure 24.5. Since $\mathcal{N}(\varepsilon)d\varepsilon$ represents by definition the number of electrons within $d\varepsilon$, the integral of $\mathcal{N}(\varepsilon)d\varepsilon$ over the interval in which the electron population is affected gives an approximate number of excited electrons N_X. This integral can be represented by the shaded area in the upper diagram in Figure 24.5. Since we know from the earlier arguments that the affected range of the energy is of the order of $k_B T$ ($\ll \mu_0$), we can estimate N_X as

$$N_X = \text{shaded area in the upper diagram} \cong \mathcal{N}(\mu_0) k_B T \tag{24.39}$$

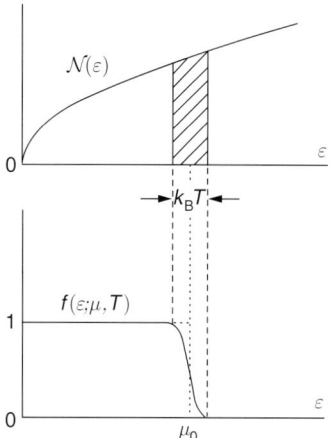

Fig. 24.5 The density of states in energy, $N(\varepsilon)$, and the Fermi distribution function $f(\varepsilon)$ are drawn as a function of the kinetic energy ε. The change in f is appreciable only near the Fermi energy μ_0 if $k_B T \ll \mu_0$. The shaded area represents approximately the number of thermally excited electrons.

where $\mathcal{N}(\mu_0)$ is the density of states at $\varepsilon = \mu_0$. From (24.36) and (24.16), we obtain

$$\mathcal{N}(\mu_0) = V \frac{2^{1/2} m^{3/2}}{\pi^2 \hbar^3} \mu_0^{1/2} = \frac{3N}{2\mu_0} \tag{24.40}$$

Using this expression, we get from (24.39)

$$N_X = \left(\frac{3}{2}\right) \frac{N k_B T}{\mu_0} \tag{24.41}$$

The electrons affected will move up with the extra energy of the order of $k_B T$ per particle. Therefore, the change in the total energy ΔE will approximately be

$$\Delta E = N_X \times k_B T = \left(\frac{3}{2}\right) \frac{N(k_B T)^2}{\mu_0} \tag{24.42}$$

Differentiating this equation with respect to T, we obtain

$$C_V = \frac{\partial}{\partial T} \Delta E = 3 N_0 k_B^2 \frac{T}{\mu_0} = 3R \frac{T}{T_F} \quad (\mu_0 \equiv k_B T_F, \; R \equiv N_0 k_B) \tag{24.43}$$

for the *molar heat capacity*.

This expression indicates that the molar electronic heat capacity at room temperature ($T = 300$ K) is indeed small:

$$C_V = 3R \frac{300}{80\,000} = 0.011 R$$

It is stressed that (24.43) was obtained because the number of thermally excited electrons N_X is much less than the total number of electrons N [see (24.41)]. We also note that the *electronic heat capacity is linear in temperature*.

24.4
Sommerfeld's Calculations

Historically, *Sommerfeld* [1] first applied the Fermi–Dirac statistics to the conduction electrons and calculated the electronic heat capacity. His calculations resolved the heat-capacity paradox (the absence of the electronic contribution at room temperature). In this section we calculate the heat capacity quantitatively.

The *heat capacity at constant volume* C_V can be calculated by differentiating the internal energy E with respect to the temperature T:

$$C_V = \left(\frac{\partial E}{\partial T}\right)_V \tag{24.44}$$

The internal energy density for free electrons given by (24.28) can be expressed as

$$\frac{E(T, V)}{V} = \frac{2^{1/2} m^{3/2}}{\pi^2 \hbar^3} \int_0^\infty d\varepsilon \, \varepsilon^{3/2} f(\varepsilon; \mu, T) \tag{24.45}$$

Here, the *chemical potential* μ is related to the number density n by (24.37):

$$n = \frac{2^{1/2} m^{3/2}}{\pi^2 \hbar^3} \int_0^\infty d\varepsilon \, \varepsilon^{1/2} f(\varepsilon) \tag{24.46}$$

The integrals on the rhs of (24.45) and (24.46) may be evaluated as follows. Let us assume

$$\alpha \equiv \beta \mu \gg 1 \tag{24.47}$$

Next, we consider

$$F(x) \equiv \frac{1}{\exp(x - \alpha) + 1}, \quad -\frac{dF}{dx} \equiv -F'(x) = \frac{\exp(x - \alpha)}{[\exp(x - \alpha) + 1]^2} \tag{24.48}$$

whose behaviors are shown in Figure 24.6. We note that $-dF/dx$ is a sharply peaked function near $x = \alpha$. Let us take the integral:

$$I \equiv \int_0^\infty dx \, F(x) \frac{d\phi}{dx} \tag{24.49}$$

where $\phi(x)$ is a certain function of x. By integrating by parts and using (24.47) and (24.48), we obtain

$$I = F(\infty)\phi(\infty) - F(0)\phi(0) - \int_0^\infty dx \, \phi(x) \frac{dF}{dx}$$

$$= -\phi(0) - \int_0^\infty dx \, \phi(x) \frac{dF}{dx} \tag{24.50}$$

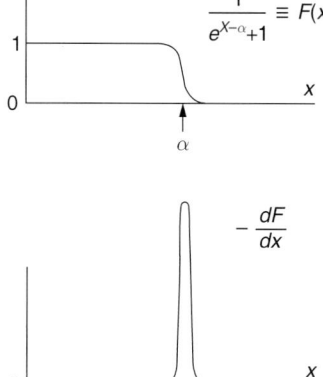

Fig. 24.6 The function $F(x)$ and $-dF/dx$ are shown for $x \geq 0$ and $\alpha \gg 1$.

We expand the function $\phi(x)$ in a Taylor series at $x = \alpha$

$$\phi(x) = \phi(\alpha) + (x - \alpha)\phi'(\alpha) + \frac{1}{2}(x - \alpha)^2 \phi''(\alpha) + \ldots \qquad (24.51)$$

introduce it in (24.50), and then integrate term by term. If $\alpha \gg 1$, then we obtain

$$-\int_0^\infty dx (x-\alpha)^n \frac{dF(x)}{dx} = \int_{-\alpha}^\infty dy\, y^n \frac{e^y}{(e^y+1)^2}$$

$$\cong \int_{-\infty}^\infty dy\, y^n \frac{e^y}{(e^y+1)^2} \equiv J_n \qquad (24.52)$$

The definite integrals J_n vanish for odd n since the integrands are odd functions of y. For small even n, we obtain

$$J_0 = 1$$

$$J_2 = -2\int_{-\infty}^\infty dy\, y^2 \frac{d}{dy}\frac{1}{e^y+1} = -2\int_0^1 dz\, \frac{\ln z}{z+1} = \frac{\pi^2}{3} \qquad (24.53)$$

Using (24.49) through (24.53), we obtain

$$I \equiv \int_0^\infty dx\, F(x)\frac{d\phi}{dx} = \phi(\alpha) - \phi(0) + \frac{\pi^2}{6}\phi''(\alpha) \qquad (24.54)$$

Equation (24.54) is useful if $\alpha \gg 1$ and $\phi(x)$ is a slowly varying function at $x = \alpha$. We apply (24.54) to evaluate the ε-integral in (24.46)

$$\int_0^\infty d\varepsilon\, \varepsilon^{1/2} f(\varepsilon) = \beta^{3/2} \int_0^\infty dx\, x^{3/2} F(x) \quad (\beta \varepsilon \equiv x)$$

24.4 Sommerfeld's Calculations

We may choose

$$\phi(x) = \frac{2}{3}x^{3/2}, \quad \frac{d\phi(x)}{dx} = x^{1/2}, \quad \frac{d^2\phi(x)}{dx^2} = \frac{1}{2}x^{-1/2}$$

and obtain

$$\int_0^\infty d\varepsilon\, \varepsilon^{3/2} f(\varepsilon) = \frac{2}{3\beta^{3/2}}(\beta\mu)^{3/2} + \frac{1}{\beta^{1/2}}\frac{\pi^2}{6}\frac{1}{2}(\beta\mu)^{-1/2}$$

$$= \frac{2}{3}\mu^{3/2}\left(1 + \frac{\pi^2}{8}\beta^{-2}\mu^{-2}\right) \qquad (24.55)$$

Using (24.54) we obtain from (24.46)

$$n = \frac{2^{1/2} m^{3/2}}{\pi^2 \hbar^3}\left(1 + \frac{\pi^2}{8}\beta^{-2}\mu^{-2}\right) \qquad (24.56)$$

Similarly, we can calculate the ε-integral in (24.45) and obtain

$$e \equiv \frac{E(T, V)}{V} = \frac{2^{3/2} m^{3/2}}{5\pi^2 \hbar^3}\mu^{5/2}\left(1 + \frac{5\pi^2}{8}\beta^{-2}\mu^{-2}\right) \qquad (24.57)$$

The chemical potential μ, in general, depends on density n and temperature T: $\mu = \mu(n, T)$. This relation can be obtained from (24.56). If we substitute $\mu(n, T)$ so obtained into (24.57), we can regard the internal energy density as a function of n and T. By subsequently differentiating this function with respect to T at fixed n, we can obtain the heat capacity at constant volume. We may also calculate the heat capacity by taking another route. The Fermi energy μ_0 is given by $\hbar^2(3\pi^2 n)^{2/3}/2m$ [see (24.16)], which depends only on n. Solving (24.56) for μ and expressing the result in terms of μ_0 and T, we obtain (Problem 24.4.1).

$$\mu = \mu_0\left[1 - \frac{\pi^2}{12}\left(\frac{k_B T}{\mu_0}\right)^2 + \ldots\right] \qquad (24.58)$$

Introducing this expression into (24.57), we obtain (Problem 24.4.2)

$$e = \frac{3}{5}n\mu_0\left[1 + \frac{5\pi^2}{12}\left(\frac{k_B T}{\mu_0}\right)^2\right] \qquad (24.59)$$

Differentiating this expression with respect to T at constant μ_0, we obtain

$$\boxed{C_V = \frac{1}{2}\pi^2 N k_B \frac{T}{T_F}} \qquad (24.60)$$

where $T_F \equiv \varepsilon_F/k_B$ is the *Fermi temperature*. Here, we see that the heat capacity C_V for degenerate electrons is greatly reduced by the factor T/T_F ($\ll 1$) compared with the ideal-gas heat capacity $3Nk_B/2$. Also note that the heat capacity changes linearly with temperature T. These findings agree with the results of our previous qualitative calculations in Section 24.3.

At normal experimental temperatures the lattice contribution to the heat capacity is much greater than the electronic contribution. Therefore, the experimental verification of the linear-T law must be done at very low temperatures, where the contribution of the lattice vibration becomes negligible. In this low-temperature region, the measured molar heat capacity should rise linearly with the temperature T. By comparing the slope with

$$\frac{C_V}{T} = \frac{\pi^2}{2} \frac{R}{T_F} \qquad (24.61)$$

[using (24.60)], we can find the Fermi temperature T_F numerically. Since this temperature T_F is related to the effective mass m^* by

$$k_B T_F = \left(\frac{\hbar^2}{2m^*}\right)(3\pi^2 n)^{2/3} \qquad (24.62)$$

[using (24.59) and (24.60)], we can obtain the numerical value for the effective mass m^* of the conduction electron. Other ways of finding the m^*-value are through the transport and optical properties of conductors, which will be discussed later.

Problem 24.4.1

Use (24.56) to verify (24.58).

HINT: Assume $\mu = \mu_0[1 + A(k_B T/\mu_0)^2]$ and find the constant A.

Problem 24.4.2

Verify (24.59).

Reference

1 Sommerfeld, A. (1928) Z. Phys., **47**, 1.

25
The Bose–Einstein Condensation

The liquid helium undergoes a superfluid phase change at 2.2 K. This is interpreted in terms of the Bose–Einstein condensation in this chapter.

25.1
Liquid Helium

Helium is the only substance in Nature that does not solidify by a lowering of the temperature under normal atmospheric pressure. There exist two main isotopes, He^4, the most abundant, and He^3. Interactions between monatomic molecules are practically identical for any pair, He^4–He^4, He^3–He^3, He^3–He^4. These interactions are mainly determined by the electronic structure that both isotopes share. The potential has a shallow attractive well such that two He^3 atoms may form a bound state with a very small binding energy. For He^4, the bound sate has a 20% greater binding energy. This is due to the difference in mass. Quantum-mechanical calculations of the ground-state energy involve the total Hamiltonian, that is, the potential energy plus kinetic energy, which contains the mass difference. The boiling points for liquid He^3 and He^4 are 3.2 and 4.2 K, respectively. These values do reflect the binding energies of the molecules.

Liquid He^4 undergoes the so-called λ-transition at 2.2 K into a *superfluid phase* whose flow properties are quite different from those of the ordinary fluid. For example, superfluid in a beaker creeps around the wall and drips down [1]. See Figure 25.1. On the other hand, liquid He^3 behaves quite differently. Very recently, a phase transition was discovered for this liquid at a temperature of 0.002 K [2]. Its superfluid phase, however, is quite different from the superfluid phase of liquid He^4.

Why does there exist such a difference between liquid He^3 and He^4? Isotope He^4 has a nucleus (α-particle) consisting of two protons and two neutrons and possessing zero nuclear spin angular momentum, and two electrons orbiting around the nucleus with vanishing angular momentum in the ground state. He^4 atoms, therefore, are bosons according to the spin-statistics theorem. Isotope He^3 has a nucleus, consisting of two protons and one neutron and possessing spin of magnitude $\hbar/2$, and accordingly they are *fermions*. This difference in statistics generates a fundamental difference in the macroscopic properties of quantum fluids.

Mathematical Physics. Shigeji Fujita and Salvador V. Godoy
Copyright © 2010 WILEY-VCH Verlag GmbH & Co. KGaA, Weinheim
ISBN: 978-3-527-40808-5

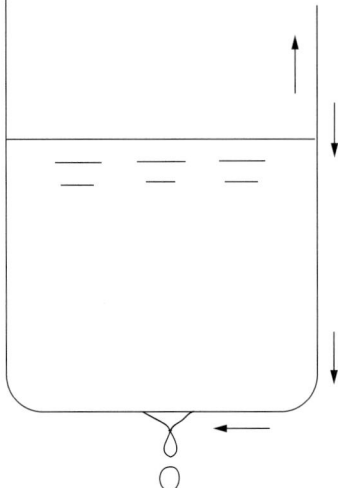

Fig. 25.1 Superfluid He in a beaker creeps around the wall and drips down.

25.2
The Bose–Einstein Condensation of Free Bosons

Let us consider a system of free bosons characterized by the Hamiltonian

$$H = \sum_{j=1}^{N} \frac{p_j^2}{2m} \tag{25.1}$$

The system in equilibrium has the momentum distribution characterized by the *Bose distribution function*

$$f_B(\varepsilon_p) = \frac{1}{e^{\beta(\varepsilon_p - \mu)} - 1}, \quad \varepsilon_p \equiv \frac{p^2}{2m} \tag{25.2}$$

which is subject to the normalization condition:

$$\frac{1}{V} \sum_p f_B(\varepsilon_p) = n \quad \text{(number density)} \tag{25.3}$$

Hereafter, we drop the subscript B.

By definition, the Bose distribution function $f(\varepsilon_p)$ represents the relative probability of finding a particle with a momentum **p**. Therefore, the function $f(\varepsilon_p)$ must be nonnegative:

$$f(\varepsilon_p) = \frac{1}{e^{\beta(\varepsilon_p - \mu)} - 1} \geq 0 \tag{25.4}$$

To insure this property for all $\varepsilon_p \geq 0$ the chemical potential μ must be nonpositive:

$$\mu \leq 0 \tag{25.5}$$

or else the Bose distribution $f(\varepsilon_p)$ would become negative for $0 < \varepsilon_p < \mu$. By the way, such restriction on the chemical potential μ does not apply to the Fermi or Boltzmann distribution function. [Why? Problem 25.2.1.]

In the bulk limit, the momentum eigenvalues form a continuous spectrum. For the moment, let us replace the sum over momentum states in (25.3) by the momentum integral:

$$n = \lim \frac{1}{V} \int d^3p \, \frac{dn}{d^3p} f(\varepsilon_p) \tag{25.6}$$

The density of states dn/d^3p for frees particles (without spin) was calculated in Section 24.2, and is given by

$$\frac{dn}{d^3p} = \frac{V}{(2\pi\hbar)^3} \tag{25.7}$$

Introducing this expression in (25.6), we obtain

$$n = \frac{1}{(2\pi\hbar)^3} \int d^3p \, f(\varepsilon_p) \tag{25.8}$$

The momentum-space integration may be carried out in the spherical polar coordinates (p, θ, φ) with $d^3p = p^2 dp \sin\theta \, d\theta \, d\varphi$ as the momentum-volume element. After performing the angular integration, we rewrite the result in the form of an energy integral and obtain

$$n = \frac{M^{1/3}}{2^{1/2} \pi^2 \hbar^3} \int_0^\infty d\varepsilon \, \varepsilon^{1/2} \frac{1}{e^{\beta(\varepsilon+|\mu|)} - 1} \equiv \frac{M^{1/3}}{2^{1/2} \pi^2 \hbar^3} F(\beta, |\mu|) \tag{25.9}$$

This result can also be obtained by finding the density of states in energy, $\mathcal{N}(\varepsilon)$, more directly as

$$\mathcal{N}(\varepsilon) = V \frac{M^{3/2}}{2^{1/2} \pi^2 \hbar^3} \varepsilon^{1/2} \tag{25.10}$$

Note that, because of the lack of spin degeneracy, this expression is one half of (14.36).

Let us now consider the ε-integral in (25.9):

$$F(\beta, |\mu|) \equiv \int_0^\infty d\varepsilon \, \varepsilon^{1/2} \frac{1}{e^{\beta(\varepsilon+|\mu|)} - 1} \tag{25.11}$$

which is a function of β and $|\mu|$. For a fixed β, the function F is a decreasing function of $|\mu|$ with a maximum occurring at $\mu = 0$:

$$F(\beta, |\mu|) \leq F(\beta, 0) \tag{25.12}$$

The maximum value for F can be evaluated as follows:

$$F(\beta, 0) = \int_0^\infty d\varepsilon \, \varepsilon^{1/2} \frac{1}{e^{\varepsilon\beta} - 1}$$

$$= \beta^{-3/2} \int_0^\infty dx \, x^{1/2} \frac{1}{e^x - 1} \quad [\varepsilon\beta = x] \tag{25.13}$$

The x-integral has the numerical value

$$\int_0^\infty dx \, x^{1/2} \frac{1}{e^x - 1} = \int_0^\infty dx \, x^{1/2} (e^{-x} + e^{-2x} + \ldots) = 1.306 \sqrt{\pi} \tag{25.14}$$

We then see that the integral on the rhs of (25.9) has an upper limit. The number density n on the lhs could, of course, be increased without limit. Something must have gone wrong in our calculations.

A closer look at (25.2) shows that the function $f(\varepsilon)$ blows up in the neighborhood of $\varepsilon = 0$ if $\mu = 0$. This behavior therefore violates the validity condition for the conversion of the sum-over-states into an integral. In such a case, we must proceed more carefully.

Let us go back to (25.3) and break the sum into two:

$$\frac{N}{V} = \frac{N_0}{V} + \frac{1}{V} \sum_{\mathbf{p}(\varepsilon_p > 0)} f(\varepsilon_p) \tag{25.15}$$

where N_0 is the number of zero-momentum bosons and is given by

$$N_0 \equiv \frac{1}{e^{\beta|\mu|} - 1} \tag{25.16}$$

This number N_0 can be made very large by choosing very small $\beta |\mu|$. For example, to have $N_0 = 10^{20}$, we may choose

$$\beta |\mu| = \ln(1 + N_0^{-1}) = \ln(1 + 10^{-20}) \simeq 10^{-20}$$

In fact, N_0 can be increased without limit. We further note that, because the density of sates in energy is proportional to $\varepsilon^{1/2}$, see (25.10), the contribution of zero-momentum bosons is not included in the ε-integral. We therefore should write the normalization condition in the bulk limit as follows:

$$n = n_0 + \frac{M^{3/2}}{2^{1/2} \pi^2 \hbar^3} \int_0^\infty d\varepsilon \, \varepsilon^{1/2} \frac{1}{e^{\beta(\varepsilon + |\mu|)} - 1} \tag{25.17}$$

The two terms in (25.17) represent the density $n_0 \equiv N_0/V$ of zero-momentum bosons and that of nonzero-momentum bosons. The term $n_0 \equiv N_0/V$ is important only when the number of zero-momentum bosons, N_0, is a significant fraction of the total number of bosons, N. The possibility of such an unusual state, called the *Bose–Einstein condensation* (BEC), was first recognized by Einstein in 1925 [3]. [Such possibilities exist neither for fermions nor classical particles.]

For low densities or high temperatures, the density of zero-momentum bosons n_0 is negligible against the (total) number density n. By raising the density or by lowering the temperature, the system undergoes a sharp transition into a state in which n_0 becomes a significant fraction of n. At absolute zero all bosons have zero momentum. The sharp change in state resembles the gas-to-liquid condensation but the BEC occurs in the momentum space. Further features of the BEC will be discussed in the following section.

Problem 25.2.1

Let us consider a system of free fermions. Find, from the normalization, the range of the chemical potential μ that appears in the Fermi distribution function:

$$f_F(\varepsilon) = [e^{\beta(\varepsilon-\mu)} + 1]^{-1}$$

Problem 25.2.2

Let us consider a system of free bosons moving in two dimensions.
1. Does this system undergo transition into a condensed state at low temperatures as in the three-dimensional case?
2. Discuss the heat capacity of the system in all temperature ranges. Calculate explicitly the heat capacity at both low- and high-temperature limits.

25.3
Bosons in Condensed Phase

At absolute zero, all bosons have zero momentum. As the temperature is raised, bosons with nonzero momenta emerge. The number of those excited bosons, given by the volume V times the second term of the rhs of (25.15) can be represented by

$$N_{\varepsilon>0} \equiv \sum_{p(\varepsilon_p>0)} f(\varepsilon_p)$$

$$\cong \frac{VM^{3/2}}{2^{1/2}\pi^2\hbar^3} \int_0^\infty d\varepsilon \varepsilon^{1/2} \frac{1}{e^{\beta(\varepsilon+|\mu|)} - 1} = \frac{VM^{3/2}}{2^{1/2}\pi^2\hbar^3} F(\beta, |\mu|) \quad (25.18)$$

Throughout the condensed phase, the chemical potential μ has a very small absolute value [Problem 25.3.1]. Since the function $F(\beta, |\mu|)$ is a slowly varying function of $|\mu|$ for $|\mu| \ll 1$, we can approximate (25.18) by its value at $\mu = 0$:

$$N_{\varepsilon>0} \cong \frac{VM^{3/2}}{2^{1/2}\pi^2\hbar^3} F(\beta,0) = \frac{1.306\sqrt{\pi}\,VM^{3/2}}{\sqrt{2}\pi^2\hbar^3}(k_B T)^{3/2} \quad (25.19)$$

where we used (25.13) and (25.14). Here, we see that the number of bosons in the excited states with positive energies grows like $T^{3/2}$ as the temperature T is raised. This number may eventually reach the total number N, as T is raised to a critical temperature T_0. At and above the critical temperature T_0 practically all bosons are in excited states. This temperature T_0 can be obtained from

$$N = \frac{1.306\sqrt{\pi}}{\sqrt{2}\pi^2} \frac{VM^{3/2}}{\hbar^3}(k_B T_0)^{3/2}$$

$$= \frac{1}{6.032}\frac{VM^{3/2}}{\hbar^3}(k_B T_0)^{3/2} \quad (25.20)$$

Solving for T_0, we obtain

$$T_0 = 3.31\frac{\hbar^2}{M k_B} n^{2/3}, \quad (6.031)^{2/3} = 3.31 \quad (25.21)$$

Using this relation, we can rewrite (25.19) in the form:

$$N_{\varepsilon>0} = N(T/T_0)^{3/2}, \quad T \leq T_0 \quad (25.22)$$

The number of zero-momentum bosons, N_0, can be obtained by subtracting this number from the total number N:

$$N_0 \equiv N - N_{\varepsilon>0} = N\left[1-(T/T_0)^{3/2}\right], \quad T \leq T_0 \quad (25.23)$$

Here, we see that N_0 is in fact a finite fraction of the total number N. The number of bosons in excited states and in the ground state, $N_{\varepsilon>0}$ and N_0, are plotted against temperature in Figure 25.2.

The thermal energy E of the system comes only from the excited bosons. The average energy per unit volume can be calculated as follows:

$$\frac{E}{V} = \lim \frac{1}{V} \sum_{p(\varepsilon_p>0)} \varepsilon_p f(\varepsilon_p)$$

$$= \frac{M^{3/2}}{2^{1/2}\pi^2\hbar^3} \int_0^\infty d\varepsilon\, \varepsilon^{3/2} \frac{1}{e^{\beta(\varepsilon+|\mu|)}-1} \quad (25.24)$$

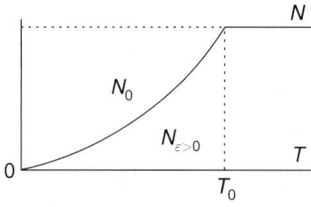

Fig. 25.2 The number of condensed bosons, N_0, below the critical temperature T_0, forms a finite fraction of the total number N.

The ε-integral here is a slowly varying function of $|\mu|$ for $|\mu| \ll 1$. We may therefore approximate it by its value at $\mu = 0$:

$$\int_0^\infty d\varepsilon\, \varepsilon^{3/2} \frac{1}{e^{\beta(\varepsilon+|\mu|)}-1} \cong \int_0^\infty d\varepsilon\, \varepsilon^{3/2} \frac{1}{e^{\beta\varepsilon}-1}$$

$$= \beta^{-5/2} \int_0^\infty dx\, x^{3/2} \frac{1}{e^x - 1} \quad [x = \beta\varepsilon] \tag{25.25}$$

The x-integral is numerically equal to 1.342:

$$\int_0^\infty dx\, x^{3/2} \frac{1}{e^x - 1} = 1.342 \tag{25.26}$$

Using (25.25) and (25.26), we obtain

$$\frac{E}{V} = \frac{M^{3/2}}{2^{1/2}\pi^2\hbar^3} 1.342 (k_B T)^{5/2} \tag{25.27}$$

This result can be rewritten with the aid of (25.20) as follows:

$$\frac{E}{V} = \frac{1.342}{2^{1/2}\pi^2} \frac{M^{3/2}}{\hbar^3} (k_B T_0)^{3/2} (k_B T)(T/T_0)^{3/2}$$

$$= 0.770\, n k_B T (T/T_0)^{3/2} \tag{25.28}$$

where we use (25.20). Note that the internal energy density E/V grows like $T^{5/2}$ in the condensed phase.

Differentiating (25.28) with respect to T, and writing the result for a mole of the gas, we obtain

$$C_V = \left(\frac{\partial E}{\partial T}\right)_V = \left(\frac{\partial E}{\partial T}\right)_{T_0} = \frac{5}{2}\frac{E}{T}$$

$$= 1.92\, N k_B (T/T_0)^{3/2} = 1.92\, R(T/T_0)^{3/2} \tag{25.29}$$

The behavior of the heat capacity at constant volume, C_V, versus the temperature T is shown in Figure 25.3. The molar heat capacity C_V increases like $T^{3/2}$ throughout the condensed phase and reaches its maximum value $1.92\,R$ at the critical temperature T_0. Above T_0, it gradually decreases and approaches the classical value $1.50\,R$ at high temperatures. At the transition point, the heat capacity C_V is continuous in value but its derivative jumps from a positive to a negative value.

The Bose–Einstein condensation is similar to the gas–liquid transformation in the sense that it is a sudden macroscopic change of state. It is, however, quite different in detail. As we just observed, the heat capacity is continuous at the transition point. In contrast, the heat capacity for the familiar gas–liquid transition has a discrete jump. Landau (Lev D. Landau, 1908–62, Russian) classified the Bose–

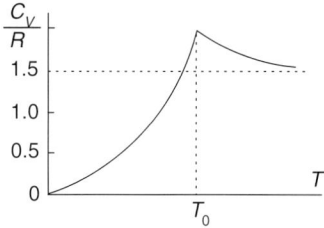

Fig. 25.3 The reduced heat capacity C_V/R exhibits a sharp cusp at the critical temperature T_0.

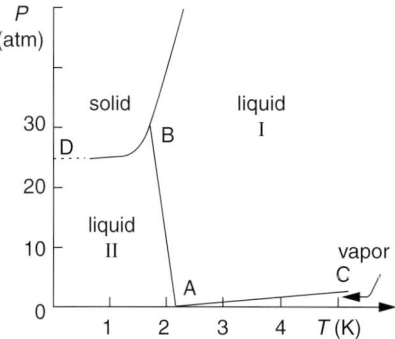

Fig. 25.4 The phase diagram of pure He^4. The line AB, called the λ-line, separates liquid He II from liquid He I. The point C represents the critical point.

Einstein condensation as a *phase transition of second order* at which the second-order derivatives of the free energy such as heat capacity and compressibility, are discontinuous at the point of transition. For a phase transition of second order, there is *no latent heat* of condensation. In contrast, a *phase transition of first order*, for which the first derivatives of the free energy such as volume, entropy, internal energy, jump between the two phases, is accompanied by a latent heat. The readers interested in more about the general theory of the phase transition, are encouraged to read chapters in the classic book on *Statistical Physics* by Landau and Lifshitz [4].

To see the relevance of the Bose–Einstein condensation to the actual liquid helium, let us look at a few properties of this substance. Figure 25.4 represents the *P–T* diagram (*phase diagram*) of pure He^4. Proceeding along the horizontal line at one atmospheric pressure the He^4 passes from gas to liquid at 4.2 K. This liquid, called the *liquid* He I, behaves like any other liquid, and has a finite viscosity. By cooling down further, the substance undergoes a sudden change at 2.18 K. Below this temperature the *liquid* He II is a superfluid. The heat capacity of liquid helium measured under its saturated vapor is shown in Figure 25.5.

Because of the similarity between this experimental curve and the inverted Greek letter λ, the transition is often called the λ-*transition*. Note that the heat capacity has an extremely sharp peak of a logarithmic type [5]. The resemblance between this curve and the heat-capacity curve of the ideal Bose gas shown in Figure 25.3, is

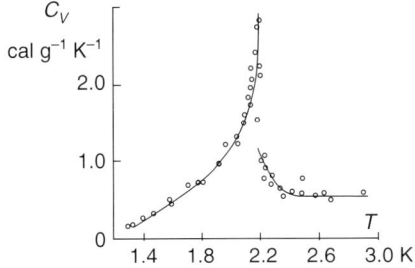

Fig. 25.5 Specific heat of liquid He^4 versus temperature. From London's book, Superfluids [3].

striking. If we calculate the value of the critical temperature T_0 from (25.21) with the use of the density of liquid helium = 0.145 g cm^{-3}, we find that $T_0 = 3.14$ K, which is remarkably close to the observed λ-transition temperature 2.18 K.

Problem 25.3.1

Assume that free bosons are in the condensed state below the critical temperature T_0. Calculate the pressure P of the system by means of the formula:

$$P = \lim \frac{1}{MV} \sum_{p\,(\text{states})} p_x^2 f(\varepsilon_p) \qquad (25.30)$$

where $f(\varepsilon_p)$ is the Bose distribution function. Does P go to zero as T approaches zero?

References

1 London, F. (1954) *Superfluids*, Vol. 2, Wiley, New York.
2 Osheroff, D.D., Richardson, R.C., and Lee, D.M. (1972) *Phys. Rev. Lett.*, **29**, 920.
3 Einstein, A. (1925) *Sitz. Berl. Acad.*, **261**, 3.
4 Landau, L. and Lifshitz, E.M. (1980) *Statistical Physics*, Part 1, Pergamon, Oxford, England, chapter 14.
5 Buckingham, M.J. and Fairbank, W.M. (1961), *The Nature of the λ-Transition*, *Progress in Law Temperature Physics*, Vol. 3, ed. by C.J. Goerter, Amsterdam, North-Holland.

26
Magnetic Susceptibility

The electron has mass m, charge $-e$, and a half spin. It has a spin magnetic moment. Pauli paramagnetism and Landau diamagnetism are discussed in this chapter.

26.1
Introduction

Let us assume that the magnetic field **B** is applied along the positive z-axis. The potential energy V of the magnetic moment μ is given by

$$V = -\mu B \cos\theta = -\mu_z B \tag{26.1}$$

where θ is the angle between the vectors $\boldsymbol{\mu}$ and **B**. The angular momentum (eigenvalues) is quantized in units of $\hbar$. The electron has a spin angular momentum **s** whose z-component can take a value equal to either $\hbar/2$ or $-\hbar/2$. Let us write

$$s'_z = \frac{1}{2}\hbar\sigma'_z \equiv \frac{1}{2}\hbar\sigma, \quad \sigma \equiv \sigma'_z = \pm 1 \tag{26.2}$$

We may assume that

$$\mu_z \propto s'_z \propto \sigma \tag{26.3}$$

We shall write this quantum relation in the form:

$$\boxed{\mu_z = \frac{1}{2}g\mu_B\sigma} \tag{26.4}$$

where

$$\mu_B \equiv \frac{e\hbar}{2mc} = 0.927 \times 10^{-20} \text{ erg gauss}^{-1} \tag{26.5}$$

called the *Bohr magneton*, has the dimensions of a magnetic moment. The constant g in (26.4) is a numerical factor of order 1, and is called a *g-factor*. If the

Mathematical Physics. Shigeji Fujita and Salvador V. Godoy
Copyright © 2010 WILEY-VCH Verlag GmbH & Co. KGaA, Weinheim
ISBN: 978-3-527-40808-5

magnetic moment of the electron is accounted for by the classical "spinning" of the classical charge around a certain axis, the g-factor should be exactly one. The experiments, however, show that this factor is 2.

In the presence of a magnetic field **B**, the electron whose spin is directed along **B**, defined as the electron with the *up-spin*, will have a lower energy than the *down-spin* electron whose spin is directed against **B**. The difference is, using (26.1) and (26.4),

$$\Delta \varepsilon = \frac{1}{2} g \mu_B B(+1) - \frac{1}{2} g \mu_B B(-1) = g \mu_B B \qquad (26.6)$$

For $B = 7000$ gauss and $g = 2$, we obtain the numerical estimate: $\Delta \varepsilon / k_B \simeq 1$ K.

If an electromagnetic wave with the frequency ν satisfying $h\nu = \Delta \varepsilon$ is applied, the electron may absorb a photon of the energy $h\nu$, and jump up to the upper energy level. The frequency corresponding to $\Delta \varepsilon / k_B = 1$ K is

$$\nu = 2.02 \times 10^{10} \text{ cycles s}^{-1} \qquad (26.7)$$

This frequency falls in the microwave region of the electromagnetic radiation spectrum.

26.2
Pauli Paramagnetism

We discuss Pauli's theory of the spin paramagnetism of a metal in this section.

Let us consider an electron moving in free space. The quantum states for the electron can be characterized by momentum **p** and spin $\sigma (= \pm 1)$. If a weak constant magnetic field **B** is applied along the positive z-axis, the energy ε associated with the quantum state (**p**, σ) is given by

$$\varepsilon = \frac{p^2}{2m} - \frac{1}{2} g \mu_B \sigma B \equiv \varepsilon(\mathbf{p}, \sigma) \qquad (26.8)$$

where the second term arises from the electromagnetic interaction, see (26.1) and (26.4). Since $g = 2$ for the electron spin, we may simplify (26.8) to

$$\varepsilon = \frac{p^2}{2m} - \mu_B B \sigma \qquad (26.9)$$

This expression shows that the electron with up-spin ($\sigma = +1$) has lower energy than the electron with down-spin ($\sigma = -1$). In other words, the spin degeneracy is removed in the presence of a magnetic field.

Let us now consider a collection of free electrons in equilibrium. At the absolute zero, the states with the lowest energies will be occupied by the electrons, the Fermi energy ε_F providing the upper limit. This situation is schematically shown in Figure 26.1, where the densities of states, $\mathcal{N}_+(\varepsilon)$ and $\mathcal{N}_-(\varepsilon)$, for electrons with

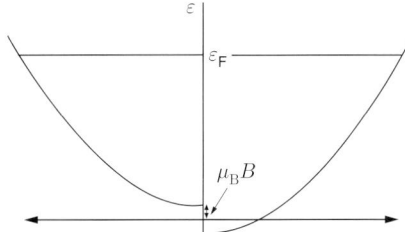

Fig. 26.1 The density of states, $\mathcal{N}_+(\varepsilon)$ and $\mathcal{N}_-(\varepsilon)$, for electrons with up- and down-spins, are drawn against the energy ε, which is measured upwards.

up- and down-spins, are drawn against the energy ε. The density of states for free electrons in the absence of the magnetic field was discussed in Chapter 24. In the absence of the field, both $\mathcal{N}_+(\varepsilon)$ and $\mathcal{N}_-(\varepsilon)$ are the same, and are given by one half of expression (24.36):

$$\mathcal{N}_0(\varepsilon) \equiv V \frac{m^{3/2}}{\sqrt{2}\pi^2 \hbar^3} \varepsilon^{1/2} \tag{26.10}$$

Because of the magnetic energy, $-\mu_B B$, the curve for the density of states $\mathcal{N}_+(\varepsilon)$ for electrons with up-spins will be displaced downward by $\mu_B B$ compared with that for zero field, and is given by

$$\mathcal{N}_+(\varepsilon) = V \frac{m^{3/2}}{\sqrt{2}\pi^2 \hbar^3} (\varepsilon + \mu_B B)^{1/2} = \mathcal{N}_0(\varepsilon + \mu_B B), \quad \varepsilon \geq -\mu_B B \tag{26.11}$$

Similarly, the curve for the *density of states*, $\mathcal{N}_-(\varepsilon)$, for electrons with down-spin is displaced upward by $\mu_B B$:

$$\mathcal{N}_-(\varepsilon) = V \frac{m^{3/2}}{\sqrt{2}\pi^2 \hbar^3} (\varepsilon - \mu_B B)^{1/2} = \mathcal{N}_0(\varepsilon - \mu_B B), \quad \varepsilon \geq \mu_B B \tag{26.12}$$

From Figure 26.1, the numbers $N_\pm$ of the electrons with up- and down-spins are given by

$$N_+ = \int_{-\mu_B B}^{\varepsilon_F} d\varepsilon \mathcal{N}_+(\varepsilon) = \int_0^{\varepsilon_F + \mu_B B} dx \mathcal{N}_0(x) \quad (x = \varepsilon + \mu_B B)$$

$$N_- = \int_{\mu_B B}^{\varepsilon_F} d\varepsilon \mathcal{N}_-(\varepsilon) = \int_0^{\varepsilon_F - \mu_B B} dx \mathcal{N}_0(x) \quad (x = \varepsilon - \mu_B B) \tag{26.13}$$

The difference $N_+ - N_-$ generates a finite magnetic moment for the system. Each electron with up-spin contributes μ_B and each electron with down-spin $-\mu_B$. Therefore, the total magnetic moment is $N_+\mu_B - N_-\mu_B$. Dividing this by volume V, we obtain, for the magnetization I,

$$I = \frac{\mu_B}{V}[N_+ - N_-]$$

$$= \frac{\mu_B}{V}\left[\int_0^{\varepsilon_F + \mu_B B} dx \mathcal{N}_0(x) - \int_0^{\varepsilon_F - \mu_B B} dx \mathcal{N}_0(x)\right]$$

$$\simeq \frac{2\mu_B^2 B}{V}\mathcal{N}_0(\varepsilon_F) \tag{26.14}$$

where we retained the term proportional to B only. Using (26.10), we can re-express this as follows:

$$I = \frac{\sqrt{2}\mu_B^2 m^{3/2}}{\pi^2 \hbar^3} \varepsilon^{1/2} B > 0 \tag{26.15}$$

The last expression shows that the magnetization is positive, and is proportional to the field B. That is, the system is paramagnetic. The *susceptibility* χ defined through the relation

$$I = \chi B \tag{26.16}$$

is given by

$$\chi = \frac{\sqrt{2}\mu_B^2 m^{3/2}}{\pi^2 \hbar^3} \varepsilon^{1/2} \tag{26.17}$$

By using the relation

$$n = \frac{2}{3}\frac{\sqrt{2}m^{3/2}}{\pi^2 \hbar^3}\varepsilon_F^{3/2} \tag{26.18}$$

we can rewrite (26.17) as

$$\chi_P = \frac{3}{2}\frac{\mu_B^2 n}{\varepsilon_F} \tag{26.19}$$

This result was first obtained by Pauli [1], and is often referred to as the *Pauli paramagnetism*.

We note that the Pauli paramagnetism is weaker than the paramagnetism of isolated atoms approximately by the factor $k_B T/\varepsilon_F$ (if this factor is small).

26.3
Motion of a Charged Particle in Electromagnetic Fields

Let us consider a particle of mass m and charge q moving in given electric and magnetic fields $(\mathbf{E}, \mathbf{B})$. In this section we shall study the motion of a charged particle classically. We are interested mainly in those situations for which the electric field

E is very small and the magnetic field B may be arbitrarily large but constant in space and time.

Let us first consider the case in which $\mathbf{E} = 0$. Newton's equation of motion for a classical particle having a charge q in the presence of a magnetic field $\mathbf{B}$ is

$$m\frac{d\mathbf{v}}{dt} = q(\mathbf{v} \times \mathbf{B}) \tag{26.20}$$

We take the dot product of this equation with $\mathbf{v}$:

$$m\mathbf{v} \cdot \frac{d\mathbf{v}}{dt} = q\mathbf{v} \cdot (\mathbf{v} \times \mathbf{B})$$

The rhs vanishes, since $\mathbf{v} \cdot (\mathbf{v} \times \mathbf{B}) = (\mathbf{v} \times \mathbf{v}) \cdot \mathbf{B} = 0$. We obtain

$$m\mathbf{v} \cdot \frac{d\mathbf{v}}{dt} = \frac{d}{dt}\left(\frac{1}{2}mv^2\right) = 0 \tag{26.21}$$

which means that the *kinetic energy* $mv^2/2$ is conserved. This result is valid regardless of how the magnetic field $\mathbf{B}$ varies in space. If the magnetic field $\mathbf{B}$ varies in time, an electric field is necessarily induced, and the above result will not hold strictly.

In the case of a constant magnetic field, we can rewrite (26.20) as

$$\frac{d\mathbf{v}}{dt} = \mathbf{v} \times \boldsymbol{\omega}_c \tag{26.22}$$

where $\boldsymbol{\omega}_c$ is the constant vector pointing along the direction of the magnetic field and having the magnitude:

$$\omega_c = \frac{qB}{m} \tag{26.23}$$

This quantity ω_c is called the *cyclotron frequency*. It is proportional to the magnetic-field strength and inversely proportional to the mass of the particle. For example, for an electron in a field of 1000 gauss, we have $\omega_c = 10^{10}\ \text{s}^{-1}$.

From (26.22) we can deduce that the motion of the electron consists of the uniform motion along the magnetic field with velocity v_z plus a circular motion with constant speed $v_\perp$, about the magnetic field, see Figure 26.2. The radius R of the circular orbit about the magnetic field, called the *cyclotron radius*, can be determined from

$$\frac{mv_\perp^2}{R} = ev_\perp B \tag{26.24}$$

(centripetal force) = (magnetic force). Solving this equation for R, we obtain

$$R = \frac{mv_\perp^2}{eB} = \frac{v_\perp}{\omega_c} \tag{26.25}$$

For the case: $B = 1000$ gauss and $v_\perp = 10^5$ cm/s, the cyclotron radius R is of the order $\sim 10^{-5}$ cm = 1000 Å. Note: The radius is inversely proportional to B. Thus, as

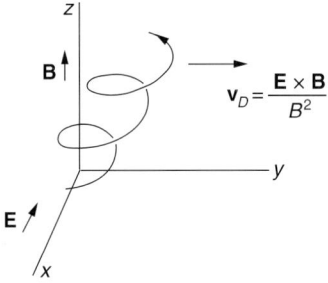

Fig. 26.2 A charged particle spirals around a slanted axis under the action of the electric **E** and magnetic **B** fields, perpendicular to each other.

the magnetic field gets greater, the electron spirals around more rapidly in smaller circles.

The magnetic field **B** does not affect the kinetic energy of the electron. The cyclotron motion describes then an *orbit of constant energy*. This feature is preserved in quantum mechanics. This fact can be used to explore the Fermi (constant-energy) surface for the conduction electrons in a metal (see Chapter 27).

The motion of an electron in static, uniform electric and magnetic fields is found to be similar to the case we have just discussed. First, if the electric field **E** is applied along the direction of **B**, the motion perpendicular to **B** is not affected so that the electron spirals around the magnetic field lines. The motion along **B** is subjected to a uniform acceleration equal to $-e\mathbf{E}/m$.

Let us now turn to the second and more interesting situation in which electric and magnetic fields are perpendicular to each other. Let us introduce $\mathbf{v}_D$ defined by

$$\mathbf{v}_D = \frac{\mathbf{E} \times \mathbf{B}}{B^2} \tag{26.26}$$

which is constant in time and has the dimension of velocity. Let us decompose the velocity **v** into two parts:

$$\mathbf{v} = \mathbf{v}' + \mathbf{v}_D \tag{26.27}$$

Substituting this into the equation of motion:

$$m\frac{d\mathbf{v}}{dt} = \text{Lorentz force} = q(\mathbf{E} + \mathbf{v} \times \mathbf{B}) \tag{26.28}$$

we obtain

$$\text{lhs} = m\frac{d}{dt}(\mathbf{v}' + \mathbf{v}_D) = m\frac{d\mathbf{v}'}{dt} \tag{26.29}$$

$$\text{rhs} = q(\mathbf{E} + \mathbf{v}' \times \mathbf{B} + \mathbf{v}_D \times \mathbf{B})$$

$$= q\left[\mathbf{E} + \frac{(\mathbf{E} \times \mathbf{B}) \times \mathbf{B}}{B^2}\right] + q(\mathbf{v}' \times \mathbf{B}) = q(\mathbf{v}' \times \mathbf{B}) \tag{26.30}$$

or

$$m\frac{d\mathbf{v}'}{dt} = q(\mathbf{v}' \times \mathbf{B}) \tag{26.31}$$

which has same form as (26.20). The motion can then be regarded as the superposition of the motion in a uniform magnetic field and a drift of the cyclotron orbit with the constant velocity $\mathbf{v}_D$ as given in (26.26). Such a motion is indicated in Figure 26.2.

The *drift velocity* $\mathbf{v}_D$ in (26.26) is perpendicular to both $\mathbf{E}$ and $\mathbf{B}$. This implies that the weak electric field will induce a macroscopic current $\mathbf{j}$ in the direction perpendicular to both $\mathbf{E}$ and $\mathbf{B}$:

$$\mathbf{j} = qn\frac{(\mathbf{E} \times \mathbf{B})}{B^2} \tag{26.32}$$

where n is the number density of the electrons. This current is called the *Hall current*. We note that drift velocity $\mathbf{v}_D$ is independent of charge and mass. This turns out to be a very important property. The measurement of the Hall effect gives information about the type of the charge carrier (electron or hole) and the number density of carriers, which will be discussed in Section 26.6.

26.4
Electromagnetic Potentials

The Lorentz force is one of the fundamental forces whose nature is most firmly understood. It is also one of the most important forces acting on electrically charged particles. The *Lorentz force* acting on a particle of charge q moving with velocity $\dot{\mathbf{r}}$ is given by

$$\mathbf{F} = q(\mathbf{E} + \dot{\mathbf{r}} \times \mathbf{B}) \tag{26.33}$$

Since the Lorentz force depends on the velocity, it is not a conservative force. However, it can be generated from a *velocity-dependent potential* U such that the correct equation of motion is obtained in the standard form with the Lagrangian defined by $L \equiv T - U$. This will be shown in the present section.

It is known in electromagnetic theory that the electric and magnetic fields $(\mathbf{E}, \mathbf{B})$ can be derived from scalar and vector potential fields $(\phi, \mathbf{A})$ through the relations:

$$\mathbf{E} = -\nabla\phi(\mathbf{r}, t) - \frac{\partial}{\partial t}\mathbf{A}(\mathbf{r}, t); \quad E_x = -\frac{\partial\phi}{\partial x} - \frac{\partial A_x}{\partial t}, \ldots \tag{26.34}$$

$$\mathbf{B} = \nabla \times \mathbf{A}; \quad B_x = \frac{\partial A_z}{\partial y} - \frac{\partial A_y}{\partial z}, \ldots \tag{26.35}$$

Let us take the function

$$U(x, y, z, \dot{x}, \dot{y}, \dot{z}) \equiv q\phi(\mathbf{r}, t) - q\dot{\mathbf{r}} \cdot \mathbf{A}(\mathbf{r}, t)$$
$$= q\phi(\mathbf{r}, t) - q(\dot{x}A_x + \dot{y}A_y + \dot{z}A_z) \tag{26.36}$$

Differentiating it with respect to x and $\dot{x}$, we obtain

$$-\frac{\partial U}{\partial x} = -q\frac{\partial \phi}{\partial x} + q\left(\dot{x}\frac{\partial A_x}{\partial x} + \dot{y}\frac{\partial A_y}{\partial x} + \dot{z}\frac{\partial A_z}{\partial x}\right) \quad (26.37)$$

$$\frac{\partial U}{\partial \dot{x}} = -qA_x$$

After the time differentiation we get

$$\frac{d}{dt}\left(\frac{\partial U}{\partial \dot{x}}\right) = -q\frac{d}{dt}A_x(x, y, z, t)$$

$$= -q\left(\frac{\partial A_x}{\partial t} + \frac{\partial A_x}{\partial x}\dot{x} + \frac{\partial A_x}{\partial y}\dot{y} + \frac{\partial A_x}{\partial z}\dot{z}\right) \quad (26.38)$$

By adding (26.37) and (26.38) together, we obtain

$$-\frac{\partial U}{\partial x} + \frac{d}{dt}\left(\frac{\partial U}{\partial \dot{x}}\right)$$

$$= q\left(-\frac{\partial \phi}{\partial x} - \frac{\partial A_x}{\partial t}\right) + q\left[\dot{y}\left(\frac{\partial A_y}{\partial x} - \frac{\partial A_x}{\partial y}\right) - \dot{z}\left(\frac{\partial A_x}{\partial z} - \frac{\partial A_z}{\partial x}\right)\right]$$

$$= qE_x + q(\dot{y}B_z - \dot{z}B_y) = q(\mathbf{E} + \dot{\mathbf{r}} \times \mathbf{B})_x \quad (26.39)$$

The quantity in the last member is just equal to the x-component of the Lorentz force in (26.33).

Let us now define the *generalized Lagrangian function L* by

$$L \equiv T - U$$

$$= \frac{m}{2}(\dot{x}^2 + \dot{y}^2 + \dot{z}^2) + q(\dot{x}A_x + \dot{y}A_y + \dot{z}A_z) - q\phi$$

$$= \frac{m}{2}\dot{\mathbf{r}}^2 + q\dot{\mathbf{r}} \cdot \mathbf{A} - q\phi \quad (26.40)$$

If we apply Lagrange's equation of the standard form

$$\frac{d}{dt}\frac{\partial L}{\partial \dot{x}} - \frac{\partial L}{\partial x} = 0 \quad (26.41)$$

we then obtain

$$0 = \frac{d}{dt}(m\dot{x}) - \frac{d}{dt}\frac{\partial U}{\partial \dot{x}} + \frac{\partial U}{\partial x} = m\ddot{x} - q(\mathbf{E} + \dot{\mathbf{r}} \times \mathbf{B})_x$$

which is in agreement with Newton's equation of motion.

Let us now define the *canonical momentum* (p_x, p_y, p_z) by

$$p_x \equiv \frac{\partial L}{\partial \dot{x}}, \quad p_y \equiv \frac{\partial L}{\partial \dot{y}}, \quad p_z \equiv \frac{\partial L}{\partial \dot{z}} \quad (26.42)$$

or in vector notation

$$\mathbf{p} = m\dot{\mathbf{r}} + q\mathbf{A} \tag{26.43}$$

Notice that the canonical momenta are distinct from the linear momenta constructed by the rule: mass × velocities.

Let us now introduce the Hamiltonian H in the standard manner. By expressing,

$$H = \dot{x} p_x + \dot{y} p_y + \dot{z} p_z - L \equiv \dot{\mathbf{r}} \cdot \mathbf{p} - L \tag{26.44}$$

in terms of the canonical variables (x, p_x, y, p_y, z, p_z), we obtain

$$H = \frac{1}{2m}\left[(p_x - qA_x)^2 + (p_y - qA_y)^2 + (p_z - qA_z)^2\right] + q\phi$$

$$= \frac{1}{2m}(\mathbf{p} - q\mathbf{A})^2 + q\phi \tag{26.45}$$

Hamilton's equations of motion for $(x, p_x, \ldots)$ are

$$\dot{x} = \frac{\partial H}{\partial p_x}, \quad \dot{y} = \frac{\partial H}{\partial p_y}, \ldots$$

$$\dot{p}_x = -\frac{\partial H}{\partial x}, \quad \dot{p}_y = -\frac{\partial H}{\partial y}, \ldots \tag{26.46}$$

Using the explicit form, (26.45), for the Hamiltonian H, we obtain for the first set of (26.46),

$$\dot{x} = \frac{\partial H}{\partial p_x} = \frac{1}{m}(p_x - qA_x), \quad \dot{y} = \frac{\partial H}{\partial p_y} = \frac{1}{m}(p_y - qA_y), \ldots$$

or

$$\dot{\mathbf{r}} = \frac{1}{m}(\mathbf{p} - q\mathbf{A}) \tag{26.47}$$

which is equivalent to (26.43). From the second set of (26.46), we obtain

$$\dot{p}_x \equiv \frac{d}{dt}[m\dot{x} + qA_x(x, y, z, t)]$$

$$= -\frac{\partial H}{\partial x} = +\frac{q}{m}(\mathbf{p} - q\mathbf{A}) \cdot \frac{\partial \mathbf{A}}{\partial x} - q\frac{\partial \phi}{\partial x} \tag{26.48}$$

The reader may verify (Problem 26.4.1) that this equation is equivalent to Newton's equation of motion:

$$m\ddot{x} = q(\mathbf{E} + \dot{\mathbf{r}} \times \mathbf{B})_x \tag{26.49}$$

Problem 26.4.1

Demonstrate the equivalence between (26.48) and (26.49).

Problem 26.4.2

Derive the Lagrangian L as a function of cylindrical polar coordinates and velocities $(\rho, \phi, z, \dot{\rho}, \dot{\phi}, \dot{z})$.

HINT: See Sections 1.8 and 6.3 for these coordinates.

26.5
The Landau States and Energies

We have so far discussed the motion of an electron using classical mechanics. Most of the qualitative features hold also in quantum mechanics. The most important quantum effect is the *quantization of the cyclotron motion*. Let us calculate the energy levels of an electron in a constant magnetic field **B**. We choose

$$(A_x, A_y, A_z) = (0, Bx, 0) \tag{26.50}$$

which yields a constant field **B** in the z-direction as can be verified from $\mathbf{B} = \nabla \times \mathbf{A}$. The Hamiltonian H then is given by

$$H = \frac{1}{2m}|\mathbf{p} + e\mathbf{A}|^2 = \frac{1}{2m}\left[p_x^2 + \left(p_y + eBx\right)^2 + p_z^2\right] \tag{26.51}$$

The Schrödinger eigenvalue equation can now be written down as:

$$-\frac{\hbar^2}{2m}\left[\frac{\partial^2}{\partial x^2} + \left(\frac{\partial}{\partial y} + \frac{ieB}{\hbar}x\right)^2 + \frac{\partial^2}{\partial z^2}\right]\psi = E\psi \tag{26.52}$$

Since the Hamiltonian H contains neither y nor z explicitly, we try a wavefunction ψ of the form

$$\psi(x, y, z) = e^{-i(k_y y + k_z z)}\varphi(x) \tag{26.53}$$

Substituting this expression into (26.52) yields the following equation for $\varphi(x)$:

$$\left[-\frac{\hbar^2}{2m}\frac{d^2}{dx^2} + \frac{1}{2}m\omega_0^2\left(x - \frac{\hbar k_y}{eB}\right)^2\right]\varphi(x) = E_1\varphi(x) \tag{26.54}$$

with

$$E_1 \equiv E - \frac{\hbar^2 k_z^2}{2m}. \tag{26.55}$$

Equation (26.54) is the wave equation for a harmonic oscillator with the angular frequency $\omega_c \equiv eB/m$ and the center of oscillation displaced from the origin by

(26.55) is the wavefunction for a harmonic oscillator with the angular frequency $\omega_c \equiv eB/m$ and the center of oscillation displaced from the origin by

$$X = \frac{\hbar k_y}{eB} \tag{26.56}$$

The energy eigenvalues, therefore, are given by

$$E_1 = \left(n + \frac{1}{2}\right)\hbar\omega_c, \quad n = 0, 1, 2, \ldots \tag{26.57}$$

Combining this with (26.55) we obtain

$$E = \left(n + \frac{1}{2}\right)\hbar\omega_c + \frac{\hbar^2 k_z^2}{2m} \tag{26.58}$$

These energy eigenvalues are called *Landau levels*. The corresponding quantum states, called the *Landau states*, are characterized by the quantum numbers (n, k_y, k_z). We note that the energies do not depend on k_y. The Landau states are quite different from the momentum eigenstates. This has significant consequences on magnetization and galvanomagnetic phenomena. The electron in a Landau state may be pictured as in a rotation with the angular frequency ω_c around the magnetic field plus a translation along the field. If a radiation having a frequency ω_c is applied, the electron may jump up from one Landau state to another by absorption of a photon of the energy equal to $\hbar\omega_c$. This generates a phenomenon of *cyclotron resonance*.

26.6
The Degeneracy of the Landau Levels

The Landau levels are highly degenerate. The *degeneracy*, that is, the number of electrons that can occupy each Landau level is

$$\frac{eBA}{2\pi\hbar} \tag{26.59}$$

where A is the sample area perpendicular to the magnetic field **B**. This is showed in this section.

Let us consider a particle moving along a straight line of length L. In classical mechanics, a dynamical state of the system, (x, p), can be represented by a point in the phase space. In quantum mechanics, the dynamical state cannot be represented by a point because of *Heisenberg's uncertainty principle*. The set of momentum eigenstates,

$$p_k = 2\pi\hbar k/L, \quad k = 0, \pm 1, \pm 2, \ldots \tag{26.60}$$

however, may be represented by the set of quantum cells in phase space as shown in Figure 26.3.

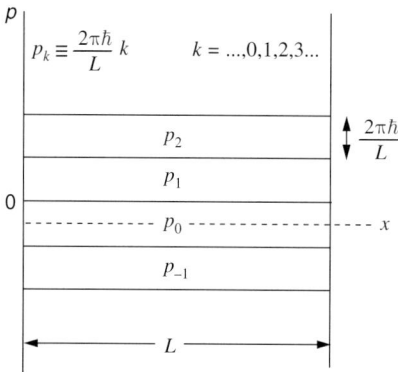

Fig. 26.3 The momentum states $\{p_k\}$ for linear motion are represented by rectangular cells of equal area $(2\pi\hbar)$ in phase space.

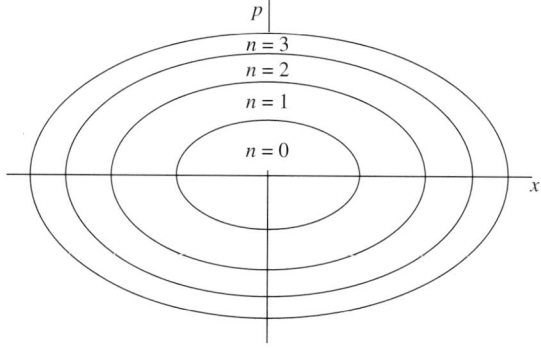

Fig. 26.4 The quantum-mechanical eigenstates for a *simple harmonic oscillator* are represented by the quantum cells of phase-space area $2\pi\hbar$.

We note that
1. each cell extends from 0 to L in position, implying that a particle can be found with an equal probability everywhere along the line from 0 to L, and
2. the area of each cell equals $2\pi\hbar = h$.

Such a representation of the quantum states in phase space is not restricted to this special case. The quantum states for a simple harmonic oscillator can be represented by the elliptical shells as shown in Figure 26.4. Each cell has an area equal to $2\pi\hbar$.

Let us now consider the case with a constant magnetic field **B**. The Hamiltonian H is given by (26.51). We can show (Problem 26.6.1) that one set of Hamilton's equations gives the correct Newton's equations of motion and the second set yields

$$m\dot{\mathbf{r}} \equiv m\mathbf{v} = \mathbf{p} + e\mathbf{A} \equiv \mathbf{\Pi} \tag{26.61}$$

Note that this *kinetic momentum* **Π** is not equal to the *canonical momentum* **p**. The component Π_z is equal to p_z but $(\Pi_x, \Pi_y) \neq (p_x, p_y)$. We now go over to quantum mechanics. Using the *quantum condition*:

$$[x, p_x] = [y, p_y] = i\hbar$$

$$[x, y] = [x, p_y] = [y, p_x] = \cdots = 0 \tag{26.62}$$

we obtain (Problem 26.6.2)

$$[\Pi_x, \Pi_y] = -ie\hbar B$$

$$[\Pi_z, \Pi_x] = [\Pi_y, \Pi_z] = 0 \tag{26.63}$$

The noncommutativity of Π_x and Π_y means that the x- and y-motion is correlated, making the motion one-dimensional (1D) harmonic-oscillator-like instead of 2D oscillator-like.

In fact, we can illustrate this behavior in more detail as follows: We introduce new canonical variables (Q, P):

$$\frac{1}{\sqrt{m}}\Pi_x \equiv P, \quad \frac{1}{eB}\sqrt{m}\Pi_y \equiv Q \tag{26.64}$$

We can then rewrite part of the Hamiltonian as

$$H_\perp \equiv \frac{1}{2m}\left(\Pi_x^2 + \Pi_y^2\right) \equiv \frac{1}{2m}\Pi_\perp^2 = \frac{1}{2}P^2 + \frac{1}{2}\omega_c^2 Q^2 \tag{26.65}$$

Using (26.63) and (26.64), we obtain

$$[Q, P] = i\hbar \tag{26.66}$$

The last two equations mean that the energy eigenvalues $E_\perp$ are the same as those of the harmonic oscillator with the cyclotron frequency ω_c. Hence, we obtain

$$E_\perp = (N_L + 1/2)\hbar\omega_c, \quad N_L = 0, 1, 2, \ldots \tag{26.67}$$

in agreement with (26.58).

After simple calculations, we can show (Problem 26.6.3) that

$$dx\, d\Pi_x\, dy\, d\Pi_y = dx\, dp_x\, dy\, dp_y \tag{26.68}$$

We can now represent the circulational part of the quantum states by a small quasiphase-space cell of the volume $(2\pi\hbar)^{-2} dx\, d\Pi_x\, dy\, d\Pi_y$. The Hamiltonian $H_\perp$ in (26.65) does not depend on the position (x, y). Assuming large normalization lengths (L_1, L_2), $A = L_1 L_2$, we can then represent the rotational Landau states by the concentric shells of the phase-space volume

$$(2\pi)\Pi_\perp \Delta\Pi_\perp \cdot L_1 L_2 (2\pi\hbar)^{-2} = A(2\pi\hbar)^{-1}\omega_c m^* \tag{26.69}$$

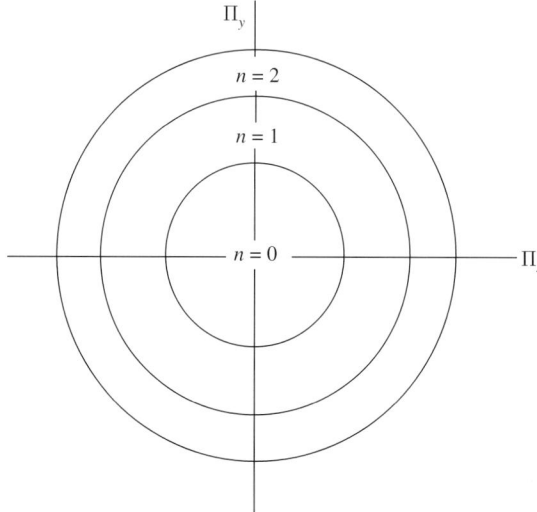

Fig. 26.5 The circulational parts of the Landau states are represented by the circular shells in the (Π_x, Π_x) space.

with the energy separation

$$\hbar\omega_c = \Delta\left(\frac{\Pi_\perp^2}{2m^*}\right) = \frac{\Pi_\perp \Delta\Pi_\perp}{m^*} \tag{26.70}$$

as shown in Figure 26.5.

The number of quantum states, dN, between the neighboring orbits is given by the phase-space volume over $(2\pi\hbar)^2$:

$$dN = \frac{L_1 L_2}{(2\pi\hbar)^2}(2\pi)\Pi_\perp \Delta\Pi_\perp$$

Using (26.70), we obtain

$$dN = \frac{eBA}{2\pi\hbar}, \quad A = L_1 L_2 \tag{26.71}$$

establishing (26.59).

In summary, as the field strength is raised the LL separation $\hbar\omega_c$ increases. Hence, the area between the neighboring orbits $2\pi\Pi_\perp\Delta\Pi_\perp$ increases in proportion to $\hbar\omega_c$ and the degeneracy dN becomes greater. This behavior is shown in Figure 26.6.

Problem 26.6.1

Obtain Hamilton's equations of motion, using the Hamiltonian given in (26.51). Show that these equations are equivalent to Newton's equations of motion.

Problem 26.6.2

Verify (26.63).

Problem 26.6.3

Verify (26.71).

26.7
Landau Diamagnetism

The electron always circulates around the magnetic flux so as to reduce the magnetic field. This is called the *Motional diamagnetism*. If we calculate this effect classically by considering the system confined to a closed volume, we obtain *zero* magnetic moments. This is known as *van Leeuwen's theorem*. We first demonstrate this theorem.

Let us take a system of free classical electrons confined in volume V. The partition function per electron is

$$Z(B) = \frac{1}{(2\pi\hbar)^3} \int_V d^3r \int d^3p \exp\left[-\frac{|\mathbf{p} + e\mathbf{A}|^2}{2mk_B T}\right] \tag{26.72}$$

We introduce *kinetic momentum* $\mathbf{\Pi} = (\Pi_x, \Pi_y, \Pi_z)$. After simple calculations we obtain

$$dx\,dy\,dz\,dp_x\,dp_y\,dp_z = dx\,dy\,dz\,d\Pi_x\,d\Pi_y\,d\Pi_z \tag{26.73}$$

Using the last two equations, we see that $Z(B)$ is equal to the classical partition function with no field;

$$Z(B) = Z(B=0) \tag{26.74}$$

This result depends on the container wall. In Figure 26.7 we show that counterclockwise currents due to the completed electron circulations are canceled out by

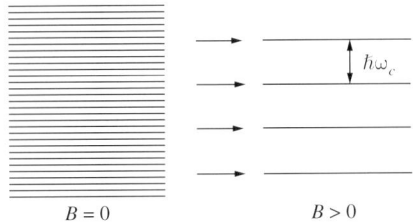

Fig. 26.6 The equidistant energy levels for 2D free electrons are bundled into the Landau levels.

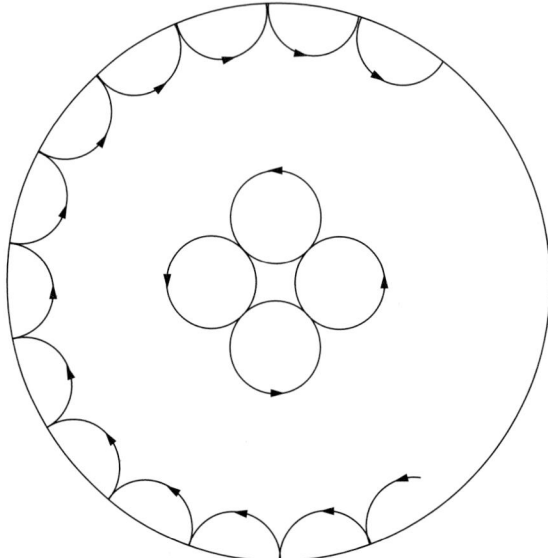

Fig. 26.7 The diamagnetic currents due to the electron circulations inside are canceled out by the currents generated near the container wall.

the clockwise currents due to the incomplete circulations near the wall. Hence, there are no magnetic moments.

In 1930, Landau [2] showed that the quantum treatment of the electron circulation yields a diamagnetic moment. We shall show this below.

Electrons obey the Fermi–Dirac statistics. Considering a system of free electrons, we define the *free energy F* as (Problems 26.7.1 and 26.7.2)

$$F = N\mu - 2k_B T \sum_i \ln\left[1 + e^{(\mu - E_i)/(k_B T)}\right] \tag{26.75}$$

where the factor 2 arises from the *spin degeneracy*. The *chemical potential* μ is determined from the condition:

$$\frac{\partial F}{\partial \mu} = 0 \tag{26.76}$$

The total magnetic moment M for the system can be found from

$$M = -\frac{\partial F}{\partial B} \tag{26.77}$$

Equation (26.76) is equivalent to the usual condition that the total number of the electrons, N, can be obtained in terms of the Fermi distribution function

$$f(E) \equiv \frac{1}{\exp[(E - \mu)/k_B T] + 1} \tag{26.78}$$

from

$$N = 2\sum_i f(E_i) \tag{26.79}$$

The Landau energy E_i, is characterized by the Landau oscillator quantum number N_L and the z-component momentum p_z. The energy E becomes continuous in the bulk limit. Let us introduce the density of states $\mathcal{N}(E) \equiv dW/dN$ such that

$$\mathcal{N}(E)dE = \text{Number of states having an energy between } E \text{ and } E + dE \tag{26.80}$$

We now write (26.75) in the form

$$F = N\mu - 2k_B T \int_0^\infty dE \frac{dW}{dE} \ln\left[1 + e^{(\mu - E)/(k_B T)}\right] \tag{26.81}$$

The *statistical weight* (number) W is the total number of states having energies less than

$$E = (N_L + 1/2)\hbar\omega_c + \frac{p_z^2}{2m} \tag{26.82}$$

Conversely, the allowed values of p_z are distributed over the range in which $|p_z|$ does not exceed

$$\{2m[E - (N_L + 1/2)\hbar\omega_c]\}^{1/2} \tag{26.83}$$

see Figure 26.8. For a fixed pair (E, N_L) the increment in the weight, dW, is given by

$$dW = eB\frac{L_1 L_2}{(2\pi\hbar)}\int dp_z \frac{L_3}{(2\pi\hbar)}$$

$$= VeB\frac{1}{(2\pi\hbar)^2} 2\{2m[E - (N_L + 1/2)\hbar\omega_c]\}^{1/2} \tag{26.84}$$

where we used (26.71); $V \equiv L_1 L_2 L_3$ is the volume of the container. After summing (26.84) with respect to N_L, we obtain

$$W(E) = A\frac{(\hbar\omega_c)^{3/2}}{(2)^{1/2}\pi} 2\sum_{N_L=0}^\infty \sqrt{\varepsilon - (2N_L + 1)\pi} \tag{26.85}$$

$$A \equiv V\frac{(2\pi m)^{3/2}}{(2\pi\hbar)^3}, \quad \varepsilon \equiv \frac{2\pi E}{\hbar\omega_c}$$

We assume high Fermi degeneracy such that

$$\mu \simeq \varepsilon_F \gg \hbar\omega_c \quad \text{for a metal} \tag{26.86}$$

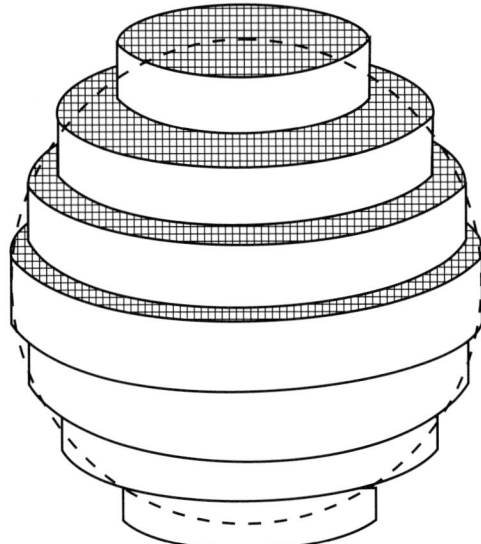

Fig. 26.8 A Fermi surface for free electrons subjected to a constant magnetic field.

The sum over N_L in (26.85) converges slowly. We use Poisson's sum formula [3, 4] and obtain, after mathematical manipulations,

$$W(E) = W_0 + W_L + W_{osc} \tag{26.87}$$

$$W_0 = A \frac{4}{3\sqrt{\pi}} E^{3/2} \tag{26.88}$$

$$W_L = -A \frac{1}{24\sqrt{\pi}} \frac{(\hbar\omega_c)^2}{E^{1/2}} \tag{26.89}$$

$$W_{osc} = A \frac{(\hbar\omega_c)^{3/2}}{(2)^{1/2}\pi^{3/2} E^{1/2}} \sum_{\nu=1}^{\infty} \frac{(-1)^\nu}{\nu^{3/2}} \sin\left(\frac{2\pi\nu E}{\hbar\omega_c} - \frac{\pi}{4}\right) \tag{26.90}$$

The detailed steps leading to (26.87) through (26.90) are given in Appendix C.

The term W_0, which is independent of B, gives the weight equal to that for a free-electron system with no field. The term W_L is negative (diamagnetic), and can generate a diamagnetic moment.

We start with (26.81), integrate by parts, and obtain

$$F = N\mu - 2\int_0^\infty dE\, W(E) f(E)$$

$$= N\mu + 2\int_0^\infty dE \frac{df}{dE} \int_0^\infty dE'\, W(E') \tag{26.91}$$

The term $-df/dE$, which can be expressed as

$$-\frac{df}{dE} = \frac{1}{4k_B T} \text{sech}^2\left(\frac{E-\mu}{2k_B T}\right) \tag{26.92}$$

has a sharp peak near $E = \mu$ if $k_B T \ll \mu$, and

$$\int_0^\infty dE \frac{df}{dE} = -1 \tag{26.93}$$

For a smoothly changing integrand in the last member of (26.91) $-df/dE$ can be regarded as a *Dirac delta function*:

$$-\frac{df}{dE} = \delta(E-\mu) \tag{26.94}$$

Using this property, (26.89) and (26.76), we obtain (Problem 26.7.4)

$$F_L = A \frac{1}{6\sqrt{\pi}} (\hbar \omega_c)^2 \varepsilon_F^{1/2} \tag{26.95}$$

Here, we set $\mu = \varepsilon_F$. This is justified since the corrections to $\mu(B, T)$ start with a B^4 term and with a T^2 term. Using (26.95) and (26.77), we obtain

$$\boxed{\chi_L = \frac{1}{V}\frac{\partial M_L}{\partial B} = -\frac{1}{V}\frac{\partial^2 F_L}{\partial B^2} = -\frac{n\mu_B^2}{2\varepsilon_F}} \tag{26.96}$$

where n is the electron density.

Comparing this result with (26.19) we observe that *Landau diamagnetism* is one third (1/3) of the *Pauli paramagnetism* in magnitude, but the calculations in this section are done with the assumption of the free-electron model. If the effective-mass (m^*) approximation is used, formula (26.96) is corrected by the factor $(m^*/m)^2$, as we can see from (26.95). Hence, the diamagnetic susceptibility for a metal is

$$\chi_L^{\text{metal}} = (m^*/m)^2 \chi_L \tag{26.97}$$

We note that the Landau susceptibility is spin independent. For a metal having a small effective mass χ_L^{metal} can be greater in magnitude than the Pauli paramagnetic susceptibility χ_P. Then, the total susceptibility expressed as

$$\chi = \chi_P + \chi_L^{\text{metal}} \tag{26.98}$$

can be negative (diamagnetic). This is observed for GaAs ($m^* = 0.07m$). The oscillatory term W_{osc} in (26.90) yields the *de Haas–van Alphen (dHVA) oscillation*.

The most frequently used means to prove the Fermi surface in conductors is to measure and analyze the dHVA data. The full theory of the dHVA oscillation turns out to require going beyond the free-electron model. The period of the oscillations $\varepsilon_F/\hbar\omega_c$ contains information about the cyclotron mass m^* but the envelope of the oscillation is controlled by the magnetotransport mass M^* of the electron dressed with two magnetic flux quanta. The interested reader is advised to examine the chapter on the dHVA oscillation in Fujita and Ito's book [5].

Problem 26.7.1

The *grand partition function* Ξ is defined by

$$\Xi \equiv \text{TR}\{\exp(\alpha N - \beta H)\} \equiv \sum_{N=0}^{\infty} e^{\alpha N} \text{Tr}_N \{\exp(\alpha N - \beta H_N)\}$$

where H_N is the Hamiltonian of the N-particle system and Tr stands for the *trace* (diagonal sum). We assume that the internal energy E, the number density n and the entropy S are given by

$$E = \langle H \rangle \equiv \text{TR}\{H \exp(\alpha N - \beta H)\}/\Xi$$

$$n = \frac{\langle N \rangle}{V} = \text{TR}\{N \exp(\alpha N - \beta H)\}/V\Xi$$

$$S = k_B (\ln \Xi + \beta E - \alpha \langle N \rangle)$$

Show that

$$F \equiv E - TS = \mu \langle N \rangle - k_B T \ln \Xi$$

where $\mu \equiv k_B T \alpha$ is the Gibbs free energy per particle (chemical potential): $G \equiv E - TS + PV = \mu \langle N \rangle$.

Problem 26.7.2

1. Evaluate the grand partition function Ξ for a free-electron system characterized by

$$H = \sum_{j=1}^{N} \frac{p_j^2}{2m}$$

2. Show that

$$n = \frac{1}{V} \frac{\partial}{\partial \alpha} \ln \Xi = \frac{2}{V} \sum_p f(\varepsilon_p)$$

$$f(\varepsilon) = \frac{1}{e^{\beta(\varepsilon-\mu)} + 1}, \quad \varepsilon_p = \frac{p^2}{2m}$$

$$E = -\frac{\partial}{\partial \beta} \ln \Xi = 2 \sum_p \varepsilon_p f(\varepsilon_p)$$

where 2 is the spin-degeneracy factor.

Problem 26.7.3

Using the free energy F in (26.75), obtain (26.78) and (26.79).

Problem 26.7.4

Verify (26.95).

References

1 Pauli, W. (1927) *Z. Phys.*, **41**, 81.
2 Landau, L.D. (1930) *Z. Phys.*, **64**, 629.
3 de Haas, W.J. and van Alphen, P.M. (1930) *Leiden Comm.*, **208d**, 212a.
4 de Haas, W.J. and van Alphen, P.M. (1932) *Leiden Comm.*, **220d** (1932).
5 Fujita, S. and Ito, K. (2007) *Quantum Theory of Conducting Matter*, Springer, New York, pp. 133–149.

27
Theory of Variations

The Euler–Lagrange equation, Fermat's principle, Hamilton's principle, and Lagrange's field equation are discussed in this chapter.

27.1
The Euler–Lagrange Equation

Let us consider two points 1 and 2 in a plane. What is the shortest path connecting the two points? We know the answer: a straight line passing through the two points, see Figure 27.1. The method by which we treat this problem is useful for treating other problems.

Any curve joining the two points (x_1, y_1) and (x_2, y_2) can be represented by an equation

$$y = y(x) \tag{27.1}$$

such that the function $y(x)$ satisfies the boundary conditions:

$$y(x_1) = y_1, \quad y(x_2) = y_2 \tag{27.2}$$

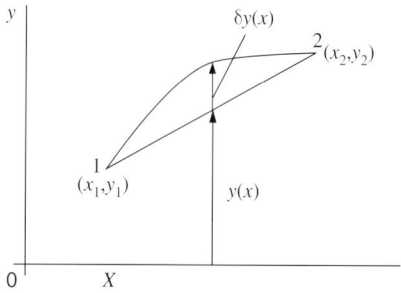

Fig. 27.1 The shortest path (straight line) between the points 1 and 2. A path close to it is also shown.

Mathematical Physics. Shigeji Fujita and Salvador V. Godoy
Copyright © 2010 WILEY-VCH Verlag GmbH & Co. KGaA, Weinheim
ISBN: 978-3-527-40808-5

We consider two neighboring points on this curve. The distance dl between them is given by

$$dl = (dx^2 + dy^2)^{1/2} = (1 + y'^2)^{1/2} dx \tag{27.3}$$

where $y' \equiv dy/dx$. The total length l of the curve is

$$l = \int_{x_1}^{x_2} (1 + y'^2)^{1/2} dx \tag{27.4}$$

The problem is then to find the function $y(x)$, which makes the length l a minimum.

We first consider a more general problem of finding the extremum (maximum or minimum) values of an integral of the form

$$I = \int_{x_1}^{x_2} f(y, y') dx \tag{27.5}$$

where $f(y, y')$ is a function of y and its first derivative y'. Consider a small variation $\delta y(x)$ in the function $y(x)$ subject to the condition that the values of y at the endpoints are unchanged so that

$$\delta y(x_1) = 0, \qquad \delta y(x_2) = 0 \tag{27.6}$$

The variation in $f(y, y')$ is, to first order,

$$\delta f = \frac{\partial f}{\partial y} \delta y + \frac{\partial f}{\partial y'} \delta y' \tag{27.7}$$

where

$$\delta y' = \frac{d}{dx} \delta y \tag{27.8}$$

The variation of the integral I must vanish:

$$\delta I = \int_{x_1}^{x_2} \left[\frac{\partial f}{\partial y} \delta y + \frac{\partial f}{\partial y'} \frac{d}{dx} \delta y \right] dx = 0 \tag{27.9}$$

for the extremum $f(x)$.

In the second term, we may integrate by parts, and the integrated term vanishes:

$$\left[\frac{\partial f}{\partial y'} \delta y \right]_{x_1}^{x_2} = 0$$

because of (27.6). Hence, we obtain

$$\delta I = \int_{x_1}^{x_2} dx \left[\frac{\partial f}{\partial y} - \frac{d}{dx} \left(\frac{\partial f}{\partial y'} \right) \right] \delta y = 0 \tag{27.10}$$

For an *arbitrary* small variation $\delta y(x)$, the bracketed quantity must then vanish:

$$\boxed{\frac{\partial f}{\partial y} - \frac{d}{dx}\left(\frac{\partial f}{\partial y'}\right) = 0} \tag{27.11}$$

This is known as the *Euler–Lagrange equation*. It is in general a second-order differential equation, whose solutions contain two constants that may be determined by the boundary conditions at x_1 and x_2.

We now go back to the shortest-path problem. We choose

$$f = (1 + y'^2)^{1/2} \tag{27.12}$$

We then have

$$\frac{\partial f}{\partial y} = 0, \quad \frac{\partial f}{\partial y'} = \frac{y'}{(1 + y'^2)^{1/2}} \tag{27.13}$$

Thus, the Euler–Lagrange equation is

$$\frac{d}{dx}\left[\frac{y'}{(1 + y'^2)^{1/2}}\right] = 0 \tag{27.14}$$

whose solution is

$$y' = \text{constant} \tag{27.15}$$

This solution has a constant slope, and represents a straight line path between the two points (x_1, y_1) and (x_2, y_2).

27.2
Fermat's Principle

In ray optics the light bends at the water surface, as shown in Figure 27.2. Snell's law states that

$$n_1 \sin \theta_1 = n_2 \sin \theta_2 \tag{27.16}$$

where n_1 and n_2 are the refractive indices for the water and the air, respectively, which have an inequality:

$$n_1 > n_2 \tag{27.17}$$

We derive Snell's law by using *Fermat's principle* that *a light ray follows the path of minimum time*.

The time dt for light to travel a small distance dl is

$$dt = dl/v \tag{27.18}$$

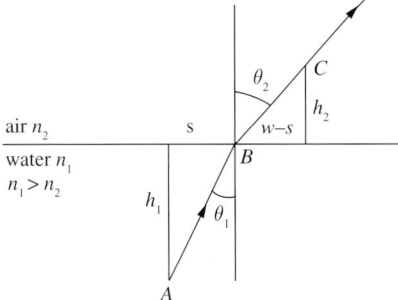

Fig. 27.2 The light bends at the water surface. Snell's law: $n_1 \sin\theta_1 = n_2 \sin\theta_2$, $n_1 > n_2$.

where v is the light speed. We assume that the speed v in a medium with refractive index n is

$$v = c/n \tag{27.19}$$

where c is the light speed in vacuum.

In the water the light travels on a straight line from A to B, which may be shown as follows. The light speed v_1 in the water is $v_1 = c/n_1$, which is constant. Then, the shortest time t, after integrating (27.18) with respect to dl, is achieved by the shortest distance traveled. That is, the minimum time problem between A and B is reduced to the minimum distance problem. The same condition holds for the air, where the light travels with the speed $v_2 = c/n_2$.

The time t required to travel for $A \to B \to C$ is from Figure 27.2

$$t = \frac{\sqrt{h_1^2 + s^2}}{c/n_1} + \frac{\sqrt{h_2^2 + (w-s)^2}}{c/n_2} \tag{27.20}$$

where h_1, h_2 and w are the distances shown in Figure 29.2. For the actual physical path, the position s must be determined by Fermat's principle. The variation δt in the time due to the variation δs must be stationary:

$$\delta t = \frac{dt}{ds}\delta s = 0 \tag{27.21}$$

Hence, we obtain.

$$\frac{dt}{ds} = \frac{n_1}{c}\frac{s}{\sqrt{h_1^2 + s^2}} - \frac{n_2}{c}\frac{w-s}{\sqrt{h_2^2 + (w-s)^2}} = 0 \tag{27.22}$$

which yields Snell's law in (27.16).

We note that if w is much greater than h_1, h_2, then Snell's law does not apply, and the light cannot pass from the water to the air. A *total reflection* occurs at the water surface.

27.3
Hamilton's Principle

The calculus of variations can be applied to the time domain. It is also easy to generalize for a multivariable function.

Let us take a function f of n variables $q_1, q_2, \ldots, q_n$ and their time derivatives $\dot{q}_1, \dot{q}_2, \ldots, \dot{q}_n$. We take a time integral

$$I = \int_{t_1}^{t_2} dt\, f(q_1, \ldots, q_n, \dot{q}_1, \ldots, \dot{q}_n) \tag{27.23}$$

We consider a small variation $\delta q_i(t)$, $i = 1, 2, \ldots, n$, subject to the condition that the values of $q_i(t)$ at the end-points are unchanged so that

$$\delta q_i(t_1) = \delta q_i(t_2) = 0 \quad i = 1, 2, \ldots, n \tag{27.24}$$

From

$$\delta I = 0 \tag{27.25}$$

we obtain n Euler–Lagrange equations:

$$\boxed{\frac{\partial f}{\partial q_i} - \frac{d}{dt}\left(\frac{\partial f}{\partial \dot{q}_i}\right) = 0, \quad i = 1, 2, \ldots, n} \tag{27.26}$$

Let us now consider a dynamical system with the Lagrangian function $L(q, \dot{q})$, where q is a generalized coordinate and $\dot{q}$ its time derivative. We define the *action integral A* by

$$A \equiv \int_{t_1}^{t_2} dt\, L(q_1, \ldots, q_n, \dot{q}_1, \ldots, \dot{q}_n) \tag{27.27}$$

Hamilton's principle of least action states that the action integral A is stationary (maximum or minimum) under arbitrary variations $(\delta q_1, \delta q_2, \ldots, \delta q_n)$, which vanish at the limits of integration, t_1, and t_2. The Euler–Lagrange equations (27.26) then are precisely the well-known Lagrange's equations of motion:

$$\frac{\partial L}{\partial q_i} - \frac{d}{dt}\left(\frac{\partial L}{\partial \dot{q}_i}\right) = 0, \quad i = 1, 2, \ldots, n \tag{27.28}$$

27.4
Lagrange's Field Equation

Earlier we saw that the transverse displacement $y(x, t) \equiv u(x, t)$ of a stretched string obeys the wave equation, see (5.11):

$$\frac{\partial^2 u(x, t)}{\partial x^2} = \frac{1}{c^2}\frac{\partial^2 u(x, t)}{\partial t^2}, \quad c^2 = \tau/\mu \tag{27.29}$$

For definitions of τ, μ and c, see Section 5.1.

The Lagrangian $\mathcal{L}$ is, from (5.51), given by

$$\mathcal{L} = \int_0^L dx \left[\frac{1}{2}\mu \left(\frac{\partial u}{\partial t}\right)^2 - \frac{1}{2}\tau \left(\frac{\partial u}{\partial x}\right)^2 \right] \tag{27.30}$$

The integrand

$$\frac{1}{2}\mu \left(\frac{\partial u}{\partial t}\right)^2 - \frac{1}{2}\tau \left(\frac{\partial u}{\partial x}\right)^2 \equiv \mathbb{L} \tag{27.31}$$

is called the Lagrangian density.

Let us consider a general case in which the Lagrangian density $\mathbb{L}$ is a functional of u, $\partial u/\partial t$, and $\partial u/\partial x$. We construct an action integral by

$$A = \int_{t_1}^{t_2} dt \int_0^L dx \mathbb{L}\left(u, \frac{\partial u}{\partial t}, \frac{\partial u}{\partial x}\right) \tag{27.32}$$

This is a generalization of (27.23), where the continuous field variables $u(x, t)$ enter instead of n discrete variables $q_1, q_2, \cdots, q_n$. We consider a small field variation

$$\delta u(x, t) \tag{27.33}$$

subject to the condition that the values of u at the end-time points are unchanged so that

$$\delta u(x_1, t_1) = \delta u(x_2, t_2) = 0 \tag{27.34}$$

From Hamilton's principle of least action, we require that

$$\delta A = 0 \tag{27.35}$$

After the variational calculation, we then obtain (Problem 27.4.1).

$$\frac{\partial \mathbb{L}}{\partial u} - \frac{\partial}{\partial t}\left(\frac{\partial \mathbb{L}}{\partial u/\partial t}\right) - \frac{\partial}{\partial x}\left(\frac{\partial \mathbb{L}}{\partial u/\partial x}\right) = 0 \tag{27.36}$$

This is called *Lagrange's field equation*. If we use the Lagrangian density for the stretched string, (27.31), we can then obtain the wave equation given in (27.29). (Problem 27.4.2).

Problem 27.4.1

Obtain Lagrange's field equation (27.36), using (27.32)–(27.35).

Problem 27.4.2

Verify the wave equation (27.29), using (27.31) and (27.36).

28
Second Quantization

The most remarkable fact about a system of fermions is that no more than one fermion can occupy a quantum particle state (Pauli's exclusion principle). For bosons no such restriction applies. That is, any number of bosons can occupy the same state. We shall discuss the second quantization formalism in which creation and annihilation operators associated with each quantum state are used. This formalism is extremely useful in treating many-boson and/or many-fermion systems. Zero-mass bosons such as photons and phonons can be created or annihilated. These dynamical processes can only be described in second quantization.

28.1
Boson Creation and Annihilation Operators

The quantum state for a system of bosons (or fermions) can most conveniently be represented by a set of occupation numbers $\{n'_a\}$, where n'_a are the numbers of bosons (or fermions) occupying the quantum particle states a. This representation is called the *occupation-number representation* or simply the *number representation*. For bosons, the possible values for n'_a are zero, one, two, or any positive integers:

$$n'_a = 0, 1, 2, \ldots \quad \text{for bosons} \tag{28.1}$$

The many-boson state can best be represented by the distribution of particles (balls) in the states (boxes) as shown in Figure 23.2.

Let us introduce operators n_a (without prime) whose eigenvalues are given by $0, 1, 2, \ldots$ Since (28.1) is meant for each and every state a independently, we assume that

$$[n_a, n_b] \equiv n_a n_b - n_b n_a = 0 \tag{28.2}$$

It is convenient to introduce complex dynamic operators η and $\eta^\dagger$ instead of directly dealing with the number operators n. We attach labels $a, b, \ldots$ to the dynamic operators η and $\eta^\dagger$ associated with the states $a, b, \ldots$ and assume that η and $\eta^\dagger$ satisfy the following *Bose commutation rules*:

$$\left[\eta_a, \eta_b^\dagger\right] = \delta_{ab}, \quad [\eta_a, \eta_b] = \left[\eta_a^\dagger, \eta_b^\dagger\right] = 0 \tag{28.3}$$

Mathematical Physics. Shigeji Fujita and Salvador V. Godoy
Copyright © 2010 WILEY-VCH Verlag GmbH & Co. KGaA, Weinheim
ISBN: 978-3-527-40808-5

Let us set

$$\eta_a^\dagger \eta_a \equiv n_a \; (= n_a^\dagger) \tag{28.4}$$

which is Hermitean.

We shall show that n_a has as eigenvalues all non-negative integers. Let n' be an eigenvalue of n (dropping the suffix a) and $|n'\rangle$ an eigenket belonging to it. By definition

$$\langle n'|\eta^\dagger \eta |n'\rangle = n' \langle n'|n'\rangle \tag{28.5}$$

Now, $\langle n'|\eta^\dagger \eta |n'\rangle$ is the squared length of the ket $\eta |n'\rangle$ and hence

$$\langle n'|\eta^\dagger \eta |n'\rangle \geq 0 \tag{28.6}$$

Also, by definition $\langle n'|n'\rangle > 0$; hence from (28.5) and (28.6), we obtain

$$n' \geq 0 \tag{28.7}$$

the case of equality occurring only if

$$\eta |n'\rangle = 0 \tag{28.8}$$

Consider now $[\eta, n] \equiv [\eta, \eta^\dagger \eta]$. We may use the following identities:

$$[A, BC] = B[A, C] + [A, B]C, \quad [AB, C] = A[B, C] + [A, C]B \tag{28.9}$$

and obtain

$$[\eta, \eta^\dagger \eta] = \eta^\dagger [\eta, \eta] + [\eta, \eta^\dagger]\eta = \eta, \quad \text{or} \quad \eta n - n\eta = \eta \tag{28.10}$$

Hence,

$$n\eta |n'\rangle = (\eta n - \eta)|n'\rangle = (n' - 1)\eta |n'\rangle \tag{28.11}$$

Now, if $\eta |n'\rangle \neq 0$, then $\eta |n'\rangle$ is, according to (28.11), an eigenket of n belonging to the eigenvalue $n'-1$. Hence, for nonzero n', $n'-1$ is another eigenvalue. We can repeat the argument and deduce that, if $n'-1 \neq 0$, $n'-2$ is another eigenvalue of n. Continuing in this way, we obtain a series of eigenvalues $n', n'-1, n'-2, \ldots$ that can terminate *only* with the value 0 because of inequality (28.7). By a similar process, we can show from the Hermitean conjugate of (28.10): $n\eta^\dagger - \eta^\dagger n = \eta^\dagger$ that the eigenvalue of n has no upper limit [Problem 28.1.1]. Hence, the eigenvalues of n are non-negative integers: $0, 1, 2, \ldots$ (Q. E. D.)

Let $|\phi_a\rangle$ be a normalized eigenket of n_a belonging to the eigenvalue 0 so that

$$n_a |\phi_a\rangle = \eta_a^\dagger \eta_a |\phi_a\rangle = 0 \tag{28.12}$$

By multiplying all these kets $|\phi_a\rangle$ together, we construct a normalized eigenket:

$$|\Phi_0\rangle \equiv |\phi_a\rangle |\phi_b\rangle \ldots \tag{28.13}$$

which is a simultaneous eigenket of all n belonging to the eigenvalues zero. This ket is called the *vacuum ket*. It has the following property:

$$\eta_a |\Phi_0\rangle = 0 \quad \text{for any } a \tag{28.14}$$

Using the Bose commutation rules (28.3) we obtain a relation (dropping suffix a)

$$\eta(\eta^\dagger)^{n'} - (\eta^\dagger)^{n'}\eta = n'(\eta^\dagger)^{n'-1} \tag{28.15}$$

which may be proved by induction (Problem 28.1.2). Multiply (28.15) by $\eta^\dagger$ from the left and operate the result to $|\Phi_0\rangle$. Using (28.14) we obtain

$$n(\eta^\dagger)^{n'}|\phi\rangle = n'(\eta^\dagger)^{n'}|\phi\rangle \tag{28.16}$$

indicating that $(\eta^\dagger)^{n'}|\phi\rangle$ is an eigenket belonging to the eigenvalue n'. The square length of $(\eta^\dagger)^{n'}|\phi\rangle$ is

$$\langle \phi | \eta^{n'}(\eta^\dagger)^{n'}|\phi\rangle = n' \langle \phi | \eta^{n'-1}(\eta^\dagger)^{n'-1}|\phi\rangle = \cdots = n'! \tag{28.17}$$

We see from (28.11) that $\eta|n'\rangle$ is an eigenket of n belonging to the eigenvalue $n'-1$. Similarly, we can show from $[n, \eta^\dagger] = \eta^\dagger$ that $\eta^\dagger|n'\rangle$ is an eigenket of n belonging to the eigenvalue $n'+1$. Thus, operator η, acting on the number eigenket, annihilates a particle, while operator $\eta^\dagger$ creates a particle. Therefore, η and $\eta^\dagger$ are called *annihilation and creation operators*, respectively. From (28.16) and (28.17) we see that if $n'_1, n'_2, \ldots$ are any non-negative integers,

$$(n'_1! n'_2! \ldots)^{-1/2} \left(\eta_1^\dagger\right)^{n'_1} \left(\eta_2^\dagger\right)^{n'_2} \ldots |\Phi_0\rangle \equiv |n'_1, n'_2, \ldots\rangle \tag{28.18}$$

is a normalized simultaneous eigenket of all the n belonging to the eigenvalues $n'_1, n'_2, \ldots$ Various kets obtained by taking different n' form a complete set of kets all orthogonal to each other.

Following Dirac [1], we postulate that the quantum states for N bosons can be represented by a *symmetric ket*

$$S\left[|\alpha_a^{(1)}\rangle |\alpha_b^{(2)}\rangle \ldots |\alpha_g^{(N)}\rangle\right] \equiv |\alpha_a \alpha_b \ldots \alpha_g\rangle_S \tag{28.19}$$

where S is the *symmetrizing operator*:

$$S \equiv \frac{1}{\sqrt{N!}} \sum_P P \tag{28.20}$$

and P are permutation operators for the particle indices $(1, 2, \ldots, N)$. The ket in (28.19) is not normalized but

$$(n_1! n_2! \ldots)^{-1/2} |\alpha_a \alpha_b \ldots \alpha_g\rangle_S \equiv |\{n\}\rangle \tag{28.21}$$

is a normalized ket representing the same state. Comparing (28.21) and (28.18), we obtain

$$|\alpha_a \alpha_b \ldots \alpha_g\rangle_S = \eta_a^\dagger \eta_b^\dagger \ldots \eta_g^\dagger |\Phi_0\rangle \tag{28.22}$$

That is, unnormalized symmetric kets $|\alpha_a \alpha_b \ldots \alpha_g\rangle_S$ for the system can be constructed by applying N creation operators $\eta_a^\dagger \eta_b^\dagger \ldots \eta_g^\dagger$ to the vacuum ket $|\Phi_0\rangle$.

So far we have tacitly assumed that the total number of bosons is fixed at N'. If this number is not fixed but is variable, we can easily extend the theory to this case. Let us introduce a Hermitean operator N defined by

$$N \equiv \sum_a \eta_a^\dagger \eta_a = \sum_a n_a \tag{28.23}$$

the summation extending over the whole set of boson states. Clearly, the operator N has eigenvalues $0, 1, 2, \ldots$, and the ket $|\alpha_a \alpha_b \ldots \alpha_g\rangle_S$ is an eigenket of N belonging to the eigenvalue N'. We may arrange kets in the order of N', that is zero-particle state, one-particle states, two-particle states, ...:

$$|\Phi_0\rangle, \quad \eta_a^\dagger |\Phi_0\rangle, \quad \eta_a^\dagger \eta_b^\dagger |\Phi_0\rangle, \quad \ldots \tag{28.24}$$

These kets are all orthogonal to each other, two kets referring to the same number of bosons are orthogonal as before, and two referring to different numbers of bosons are orthogonal because they have different eigenvalues N'. By normalizing the kets, we get a set of kets like (28.21) with no restriction on $\{n'\}$. These kets form the basic kets in a representation where $\{n_a\}$ are diagonal.

Problem 28.1.1

(a) Show (twice) that $n\eta^\dagger - \eta^\dagger n = \eta^\dagger$, by taking the Hermitian-conjugation of (28.10) *and* using (28.9). (b) Use this relation and obtain a series of eigenvalues $n', n'+1, n'+2, \ldots$, where n' is an eigenvalue of n.

Problem 28.1.2

Prove (28.15) by mathematical induction. Hint: use (28.9).

28.2
Observables

We wish to express observable physical quantities (observables) for the system of identical bosons in terms of η and $\eta^\dagger$. These observables are by postulate symmetric functions of the boson variables.

An observable can be written in the form:

$$\sum_j y^{(j)} + \sum_i \sum_j z^{(ij)} + \ldots \equiv Y + Z + \ldots \tag{28.25}$$

where $y^{(j)}$ is a function of the dynamic variables of the jth boson, $z^{(ij)}$ that of the dynamic variables of the ith and jth bosons, and so on.

28.2 Observables

We take $Y \equiv \sum_j y^{(j)}$. Since $y^{(j)}$ acts only on the ket $|a^{(j)}\rangle$ of the jth boson, we have

$$y^{(j)}\left(\left|a_{x_1}^{(1)}\right\rangle \left|a_{x_2}^{(2)}\right\rangle \cdots \left|a_{x_j}^{(j)}\right\rangle \cdots\right)$$
$$= \sum_a \left(\left|a_{x_1}^{(1)}\right\rangle \left|a_{x_2}^{(2)}\right\rangle \cdots \left|a_a^{(j)}\right\rangle \cdots\right) \left\langle a_a^{(j)}\right| y^{(j)} \left|a_{x_j}^{(j)}\right\rangle \quad (28.26)$$

The matrix element $\langle a_a^{(j)}|y^{(j)}|a_{x_j}^{(j)}\rangle \equiv \langle a_a|y|a_{x_j}\rangle$ does not depend on the particle index j. Summing (28.26) over all j and applying operator S to the result, we obtain

$$S Y \left(\left|a_{x_1}^{(1)}\right\rangle \left|a_{x_2}^{(2)}\right\rangle \cdots\right)$$
$$= \sum_j \sum_a S \left(\left|a_{x_1}^{(1)}\right\rangle \left|a_{x_2}^{(2)}\right\rangle \cdots \left|a_{x_1}^{(1)}\right\rangle \left|a_a^{(j)}\right\rangle \cdots\right) \langle a_a| y |a_{x_j}\rangle \quad (28.27)$$

Since Y is symmetric, we can replace SY by YS for the lhs. After straightforward calculations, we obtain, from (28.27),

$$Y\eta_{x_1}^\dagger \eta_{x_2}^\dagger \cdots |\Phi_0\rangle = \sum_j \sum_a \eta_{x_1}^\dagger \eta_{x_2}^\dagger \cdots \eta_{x_{j-1}}^\dagger \eta_a^\dagger \eta_{x_{j+1}}^\dagger \cdots |\Phi_0\rangle \langle a_a| y |a_{x_j}\rangle$$
$$= \sum_a \sum_b \eta_a^\dagger \sum_j \eta_{x_1}^\dagger \eta_{x_2}^\dagger \cdots \eta_{x_{j-1}}^\dagger \eta_{x_{j+1}}^\dagger \cdots |\Phi_0\rangle \delta_{bx_j} \langle a_a| y |a_b\rangle$$
$$\quad (28.28)$$

Using the commutation rules and the property (28.14) we can show that

$$\eta_b \eta_{x_1}^\dagger \eta_{x_2}^\dagger \cdots |\Phi_0\rangle = \sum_j \eta_{x_1}^\dagger \eta_{x_2}^\dagger \cdots \eta_{x_{j-1}}^\dagger \eta_{x_{j+1}}^\dagger \cdots |\Phi_0\rangle \delta_{bx_j} \quad (28.29)$$

(Problem 28.2.1). Using this relation, we obtain from (28.28)

$$Y\eta_{x_1}^\dagger \eta_{x_2}^\dagger \cdots |\Phi_0\rangle = \sum_a \sum_b \eta_a^\dagger \eta_b \langle a_a| y |a_b\rangle \left(\eta_{x_1}^\dagger \eta_{x_2}^\dagger \cdots |\Phi_0\rangle\right) \quad (28.30)$$

Since the kets $\eta_{x_1}^\dagger \eta_{x_2}^\dagger \cdots |\Phi_0\rangle$ form a complete set, we obtain

$$Y = \sum_a \sum_b \eta_a^\dagger \eta_b \langle a_a| y |a_b\rangle \quad (28.31)$$

In a similar manner Z in (28.25) can be expressed by [Problem 28.2.2

$$Z = \sum_a \sum_b \sum_c \sum_d \eta_a^\dagger \eta_b^\dagger \eta_d \eta_c \langle a_a a_b| y |a_c a_d\rangle \quad (28.32)$$

$$\langle a_a a_b| y |a_c a_d\rangle \equiv \left\langle a_a^{(1)}\right| \left\langle a_b^{(2)}\right| z^{(12)} \left|a_d^{(2)}\right\rangle \left|a_c^{(1)}\right\rangle \quad (28.33)$$

Problem 28.2.1

Prove (28.29). Hint: Start with cases of one- and two-particle-state kets.

Problem 28.2.2

Prove (28.32) by following those steps similar to (28.27)–(28.31).

28.3
Fermions Creation and Annihilation Operators

In this section we treat a system of identical fermions in a parallel manner.

The quantum states for fermions, by postulate, are represented by *Antisymmetric kets*:

$$|a_a a_b \ldots a_g\rangle_A \equiv A\left(\left|a_a^{(1)}\right\rangle\left|a_b^{(2)}\right\rangle \ldots \left|a_g^{(N)}\right\rangle\right) \tag{28.34}$$

where

$$A \equiv \frac{1}{\sqrt{N!}} \sum_P \delta_P P \tag{28.35}$$

is the *antisymmetrizing operator*, with δ_P being +1 or −1 according to whether P is even or odd. Each antisymmetric ket in (28.34) is characterized such that it changes its sign if an odd permutation of particle indices is applied to it, and the fermion states $a, b, \ldots, g$ are all different. Just as for a boson system, we can introduce observables $n_1, n_2, \ldots$ each with eigenvalues 0 or 1, representing the number of fermions in the states $\alpha_1, \alpha_2, \ldots$, respectively.

We can also introduce a set of linear operators $(\eta, \eta^\dagger)$, one pair $(\eta_a, \eta_a^\dagger)$ for each state α_a, satisfying the *Fermi anticommutation rules*:

$$\{\eta_a, \eta_b^\dagger\} \equiv \eta_a \eta_b^\dagger + \eta_b^\dagger \eta_a = \delta_{ab}, \quad \{\eta_a, \eta_b\} = \{\eta_a^\dagger, \eta_b^\dagger\} = 0 \tag{28.36}$$

The number of fermions in the state α_a is again represented by

$$n_a = \eta_a^\dagger \eta_a = n_a^\dagger \tag{28.37}$$

Using (28.36), we obtain

$$n_a^2 = \eta_a^\dagger \eta_a \eta_a^\dagger \eta_a = \eta_a^\dagger (1 - \eta_a^\dagger \eta_a) \eta_a = \eta_a^\dagger \eta_a = n_a$$

or

$$n_a^2 - n_a = 0 \tag{28.38}$$

If an eigenket of n_a belonging to the eigenvalue n_a' is denoted by $|n_a'\rangle$, (28.38) yields

$$(n_a^2 - n_a)|n_a'\rangle = (n_a'^2 - n_a')|n_a'\rangle = 0 \tag{28.39}$$

Since $|n'_a\rangle \neq 0$, we obtain $n'_a(n'_a - 1) = 0$, meaning that the eigenvalues n'_a are either 0 or 1 as required:

$$n'_a = 0 \text{ or } 1 \tag{28.40}$$

Similarly to the case of bosons, we can show that

$$|a_a a_b \ldots a_g\rangle_A = \eta^\dagger_a \eta^\dagger_b \ldots \eta^\dagger_g |\Phi_0\rangle \tag{28.41}$$

which is normalized to unity.

Observables describing the system of fermions can be expressed in terms of operators η and $\eta^\dagger$, and the results have the same form (28.31) and (28.32) as for the case of bosons.

In summary, both states and observables for a system of identical particles can be expressed in terms of creation and annihilation operators. This formalism, called the *second quantization*, has some notable advantages over the usual Schrödinger formalism. First, the permutation-symmetry property of the quantum particles is represented simply in the form of Bose commutation (or Fermi anticommutation) rules. Second, observables in second quantization are defined for an arbitrary number of particles so that the formalism may apply to systems in which the number of particles is not fixed, but variable. Third, and most importantly, all relevant quantities (states and observables) can be defined referring only to the single-particle states. This property allows one to describe the motion of the many-body system in the 3D space. This is a natural description since all particles in nature move in 3D. In fact, relativistic quantum field theory can be developed only in second quantization.

28.4
Heisenberg Equation of Motion

In the Schrödinger picture (SP), the energy eigenvalue equation is

$$H|E\rangle = E|E\rangle \tag{28.42}$$

where H is the Hamiltonian and E the eigenvalue. In the position representation this equation is written as

$$H\left(x_1, -i\hbar\frac{\partial}{\partial x_1}, x_2, -i\hbar\frac{\partial}{\partial x_2}, \ldots\right)\Psi(x_1, x_2, \ldots) = E\Psi \tag{28.43}$$

where Ψ is the wavefunction for the system. We considered here a one-dimensional motion for conceptional and notational simplicity. [For a three-dimensional motion, (x, p) should be replaced by $(x, y, z, p_x, p_y, p_z) = (\mathbf{r}, \mathbf{p})$.] If the number of electrons N is large, the wavefunction Ψ contains many electron variables $(x_1, x_2, \ldots)$. This complexity needed in dealing with many electron variables can be avoided if we use the second quantization formulation and the Heisenberg picture (HP), which will be shown in this section.

If the Hamiltonian H is the sum of single-particle Hamiltonians:

$$H = \sum_j h^{(j)} \tag{28.44}$$

this Hamiltonian H can be represented by

$$H = \sum_a \sum_b \langle a | h | b \rangle \eta_a^\dagger \eta_b \equiv \sum_a \sum_b h_{ab} \eta_a^\dagger \eta_b \tag{28.45}$$

where $\eta_a (\eta_a^\dagger)$ are annihilation (creation) operators associated with particle-state a and satisfying the Fermi anticommutation rules in (28.36).

In the HP a variable $\xi(t)$ changes in time, following the *Heisenberg equation of motion*:

$$-i\hbar \frac{d\xi(t)}{dt} = [H, \xi] \equiv H\xi - \xi H \tag{28.46}$$

Setting $\xi = \eta_a^\dagger$, we obtain

$$-i\hbar \frac{d\eta_a^\dagger}{dt} = [H, \eta_a^\dagger] \tag{28.47}$$

whose Hermitian conjugate is given by

$$i\hbar \frac{d\eta_a}{dt} = ([H, \eta_a^\dagger])^\dagger = -[H, \eta_a] \tag{28.48}$$

By the quantum postulate the physical observable ξ is Hermitian: $\xi^\dagger = \xi$. Variables η_a and $\eta_a^\dagger$ are *not* Hermitian, but both obey the *same* Heisenberg equation of motion.

We introduce (28.45) into (28.47), and calculate the commutator $[H, \eta_a^\dagger]$. In such a commutator calculation the following identities:

$$[A, BC] = \{A, B\}C - B\{A, C\}$$

$$[AB, C] = A\{B, C\} - \{A, C\}B \tag{28.49}$$

are very useful. Note: The negative signs on the right-hand terms in (28.49) occur when the cyclic order is destroyed. We obtain from (28.47) and (28.48)

$$-i\hbar \frac{d\eta_a^\dagger}{dt} = \sum_c \sum_b h_{cb}[\eta_c^\dagger \eta_b, \eta_a^\dagger]$$

$$= \sum_c \sum_b h_{cb} \eta_c^\dagger \{\eta_b, \eta_a^\dagger\} = \sum_c h_{ca} \eta_c^\dagger \tag{28.50}$$

or

$$i\hbar \frac{d\eta_a}{dt} = \sum_c h_{ac} \eta_c \tag{28.51}$$

Equation (28.50) means that the change of the one-body operator $\eta_a^\dagger$ is determined by the one-body Hamiltonian h. This is the major advantage of working in the HP. Equations (28.50) and (28.51) are valid for *any* single-particle states $\{a\}$.

In the field operator language (28.51) reads

$$i\hbar \frac{\partial \psi(\mathbf{r}, t)}{\partial t} = h\left(\mathbf{r}, -i\hbar \frac{\partial}{\partial \mathbf{r}}\right) \psi(\mathbf{r}, t) \qquad (28.52)$$

which is formally identical to the Schrödinger equation of motion for a particle.

If the system Hamiltonian H contains an interparticle interaction

$$V = \frac{1}{2} \int d^3 r \int d^3 r' \, v(\mathbf{r} - \mathbf{r}') \psi^\dagger(\mathbf{r}, t) \psi^\dagger(\mathbf{r}', t) \psi(\mathbf{r}', t) \psi(\mathbf{r}, t) \qquad (28.53)$$

the evolution equation for $\psi(\mathbf{r}, t)$ is nonlinear (Problem 28.4.2):

$$i\hbar \frac{\partial \psi(\mathbf{r}, t)}{\partial t} = h\left(\mathbf{r}, -i\hbar \frac{\partial}{\partial \mathbf{r}}\right) \psi(\mathbf{r}, t)$$
$$+ \int d^3 r' \, v(\mathbf{r} - \mathbf{r}') \psi^\dagger(\mathbf{r}', t) \psi(\mathbf{r}', t) \psi(\mathbf{r}, t) \qquad (28.54)$$

In quantum field theory the basic dynamical variables are particle-field operators. The quantum statistics of the particles are given by the Bose commutation or the Fermi anticommutation rules satisfied by the field operators. The evolution equations of the field operators are intrinsically nonlinear when the interparticle interaction is present.

Problem 28.4.1

Verify that the equation of motion (28.50) holds for bosons.

Problem 28.4.2

Use (28.46) to verify (28.54).

Reference

1 Dirac, P.A.M. (1928) *Proc. Roy. Soc.* (London), **117**, 610.

29
Quantum Statistics of Composites

The Ehrenfest–Oppenheimer–Bethe (EOB) rule with respect to the center-of-mass motion is that a composite particle moves as a boson (fermion) if it contains an even (odd) number of elementary fermions. This rule is proved in this chapter. There exist no elementary bosons in nature, which is shown.

29.1
Ehrenfest–Oppenheimer–Bethe's Rule

Experiments indicate that every quantum particle in nature moves either as a boson or as a fermion [1]. This statement, applied to elementary particles, is known as the *quantum-statistical postulate* (or *principle*). Bosons (fermions), by definition, can (cannot) multiply occupy one and the same quantum-particle state. Spin and isospin (charge), which are part of particle-state variables are included in the definition of the state. Electrons (e) and nucleons (protons p, neutrons n) are examples of elementary fermions [1, 2]. Composites such as deuterons (p, n) tritons ($p, 2n$), hydrogen H (p, e) are indistinguishable and obey quantum statistics. According to the Ehrenfest–Oppenheimer–Bethe (EOB) rule [3, 4] a composite is fermionic (bosonic) if it contains an odd (even) number of elementary fermions. Let us review the arguments leading to EOB's rule as presented in Bethe–Jackiw's book [4]. Take a composite of two identical fermions and study the symmetry of the wavefunction for two composites, which has four particle-state variables, two for the first composite and two for the second one. Imagine that the exchange between the two composites is carried out particle by particle. Each exchange of fermions (bosons) changes the wavefunction by the factor –1 (+1). In the present example, the sign changes twice and the wavefunction is therefore unchanged. If a composit contains different types of particles, as in the case of H, the symmetry of the wavefunction is deduced by the interchange within each type. We shall see later that these arguments are incomplete. We note that Feynman used these arguments to deduce that Cooper pairs [5] are bosonic [6]. The symmetry of the many-particle wavefunction and the quantum statistics for elementary particles are one-to-one [1]. A set of elementary fermions (bosons) can be described in terms of creation and annihilation operators satisfying the Fermi anticommutation (Bose commutation)

Mathematical Physics. Shigeji Fujita and Salvador V. Godoy
Copyright © 2010 WILEY-VCH Verlag GmbH & Co. KGaA, Weinheim
ISBN: 978-3-527-40808-5

rules. But no one-to-one correspondence exists for composites since composites by construction have extra degrees of freedom. Wavefunctions and second-quantized operators are important auxiliary quantum variables but they are not observables in Dirac's sense [1]. We must examine the observable occupation numbers for the study of the quantum statistics of composites. In the present chapter we shall show that EOB's rule applies to the center-of-mass (CM) motion of the composites.

29.2
Two-Particle Composites

Let us consider two-particle composites. There are four important cases represented by (A) electron–electron (pairon), (B) electron–proton (hydrogen H), (C) nucleon–pion, and (D) boson–boson.

(A) Identical fermion composite.

Second-quantized operators for a pair of electrons are defined by [8]

$$B_{12}^\dagger \equiv B_{k_1 k_2}^\dagger \equiv c_{k_1}^\dagger c_{k_2}^\dagger \equiv c_1^\dagger c_2^\dagger, \quad B_{34} = c_4 c_3 \tag{29.1}$$

where $c_{k_1}^\dagger (c_{k_1}) \equiv c_1^\dagger (c_1)$ are creation (annihilation) operators (spins indices omitted) satisfying the Fermi anticommutation rules:

$$\{c_1, c_2^\dagger\} \equiv c_1 c_2^\dagger + c_2^\dagger c_1 = \delta_{k_1 k_2}, \quad \{c_1, c_2\} = 0 \tag{29.2}$$

The commutators among B and $B^\dagger$ can be computed by using (29.2) and are given by [8]

$$[B_{12}, B_{34}] \equiv B_{12} B_{34} - B_{34} B_{12} = 0, \quad (B_{12})^2 = 0 \tag{29.3}$$

$$[B_{12}, B_{34}^\dagger] = \begin{cases} 1 - n_1 - n_2 & \text{if } k_1 = k_3, k_2 = k_4 \\ c_2 c_4^\dagger & \text{if } k_1 = k_3, k_2 \neq k_4 \\ c_1 c_3^\dagger & \text{if } k_1 \neq k_3, k_2 = k_4 \\ 0 & \text{otherwise} \end{cases} \tag{29.4}$$

where

$$n_j = c_j^\dagger c_j \quad (j = 1, 2) \tag{29.5}$$

represent the number operators for electrons. Using (29.1)–(29.5) and

$$n_{12} \equiv B_{12}^\dagger B_{12} \tag{29.6}$$

we obtain

$$n_{12}^2 = B_{12}^\dagger \left(1 - n_1 - n_2 + B_{12}^\dagger B_{12}\right) B_{12} = n_{12} \tag{29.7}$$

29.2 Two-Particle Composites

If $|n'_{12}\rangle$ is the eigenstate of n_{12}, we obtain

$$\left(n'^2_{12} - n_{12}\right)|n'_{12}\rangle = \left(n'^2_{12} - n'_{12}\right)|n'_{12}\rangle = 0, \quad |n'_{12}\rangle \neq 0$$

yielding

$$n'_{12} = 0 \text{ or } 1 \tag{29.8}$$

Let us now introduce the relative and net (or CM) momenta (**k**, **q**) such that

$$\mathbf{k} \equiv \frac{1}{2}(\mathbf{k}_1 - \mathbf{k}_2), \quad \mathbf{q} \equiv \mathbf{k}_1 + \mathbf{k}_2; \quad \mathbf{k}_1 = \mathbf{k} + \frac{1}{2}\mathbf{q}, \quad \mathbf{k}_2 = -\mathbf{k} + \frac{1}{2}\mathbf{q} \tag{29.9}$$

We may alternatively represent the pair operators by

$$B'_{kq} \equiv c_{-k+\frac{1}{2}q} c_{k+\frac{1}{2}q} \equiv B_{12}, \quad B'^{\dagger}_{kq} \equiv c^{\dagger}_{k+\frac{1}{2}q} c^{\dagger}_{-k+\frac{1}{2}q} \tag{29.10}$$

The prime on B_{kq} will be dropped hereafter. From (29.8) we deduce that the number operator in the k–q representation.

$$n_{kq} \equiv B^{\dagger}_{kq} B_{kq} \tag{29.11}$$

has eigenvalues 0 or 1:

$$n'_{kq} = 0 \text{ or } 1 \tag{29.12}$$

The total number of a system of pairons, N, is represented by

$$N \equiv \sum_{k_1}\sum_{k_2} n_{12} = \sum_{k}\sum_{q} n_{kq} = \sum_{q} n_q \tag{29.13}$$

where

$$n_q \equiv \sum_{k} n_{kq} = \sum_{k} B^{\dagger}_{kq} B_{kq} \tag{29.14}$$

represents the number of pairons having net momentum **q**. From (29.12)–(29.14) we see that the eigenvalues of the number operator n_q can be non-negative integers. To explicitly see this property, we introduce

$$B_q \equiv \sum_{k} B_{kq} \tag{29.15}$$

and obtain, after using (29.2)–(29.5) and (29.10),

$$[B_q, n_q] = \sum_{k}(1 - n_{k+\frac{1}{2}q} - n_{-k+\frac{1}{2}q}) B_{kq} = B_q, \quad \left[n_q, B^{\dagger}_q\right] = B^{\dagger}_q \tag{29.16}$$

Although the occupation number n_q is not connected with B_q as $n_q \neq B^{\dagger}_q B_q$, the eigenvalues n'_q of n_q satisfying (29.16) can be shown straightforwardly to yield [1]

$$n'_q = 0, 1, 2, \ldots \tag{29.17}$$

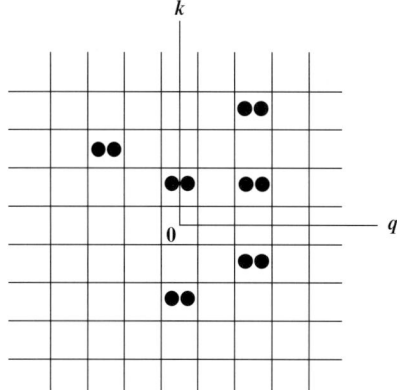

Fig. 29.1 The number representation of many electron pairs in the (k, q) space.

with the eigenstates

$$|0\rangle \,, \quad |1\rangle = B_q^\dagger |0\rangle \,, \quad |2\rangle = B_q^\dagger B_q^\dagger |0\rangle \,, \ldots \tag{29.18}$$

This is important. We illustrate it by taking a one-dimensional motion. The pairon occupation-number states may be represented by drawing quantum cells in the (k, q) space. From (29.12) the number n'_{kq} are limited to 0 or 1, see Figure 7.1. The number of pairons characterized by net momentum q only, n'_q, is the sum of the numbers of pairs at column q, and clearly it is zero or a positive integer.

In summary, pairons with both $\mathbf{k}$ and $\mathbf{q}$ specified are subject to the Pauli exclusion principle, see (29.12). Yet, the occupation numbers n'_q of the pair having CM momentum $\mathbf{q}$ are $0, 1, 2, \ldots$, see (29.17). Note that our results (29.8), (29.12) and (29.17) are obtained by using the pair commutators (29.3) and (29.4). Further note that our result (29.17) does not follow from examination of the symmetry of the wavefunction, the symmetry arising from (29.3) only. Equation (29.4) is needed to prove (29.16). The fact that the quantum statistics depends on whether we specify $(\mathbf{k}, \mathbf{q})$ or $\mathbf{q}$ alone, arises because a composite by construction has more degrees of freedom. Only with respect to the CM motion are the Cooper pairs bosonic and can multiply occupy the same momentum state $\mathbf{q}$. We say in short that the pair moves as a boson.

(B) Different-fermion composite.

The quantum state for two *distinguishable* particles (1,2) can be represented by

$$\left|\mathbf{k}_a^{(1)}, \mathbf{k}_b^{(2)}\right\rangle \equiv \left|\mathbf{k}_a^{(1)}\right\rangle \left|\mathbf{k}_b^{(2)}\right\rangle \tag{29.19}$$

We may represent the state $|\mathbf{k}_a^{(j)}\rangle$ for the particle j by specifying a set of occupation numbers $(n_a'^{(j)}, n_b'^{(j)}, \ldots)$ with the restriction that each $n_a'^{(j)}$ can take on a value either 0 or 1 and only one member of the set takes the value 1. These numbers $n_a'^{(j)}$ can be represented by [10]

$$n_a^{(j)} \equiv \eta_a^{(j)\dagger} \eta_a^{(j)} \tag{29.20}$$

where creation (annihilation) operators $\eta_a^{(j)\dagger}$ ($\eta_a^{(j)}$) satisfy the Fermi anticommutation rules (29.2). The quantum state for a many-electron–many-proton system may be represented by a generalization of (29.19), the direct product of a many-electron (antisymmetric) state and a many-proton (antisymmetric) state. Such states can be described in terms of second-quantized operators c's (electrons) and a's (protons), both satisfying the anticommutation rules (29.2). We postulate that observables for different particles commute:

$$\left[n_a^{(1)}, n_b^{(2)}\right] = 0 \tag{29.21}$$

based on which we may choose such that c's and a's anticommute with each other:

$$\{c, a^\dagger\} = \{c, a\} = 0 \tag{29.22}$$

Pair operators are defined by

$$B_{12}^\dagger \equiv a_1^\dagger c_2^\dagger, \quad B_{34} \equiv c_4 a_3 \tag{29.23}$$

We study the number operator $n_q^{(H)}$ defined in the form (29.14) and show by means of (29.16) that the eigenvalues of $n_q^{(H)}$ are $0, 1, 2, \ldots$. That is, hydrogens H move as bosons.

(C) Fermion–boson composite.

We define pair operators in the form (29.23) with boson operators (b's) satisfying the Bose commutation rules:

$$\left[b_1, b_2^\dagger\right] \equiv b_1 b_2^\dagger - b_2^\dagger b_1 = \delta_{k_1 k_2}, \quad [b_1, b_2] = 0 \tag{29.24}$$

Fermion and boson operators mutually anticommute, which is in accord with (29.21). We obtain

$$\{B_{12}, B_{34}\} \equiv B_{12} B_{34} + B_{34} B_{12} = 0 \tag{29.25}$$

$$\{B_{12}, B_{34}^\dagger\} = \begin{cases} 1 + n_1 - n_2 & \text{if } k_1 = k_3, k_2 = k_4 \\ c_2 c_4^\dagger & \text{if } k_1 = k_3, k_2 \neq k_4 \\ b_1 b_3^\dagger & \text{if } k_1 \neq k_3, k_2 = k_4 \\ 0 & \text{otherwise} \end{cases} \tag{29.26}$$

We define n_q and B_q as in (29.14) and (29.15), use (29.26) and obtain

$$[n_q, B_q] + B_q + \sum_k n_{k+\frac{1}{2}q} B_{kq} = 0 \tag{29.27}$$

$$[n_q, B_q^\dagger] - B_q^\dagger - \sum_k B_{kq}^\dagger n_{k+\frac{1}{2}q} = 0 \tag{29.28}$$

The vacuum state, $|0\rangle$, satisfying

$$a_k |0\rangle = b_k |0\rangle = 0, \quad \text{(all k)} \tag{29.29}$$

is defined. One-pair states $\left|n'_q = 1\right\rangle$ are constructed by

$$\left|n'_q = 1\right\rangle \equiv |1\rangle = B_q^\dagger |0\rangle, \quad n_q |1\rangle = |1\rangle \tag{29.30}$$

The two states ($|0\rangle$, $|1\rangle$) are the only pair-number states at q that can be constructed without violating the restriction imposed by (29.28). In fact, applying (29.28) to $|1\rangle$ we obtain

$$\left(\left[n_q, B_q^\dagger\right] - B_q^\dagger - \sum_k B_{kq}^\dagger n_{k+\frac{1}{2}q}\right)|1\rangle = \left(n_q B_q^\dagger - 2B_q^\dagger - \sum_k B_{kq}^\dagger\right)|1\rangle$$

$$= \left(n_q B_q^\dagger - 3B_q^\dagger\right)|1\rangle = 0$$

or

$$n_q B_q^\dagger |1\rangle = 3 B_q^\dagger |1\rangle \tag{29.31}$$

indicating that no two-pair state can be constructed in a regular manner. That is, $|2\rangle \neq B_q^\dagger |1\rangle$. Hence fermion–boson composites move as fermions.

(D) Identical boson composite.

We introduce pair operators:

$$B_{12}^\dagger \equiv a_{k_1}^\dagger a_{k_2}^\dagger \equiv a_1^\dagger a_2^\dagger, \quad B_{12} \equiv a_2 a_1 \tag{29.32}$$

We compute commutators among B and $B^\dagger$ and obtain

$$[B_{12}, B_{34}] = 0 \tag{29.33}$$

$$\left[B_{12}, B_{34}^\dagger\right] = \begin{cases} 1 + n_1 + n_2 & \text{if } k_1 = k_3, k_2 = k_4, k_2 \neq k_3, k_1 \neq k_4 \\ 2 + 4n_1 & \text{if } k_1 = k_3, k_2 = k_4, k_2 = k_3, k_1 = k_4 \\ a_2 a_4^\dagger & \text{if } k_1 = k_3, k_2 \neq k_4, k_2 \neq k_3, k_1 \neq k_4 \\ a_2 a_4^\dagger + a_2 a_3^\dagger & \text{if } k_1 = k_3, k_2 \neq k_4, k_2 \neq k_3, k_1 = k_4 \\ a_2 a_4^\dagger + a_4^\dagger a_1 & \text{if } k_1 = k_3, k_2 \neq k_4, k_2 = k_3, k_1 \neq k_4 \\ a_3^\dagger a_1 & \text{if } k_1 \neq k_3, k_2 = k_4, k_2 \neq k_3, k_1 \neq k_4 \\ a_3^\dagger a_1 + a_2 a_3^\dagger & \text{if } k_1 \neq k_3, k_2 = k_4, k_2 \neq k_3, k_1 = k_4 \\ a_3^\dagger a_1 + a_4 a_1^\dagger & \text{if } k_1 \neq k_3, k_2 = k_4, k_2 \neq k_3, k_1 = k_4 \\ 0 & \text{otherwise} \end{cases} \tag{29.34}$$

Consider a pair-creation operator

$$B_q^\dagger \equiv \sideset{}{'}\sum_{k \neq 0} B_{kq}^\dagger + B_{0q}^\dagger \tag{29.35}$$

Multiplying this equation from the left by n_q and from the right by $|\Phi_0\rangle$, we obtain

$$n_q B_q^\dagger |\Phi_0\rangle \equiv n_q \left(\sideset{}{'}\sum_{k \neq 0} B_{kq}^\dagger + B_{0q}^\dagger\right)|\Phi_0\rangle = \left(\sideset{}{'}\sum_{k \neq 0} B_{kq}^\dagger + 2 B_{0q}^\dagger\right)|\Phi_0\rangle \tag{29.36}$$

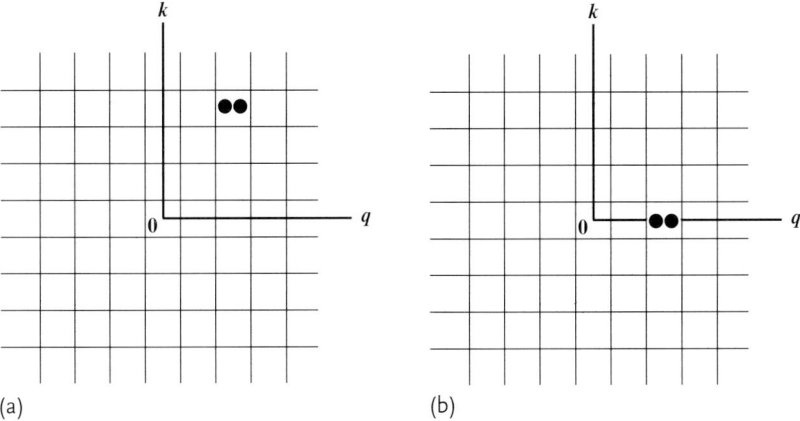

Fig. 29.2 (a) State $B_{kq}^\dagger |\Phi_0\rangle$, $k \neq 0$. (b) State $B_{0q}^\dagger |\Phi_0\rangle$.

which indicates that $B_q^\dagger |\Phi_0\rangle$ is not the eigenstate of n_q. This is significant. The state corresponding to $B_{kq}^\dagger |\Phi_0\rangle$, $k \neq 0$ in one dimension is shown in Figure 7.2(a). The corresponding occupation number $n_{kq} = B_{kq}^\dagger B_{kq}$ has the eigenvalue one since

$$n_{kq} B_{kq}^\dagger |\Phi_0\rangle = a_{k+\frac{1}{2}q}^\dagger a_{-k+\frac{1}{2}q}^\dagger a_{-k+\frac{1}{2}q} a_{k+\frac{1}{2}q} a_{k+\frac{1}{2}q}^\dagger a_{-k+\frac{1}{2}q}^\dagger |\Phi_0\rangle$$
$$= B_{kq}^\dagger |\Phi_0\rangle, \quad k \neq 0 \tag{29.37}$$

For **k** = 0 a straightforward calculation gives

$$n_{0q} B_{0q}^\dagger |\Phi_0\rangle = a_{\frac{1}{2}q}^\dagger a_{\frac{1}{2}q}^\dagger a_{\frac{1}{2}q} a_{\frac{1}{2}q} a_{\frac{1}{2}q}^\dagger a_{\frac{1}{2}q}^\dagger |\Phi_0\rangle = 2 B_{0q}^\dagger |\Phi_0\rangle \tag{29.38}$$

Thus, the operator n_{0q} has the eigenvalue 2. The state $B_{0q}^\dagger |\Phi_0\rangle$ is represented by Figure 7.2(b). These last two results generate (29.36). In the presence of the double occupancy at $k = 0$, we find no one-pair number state. This anomaly does not occur when dealing with elementary fermions since the double occupancy is excluded by Pauli's principle.

29.3
Discussion

In 1940 Pauli established the spin-statistics theorem [2]: half-integral spin elementary particles are fermions, while integral spin particles are bosons. He derived it by applying general principles of quantum theory and relativity to elementary particles. Just as elementary particles, composites are experimentally found to be indistinguishable and move either as bosons or as fermions (quantum-statistical principle). This can be understood simply if the CM of a composite moves, following the same general principles, and if the spin-statistics theorem is applied. We take Democritos' atomistic view: every matter is composed of massive "atoms" (elementary particles). (Massless quantum particles such as photons, neutrinos will not be considered hereafter.) We saw in case (**D**) that no one-pair state for the identical boson pair can be constructed. Hence, this composite moves neither as a boson nor as a fermion, in violation of the quantum-statistical principle. The arguments quoted earlier for the EOB's rule fail in this case. Electrons and nucleons have half-spins while pions have zero spin. Hence the other three cases (**A**)–(**C**) are in accord with the spin-statistics theorem and also with the EOB rule.

In our derivation we omitted consideration of spin, isospin, ... We now discuss this point. Following Dirac [1], we define the indistinguishability of a system of identical elementary particles in terms of the permutation symmetry:

$$[P, H] = [P, \xi] = 0, \quad \text{(all } \xi \text{ and all } P\text{)} \tag{29.39}$$

$$[P, \rho] = 0 \tag{29.40}$$

where

$$H = H(\eta_1, \eta_2, \ldots, \eta_N) \tag{29.41}$$

is the Hamiltonian of N particle variables η containing position, momentum and other quantum variables such as spin, isospin, ...; ξ is a system-dynamical function such as the center-of mass and the total momentum; P's are permutation operators of N particle indices. The density operator ρ is defined in the form:

$$\rho \equiv \sum_\nu |\nu\rangle P_\nu \langle \nu|, \quad \sum_\nu P_\nu = 1, \quad P_\nu = \text{probability} \tag{29.42}$$

where $|\nu\rangle$ are symmetric (antisymmetric) kets for bosons (fermions). The particle state is characterized by momentum $\mathbf{k}$, spin-component σ, isospin-component τ, ... The state may equivalently be represented by the continuous position, $\mathbf{r}, \sigma, \tau, \ldots$. The set of momenta, $\{\mathbf{k}\}$, is infinite since the position conjugate to the momentum is a continuous variable. Dirac's relativistic wave equation [1] indicates that the electron and antiparticle (positron) has spin 1/2. Pauli's spin-statistics theorem [2] originates in the relativistic quantum motion of the particles in the ordinary three-dimensional space [2]. In contrast, other sets $\{\sigma\}, \{\tau\}, \ldots$ are all finite, and hence these variables play secondary roles in quantum statistics. This is so because the quantum statistics of the particles must be defined with the condition that there are an infinite set of particle states. In fact, if there were only one state, neither symmetric nor antisymmetric states could be constructed. If there were only two states, no antisymmetric states for three or more particles could be constructed. Limiting the number of particles in the theory is unnatural.

We have studied the eigenvalues of the pair-number operators (n_{12}, n_{kq}, n_q), which are observables in Dirac's sense [1]. All of our results are obtained without introducing the Hamiltonian. Hence, the results are valid independently of any interaction and energy (bound or unbound). This is significant, and is supported by the following arguments. We consider a system of interacting particles and write the Hamiltonian H in the form

$$H = H_0 + \lambda V \tag{29.43}$$

where H_0 is the sum of the single-particle Hamiltonian:

$$H_0 = \sum_{j=1}^{N} h_0(\eta_j) \tag{29.44}$$

and V is an interaction Hamiltonian and λ a coupling constant. For $\lambda = 0$ the quantum statistics is postulated. Consider now a continuous limit:

$$\lambda \to 1 \tag{29.45}$$

No continuous limit can change discrete (permutation in our case) symmetry. Hence, the quantum statistics arising from the particle-permutation symmetry and relativistic quantum dynamics is unchanged in the limit (29.45). Such a demonstration can be extended to the case of an interaction Hamiltonian with other particle fields. All experiments appear to support our view: independence of the statistics upon interaction.

We saw in (A) that the CM motion of a pairon is bosonic, while its motion with both (**k**, **q**) specified is fermionic. This means that the fermionic nature of the constituents (electrons) is important for the total description of a composite (pairon). This is a general character of any composite. In fact Bardeen, Cooper and Schrieffer, in their historic paper on superconductivity [8], used the fermionic property (29.3) to construct the ground-state of a BCS system, the state of the Cooper pairs bosonically condensed all at zero CM momentum. By assuming the spin-statistics theorem for composites Feynman argued that the pairons move as bosons [10], and proceeded to derive the Josephson equations [11, 12]. Both fermionic and bosonic properties of the pairons were used in the total description of superconductivity [13–15].

Let us now consider a three-identical fermion composite. Triplet operators $(T, T^\dagger)$ are defined by

$$T^\dagger_{123} \equiv c_1^\dagger c_2^\dagger c_3^\dagger, \quad T_{123} = c_3 c_2 c_1 \tag{29.46}$$

If any two of the momenta (k_1, k_2, k_3) are the same, Ts vanish due to Pauli's exclusion principle. We shall show that the CM motion of the triplet is fermionic. Decompose the triplet into a system of a two-fermion composite and a fermion. The CM motion of the pair composite is bosonic according to our study in case (A). Applying the result in case (C) to the system, we then deduce that the CM motion of the triplet is fermionic. The above line of argument can be extended to the case of an N-nucleon system. First, eliminate the multioccupancy states. Second, split it into a system of $(N-1)$-nucleon composite and a nucleon. Third, apply the arguments in either (B) or (C), and deduce that the addition of one nucleon changes the quantum statistics. Next, we consider an atom composed of a nucleus and one electron. By the same argument the addition of the electron changes the quantum statistics. Further addition of an electron generates the change in statistics.

In summary, the quantum statistics for the CM motion of *any* composite is determined by the total number of the (constituting) elementary fermions. If this number is odd (even), the composite moves as a fermion (boson). Composites may contain no bosons. The EOB rule with respect to the CM motion of a composite follows directly from the commutation relations (29.3) and (29.4) and their generalizations. We stress that this rule cannot be derived from the arguments based on the symmetry of a composite wavefunction equivalent to the symmetry property of the product of the creation operators alone. The quantum statistics of the constituent particles must be treated separately. For example, the CM of hydrogen molecules $(2e, 2p)$ move as bosons. But ortho- and para-hydrogens have different internal structures and behave differently because the quantum statistics of the two constituting protons play a role [16, 17].

Experiments show that photons are bosons. A photon in a vacuum runs with the light speed and cannot stop. Hence the photon does not have the position variable as a quantum observable. In this respect, it is essentially different from other elementary fermions such as the electron and nucleon. Pions (π), and kaons (K) are experimentally found to be massive bosons. As we saw in Section 8.2, no massive elementary bosons exist. These π and K must be regarded as composites. Fermi and Yang [18] regarded π as a composite of nucleon and antinucleon. In the standard model π is regarded as a composite of two quarks [19]. These theoretical approaches are in line with our theory.

In condensed-matter physics, many elementary excitations such as phonons, magnons, plasmons, and so on appear. These particles cannot travel as fast as photons and they cannot be considered as relativistic quantum particles in principle. Hence they cannot have nonzero spins. They must therefore be bosons.

References

1 Dirac, P.A.M. (1958) *Principles of Quantum Mechanics*, 4th edn, Oxford University Press, London, p. 211, pp. 136–138, p. 37, pp. 253–257.
2 Pauli, W. (1940) *Phys. Rev.*, **58**, 716.
3 Ehrenfest, P. and Oppenheimer, J.R. (1931) *Phys. Rev.*, **37**, 331.
4 Bethe, H.A. and Jackiw, R. (1968) *Intermediate Quantum Mechanics*, 2nd edn, Benjamin, New York, p. 23.
5 Cooper, L.N. (1956) *Phys. Rev.*, **104**, 1189.
6 Feynman, R.P., Leighton, R.B., and Sands, M. (1965) *Feynman Lectures on Physics*, vol. III, Addison-Wesley, Reading, MA, p. 2107–2118.
7 Fujita, S. and Morabito, D.L. (1998) *Int. J. Mod. Phys. B*, **12**, 2139.
8 Bardeen, J., Cooper, L.N., and Schrieffer, J.R. (1957) *Phys. Rev.*, **108**, 1175.
9 Fujita, S. (1983) *Introduction to Non-Equilibrium Quantum Statistical Mechanics*, Krieger, Malabar, FL, pp. 9–11.
10 Feynman, R.P. (1972) *Statistical Mechanics*, Addison-Wesley, Redwood City, CA, p. 304.
11 Josephson, B.D. (1992) *Phys. Lett.*, **1**, 251.
12 Josephson, B.D. (1964) *Rev. Mod. Phys.*, **36**, 216.

13 Fujita, S. and Godoy, S. (1996) *Quantum Statistical Theory of Superconductivity*, Plenum, New York, pp. 122–124.
14 Fujita, S. (1991) *J. Supercond.*, **4**, 297.
15 Fujita, S. (1992) *J. Supercond.*, **5**, 83.
16 Dennison, D.M. (1927) *Proc. Roy. Soc.*, London, **115**, 483.
17 Hori, T. (1927) *Z. Phys.*, **44**, 834.
18 Fermi, E. and Yang, C.N. (1949) *Phys. Rev.*, **76**, 1739.
19 Gell-Mann, M. and Ne'eman, Y. (1964) *The Eightfold Way*, Benjamin, Reading, MA.

30
Superconductivity

Basic superconducting properties, occurrence of superconductors, theoretical background, and quantum statistical theory are described in this chapter.

30.1
Basic Properties of a Superconductor

Superconductivity is characterized by the following six basic properties: zero resistance, the Meissner effect, magnetic flux quantization, Josephson effects, gaps in elementary excitation energy spectra, and a sharp phase change. We shall briefly describe these properties in this section.

30.1.1
Zero Resistance

The phenomenon of superconductivity was discovered in 1911 by Kamerlingh Onnes [1] who measured extremely small electric resistance in mercury below a certain critical temperature T_c ($\approx$ 4.2 K). His data are reproduced in Figure 30.1. This *zero resistance* property can be confirmed by a never-decaying supercurrent ring experiment described in Section 30.1.3.

30.1.2
Meissner Effect

Substances that become superconducting at finite temperatures will be called *superconductors* in the present text. If a superconductor below T_c is placed under a weak magnetic field, it repels the magnetic field **B** completely from its interior as shown in Figure 30.2. This is called the *Meissner effect*, and it was discovered by Meissner and Ochsenfeld [2] in 1933.

The Meissner effect can be demonstrated dramatically by a floating magnet as shown in Figure 30.3. A small bar magnet above T_c simply rests on a superconductor dish. If the temperature is lowered below T_c, then the magnet will float as indicated. The gravitational force exerted on the magnet is balanced by the mag-

Mathematical Physics. Shigeji Fujita and Salvador V. Godoy
Copyright © 2010 WILEY-VCH Verlag GmbH & Co. KGaA, Weinheim
ISBN: 978-3-527-40808-5

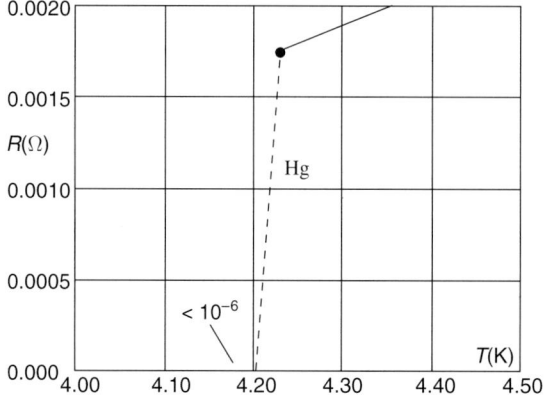

Fig. 30.1 Resistance versus temperature, after Kamerling-Onnes [1].

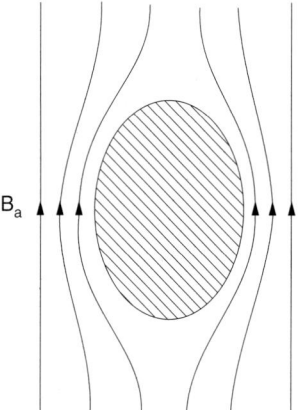

Fig. 30.2 A superconductor expels a weak magnetic field.

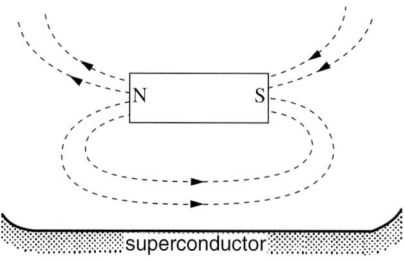

Fig. 30.3 A floating magnet.

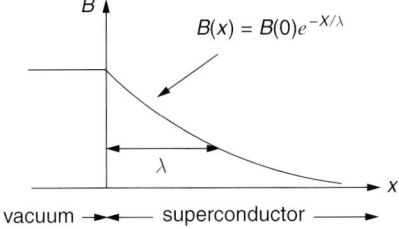

Fig. 30.4 Penetration of the magnetic field into a superconductor slab.

netic pressure (part of electromagnetic stress tensor) due to the inhomogeneous magnetic field (B-field) surrounding the magnet, which is represented by the magnetic flux lines.

Later, more refined experiments revealed that the small magnetic field penetrates into a very thin surface layer of the superconductor. Consider the boundary of a semi-infinite slab. When an external field is applied parallel to the boundary, the B-field falls off exponentially:

$$B(x) = B(0)e^{-x/\lambda} \qquad (30.1)$$

as indicated in Figure 30.4. Here, λ is called a *penetration depth*, which is of the order of 500 Å in most superconductors at lowest temperatures. Its small value on a macroscopic scale allows us to describe the superconductor as being perfectly diamagnetic. The penetration depth λ plays a very important role in the description of the magnetic properties.

30.1.3
Ring Supercurrent and Flux Quantization

Let us take a ring-shaped cylindrical superconductor. If a weak magnetic field **B** is applied along the ring axis and the temperature is lowered below T_c, then the field is expelled from the ring due to the Meissner effect. If the field is slowly reduced to zero, part of the magnetic flux lines may be trapped as shown in Figure 30.5. The magnetic moment generated is found to be maintained by a *never-decaying* supercurrent flowing around the ring [3].

More delicate experiments [4, 5] showed that the *magnetic flux* enclosed by the ring is *quantized* as

$$\Phi = n\Phi_0, \quad n = 0, 1, 2, \ldots \qquad (30.2)$$

$$\Phi_0 = \frac{h}{2e} = \frac{\pi\hbar}{e} = 2.07 \times 10^{-7} \text{ gauss cm}^2 \qquad (30.3)$$

Φ_0 is called a *flux quantum*. The experimental data obtained by Deaver and Fairbank [4] is shown in Figure 30.6. The superconductor exhibits a quantum state described by a quantum number n.

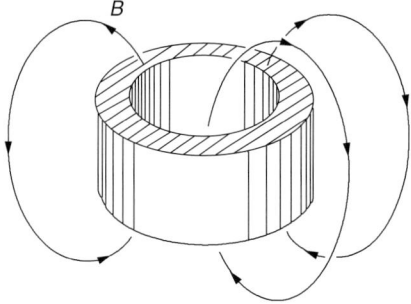

Fig. 30.5 A set of magnetic flux lines are trapped in the ring.

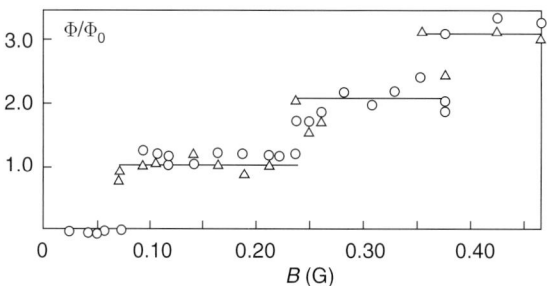

Fig. 30.6 The magnetic flux quantization, after Deaver and Fairbank [4]. The two sets of data are shown as △ and ○.

30.1.4
Josephson Effects

Let us take two superconductors (S_1, S_2) separated by an oxide layer of thickness on the order of 10 Å, called a *Josephson junction*. We use this system as part of a circuit including a battery as shown in Figure 30.7. Above T_c, the two superconductors, S_1 and S_2, and the junction I all show potential drops. If the temperature is lowered beyond T_c, the potential drops in S_1 and S_2 disappear because of zero resistance. The potential drop across the junction I also disappears! In other words, the supercurrent runs through the junction I with no energy loss. Josephson predicted [6], and later experiments [7] confirmed, this *Josephson tunneling* or *dc Josephson effect*.

Let us take a closed-loop superconductor containing two similar Josephson junctions and make a circuit as shown in Figure 30.8. Below T_c, the supercurrent I branches out into I_1 and I_2. We now apply a magnetic field **B** perpendicular to the loop. The magnetic flux can go through the junctions, and the field can be changed continuously. The total current I is found to have an oscillatory component:

$$I = I^{(0)} \cos(\pi \Phi / \Phi_0), \quad (I^{(0)} = \text{constant}) \tag{30.4}$$

where Φ is the magnetic flux enclosed by the loop, indicating that the two supercurrents I_1 and I_2, macroscopically separated (~ 1 mm), interfere just as two laser

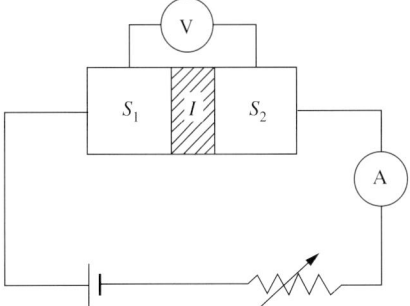

Fig. 30.7 Two superconductors, S_1 and S_2, and a Josephson junction I are connected with a battery.

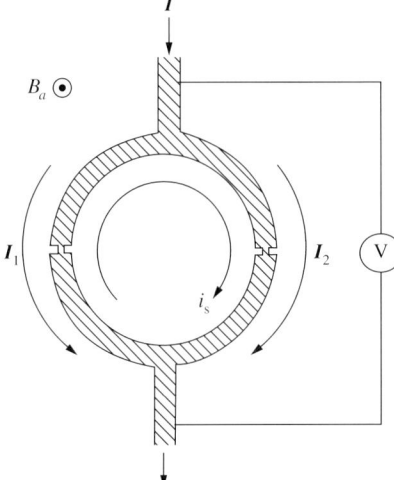

Fig. 30.8 Superconducting quantum interference device (SQUID).

beams coming from the same source. This is called *Josephson interference*. A sketch of an interference pattern [8] is shown in Figure 30.9.

The circuit in Figure 30.8 can be used to detect an extremely weak magnetic field. The device is called the *Superconducting Quantum Interference Device* (SQUID).

In true thermodynamic equilibrium, there can be no currents, super or normal. Thus, we must deal with a nonequilibrium condition when discussing the basic properties of superconductors such as zero resistance, flux quantization, and Josephson effects. All of these arise from the supercurrents that dominate the transport and magnetic phenomena. When a superconductor is used to form a circuit with a battery, and a steady state is established, all currents passing the superconductor are supercurrents. Normal currents due to the moving electrons and other charged particles do not show up because no voltage difference can be developed in a homogeneous superconductor.

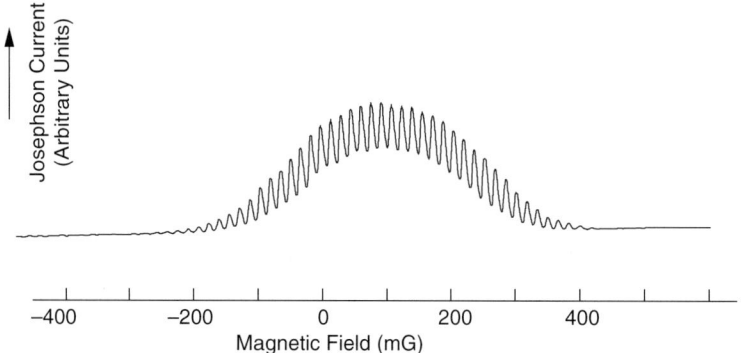

Fig. 30.9 Current versus magnetic field, after Jaklevic et al. [8].

30.1.5
Energy Gap

If a continuous band of the excitation energy is separated by a finite gap ε_g from the discrete ground-state energy level as shown in Figure 30.10, then this gap can be detected by photoabsorption [9, 10], quantum tunneling [11], heat capacity [12] and other experiments. The energy gap ε_g is found to be temperature dependent. The energy gap $\varepsilon_g(T)$ as determined from the tunneling experiments [13] is shown in Figure 30.11. The energy gap is zero at T_c, and reaches a maximum value $\varepsilon_g(0)$ as temperature approaches 0 K.

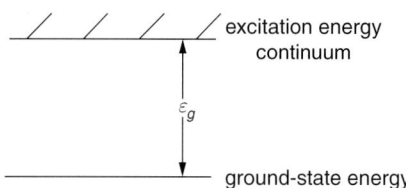

Fig. 30.10 Excitation-energy spectrum with a gap.

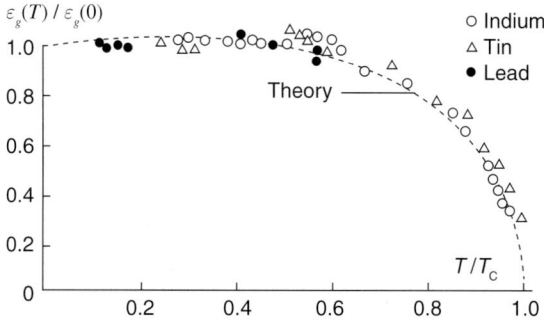

Fig. 30.11 The energy gap $\varepsilon_g(T)$ versus temperature, as determined by tunneling experiments, after Giaever and Megerle [13].

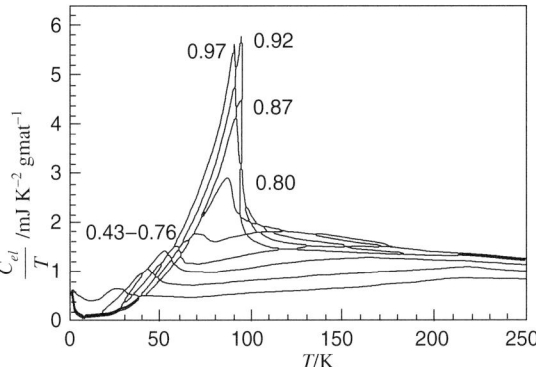

Fig. 30.12 Electronic heat capacity C_{el}/T for YBa$_2$ Cu$_3$ O$_{6+x}$ plotted against T. Loram et al. [14].

30.1.6
Sharp Phase Change

The superconducting transition is a sharp phase change. In Figure 30.12, the data of the electronic heat capacity C_{el} plotted as C_{el}/T against T as reported by Loram et al. [14], for YBa$_2$CuO$_{6+x}$ (2-D superconductor) with the x-values as shown. The data at $x = 0.92$ have the highest T_c. There is no latent heat and no discontinuity in C_{el} at T_c.

30.2
Occurrence of a Superconductor

30.2.1
Elemental Superconductors

More than 40 elements have been found to become superconducting. Table 30.1 shows the critical temperature T_c and the critical magnetic fields at 0 K, B_0.

Most nonmagnetic metals can be superconductors, with notable exceptions being monovalent metals such as Li, Na, K, Cu, Ag, and Au. Some metals can become superconductors under applied pressures and/or in thin films, and these are indicated by asterisks in Table 30.1.

30.2.2
Compound Superconductors

Hundreds of metallic compounds are found to be superconductors. A selection of compound superconductors with critical temperature T_c are shown in Table 30.2.

Compound superconductors exhibit type II magnetic behavior different from that of type I elemental superconductors. A very weak magnetic field is expelled from the body (the Meissner effect) just as by the type I superconductor. If the field

30 Superconductivity

Table 30.1 Superconductivity Parameters of the Elements. Transition temperature in K and critical magnetic field at 0 K in gauss. * denotes superconductivity in thin films or under high pressures.

1	2	3	4	5	6	7	8	9	10	11	12	13	14	15	16	17	18
Li	Be*											B	C*	N	O		Ne
Na	Mg											Al $T_c=1.18$ $B_0=105$	Si*	P	S		Ar
K	Ca	Sc	Ti 0.39 / 100	V 5.38 / 1420	Cr	Mn	Fe	Co	Ni	Cu	Zn 0.87 / 53	Ga 1.09 / 51	Ge*	As	Se*		Kr
Rb	Sr	Y*	Zr 0.54 / 47	Nb 9.20 / 1980	Mo 0.92 / 95	Tc 7.77 / 1410	Ru 0.51 / 70	Rh	Pd	Ag	Cd 3.40	In 3.40 / 293	Sn 3.72 / 309	Sb*	Te*		Xe
Cs*	Ba*	La 6.00 / 1100	Hf	Ta 4.48 / 830	W 0.01 / 1.07	Re 1.69 / 198	Os 0.65 / 65	Ir 0.14 / 19	Pt*	Au	Hg 4.15 / 412	Tl 2.39 / 171	Pb 7.19 / 803	Bi*	Po		Rn
Fr	Ra	Ac															

Ce*	Pr	Nd	Pm	Sm	Eu	Gd	Tb	Dy	Ho	Er	Tm	Yb	Lu	
Th 1.36 / 1.62	Pa 1.4	U 0.68	Np	Pu	Am	Cm	Bk	Cf	Es	Fm	Md	No	Lw	

Table 30.2 Critical temperatures of selected compounds

Compound	T_c (K)	Compound	T_c (K)
Nb_3Ge	23.0	MoN	12.0
$Nb_3(Al_{0.8}Ge_{0.2})$	20.9	V_3Ga	16.5
Nb_3Sn	18.05	V_3Si	17.1
Nb_3Al	17.5	UCo	1.70
Nb_3Au	11.5	Ti_2Co	3.44
NbN	16.0	La_3In	10.4

a Note: the critical temperature T_c tends to be higher in compounds than in elements. Nb_3Ge has the highest T_c (~ 23 K).

is raised beyond the *lower critical field* H_{c1}, the body allows a partial penetration of the field, still remaining in the superconducting state. A further field increase turns the body to a normal state upon passing the *upper critical field* H_{c2}. Between H_{c1} and H_{c2}, the superconductor is in a mixed state in which magnetic flux lines surrounded by supercurrents, called *vortices*, penetrate the body. The critical fields versus temperature are shown in Figure 30.13. The upper critical field H_{c2} can be very high ($20\,T = 2 \times 10^5$ G for Nb_3Sn). Also, the critical temperature T_c tends to be high for high-H_{c2} superconductors. These properties make compound superconductors useful for devices and magnets.

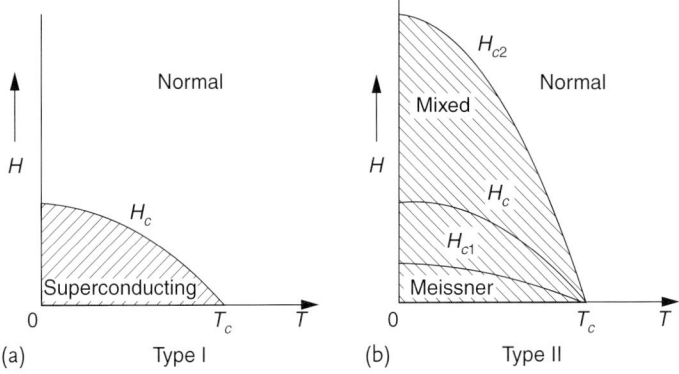

Fig. 30.13 Phase diagrams of type I and type II superconductors.

30.2.3
High-T_c Superconductors

In 1986 Bednorz and Müller [15] reported the discovery of the first *cuprate superconductors*, also called *high-temperature superconductor*, (HTSC). Since then, many investigations have been carried out on the high-T_c superconductors including YBaCuO with $T_c \sim 94$ K [16]. The boiling point of abundantly available and inexpensive liquid nitrogen (N) is 77 K. So the application potential of HTSCs, which are of type II, appears to be great. The superconducting state of these conductors is essentially the same as that of elemental superconductors.

30.3
Theoretical Survey

We review briefly the current theories.

30.3.1
The Cause of Superconductivity

At present, superconductivity in solids is believed to be caused by phonon exchange. When a phonon is exchanged between two electrons, a bound electron pair, called a *Cooper pair* (pairon) [17] is formed.

The exchange of a boson (phonon) between the two fermions (electrons) can be pictured as the *emission of a boson* by a fermion and the subsequent *absorption* of the boson by another fermion. The emission and absorption of the boson requires a *second quantization formulation* in which creation and annihilation operators are introduced. It turns out that the second quantization formulation can describe the dynamics of fermions as well. The second quantization formulation was summarized in Chapter 28.

30.3.2
The Bardeen–Cooper–Schrieffer Theory

In 1957 *Bardeen, Cooper and Schrieffer* (BCS) published a classic paper [18] that is regarded as one of the most important theoretical works in the twentieth century. Bardeen, Cooper and Schrieffer got the 1972 Nobel physics prize for this work.

We shall briefly review the BCS theory.

In spite of the Coulomb repulsion among electrons there exists a sharp Fermi surface for the normal state of a conductor, as described by the Fermi-liquid model of Landau [19, 20]. The phonon-exchange attraction can bind pairs of electrons near the Fermi surface within a distance (energy) equal to Planck's constant $\hbar$ times the Debye frequency ω_D. The bound-electron pairs, each having antiparallel spins and charge (magnitude) $2e$, are called *Cooper pairs* (pairons). Cooper pair and *pairons* both denote the same entity. When we emphasize the quasiparticle aspect rather than the two-electron composition aspect, we use the term pairon more often.

BCS started with a Hamiltonian H in the form:

$$H = \sum_{\substack{k \\ \varepsilon_k > 0}} \sum_s \varepsilon_k c^\dagger_{ks} c_{ks} + \sum_{\substack{k \\ \varepsilon_k < 0}} \sum_s |\varepsilon_k| c_{ks} c^\dagger_{ks}$$

$$+ \frac{1}{2} \sum_{k_1} \sum_{s_1} \cdots \sum_{k_4} \sum_{s_4} \langle 1, 2| U |3, 4 \rangle c^\dagger_1 c^\dagger_2 c_4 c_3 \quad (30.5)$$

where ε_{k_1} is the kinetic energy of a free electron measured relative to the Fermi energy ε_F, and $c^\dagger_{k_1 s_1}$ ($c_{k_1 s_1}$) are creation (annihilation) operators satisfying the Fermi anticommutation rules:

$$\{c_{ks}, c^\dagger_{k's'}\} = c_{ks} c^\dagger_{k's'} + c^\dagger_{k's'} c_{ks} = \delta_{k,k'} \delta_{s,s'} \quad (30.6)$$

$$\{c_{ks}, c_{k's'}\} = \{c^\dagger_{ks}, c^\dagger_{k's'}\} = 0$$

The first (second) sum on the rhs of (30.5) represents the total kinetic energy of "electrons" with positive ε_k ("holes" with negative ε_k). The matrix element $\langle 1, 2| U |3, 4 \rangle$ denotes the net interaction arising from the virtual exchange of a phonon and the Coulomb repulsion between electrons. Specifically

$$\langle 1, 2| U |3, 4 \rangle = \begin{cases} -\dfrac{V_0}{V} \delta_{k_1+k_2, k_3+k_4} \, \delta_{s_1, s_3} \, \delta_{s_2, s_4} & \text{if } |\varepsilon_m| < \hbar \omega_D \\ 0 & \text{otherwise} \end{cases} \quad (30.7)$$

where V_0 is a positive constant (energy).

Starting with the Hamiltonian in (30.5), BCS obtained an expression W for the ground-state energy

$$W = \hbar \omega_D \mathcal{N}(0) w_0 = N_0 w_0 \quad (30.8)$$

where

$$w_0 = \frac{-2\hbar \omega_D}{\exp[2/v_0 \mathcal{N}(0)] - 1}, \quad v_0 = V_0 V^{-1} \quad (30.9)$$

is the pairon ground-state energy;

$$N_0 \equiv \hbar\omega_D \mathcal{N}(0) \tag{30.10}$$

is the total number of pairons, and $\mathcal{N}(0)$ the density of states per spin at the Fermi energy. In the variational calculation of the ground-state energy BCS found that the *unpaired electrons*, often called the *quasielectrons*, not joining the ground pairons that form the supercondensate, have the energy

$$E_k = \left(\Delta^2 + \varepsilon_k^2\right)^{1/2} \tag{30.11}$$

The energy constant Δ, called the *quasielectron energy gap*, in (30.11) is greatest at 0 K and decreases to zero as temperature is raised to the critical temperature T_c. BCS further showed that the energy gap at 0 K, $\Delta(T=0) \equiv \Delta_0$ and the critical temperature T_c are related (in the weak coupling limit) by

$$2\Delta_0 = 3.53 k_B T_c \tag{30.12}$$

These findings of (30.8) to (30.12) are among the most important results obtained in the BCS theory. A large body of theoretical and experimental work followed several years after the BCS theory. By 1964 the general consensus was that the BCS theory is an essentially correct theory of superconductivity.

BCS assumed the Hamiltonian in (30.5) containing "electron" and "hole" kinetic energies. They also assumed a spherical Fermi surface. These two assumptions, however, contradict each other. If a Fermi sphere whose inside (outside) is filled with electrons is assumed, then there are "electrons" ("holes") only, see Book 1, Section 10.4. Besides this logical inconsistency, if a free-electron model having a spherical Fermi surface is assumed, then the question of why metals such as sodium (Na) and potassium (K) remain normal cannot be answered. We must incorporate the band structures of electrons more explicitly.

30.4
Quantum-Statistical Theory

In a quantum-statistical theory one starts with a reasonable Hamiltonian and derives everything from this, following step-by-step calculations. Only Heisenberg's equation of motion (quantum mechanics) Pauli's exclusion principle (quantum statistics) and Boltzmann's statistical principle (grand canonical ensemble theory) are assumed.

30.4.1
The Full Hamiltonian

BCS defined in their original work [18], see (30.5), the "electron" ("hole") as the quasiparticle having an energy higher (lower) than the Fermi energy. The "electron" ("hole") in semiconductor physics is defined as a quasiparticle possessing charge $-e$ ($+e$) that circulates counterclockwise (clockwise) when viewed from the tip of the applied magnetic field vector. We take this second definition. The "electron" ("holes") are quasiparticles generated near the Fermi surface on the positive (negative) side of the surface with the convention that the positive normal vector points in the energy-increasing direction. It then follows that the "hole" is not an antielectron in contrast with the positron. The "electron" (1) and the "hole" (2) have different effective masses:

$$m_1 = m_2 \tag{30.13}$$

Fujita and his group developed a quantum-statistical theory of superconductivity in a series of papers [21–25]. We present this theory in the present section.

In the ground state there are no currents for any system, super or normal. To describe a ring supercurrent that can run indefinitely at 0 K, we must introduce *moving pairons*, that is, pairons with finite center-of-mass (CM) momenta. Creation operators for "electron" (1) and "hole" (2) pairons are defined by

$$B_{12}^{(1)\dagger} \equiv B_{k_1 \uparrow k_2 \downarrow}^{(1)\dagger} \equiv c_1^{(1)\dagger} c_2^{(1)\dagger}, \quad B_{34}^{(2)\dagger} \equiv c_4^{(2)\dagger} c_3^{(2)\dagger} \tag{30.14}$$

(The pairon operators are denoted by Bs, which should not be confused with the magnetic field **B**.)

The number operators for "electrons" ($j=1$) and "hole" ($j=2$) are represented by

$$n_1^{(j)} \equiv c_{k_1 \uparrow}^{(j)\dagger} c_{k_1 \uparrow}^{(j)}, \quad n_2^{(j)} \equiv c_{k_2 \downarrow}^{(j)\dagger} c_{k_2 \downarrow}^{(j)} \tag{30.15}$$

Let us now introduce the relative and net momenta (**k**, **q**) such that

$$\mathbf{k} \equiv \frac{1}{2}(\mathbf{k}_1 - \mathbf{k}_2), \quad \mathbf{q} \equiv \mathbf{k}_1 + \mathbf{k}_2$$

$$\mathbf{k}_1 = \mathbf{k} + \mathbf{q}/2, \quad \mathbf{k}_2 = -\mathbf{k} + \mathbf{q}/2 \tag{30.16}$$

We can alternatively represent pairon annihilation operators by

$$B_{kq}^{\prime(1)} \equiv B_{k_1 \uparrow k_2 \downarrow}^{(1)} \equiv c_{-k+q/2\downarrow}^{(1)} c_{k+q/2\uparrow}^{(1)}, \quad B_{kq}^{\prime(2)} \equiv c_{k+q/2\uparrow}^{(2)} c_{-k+q/2\downarrow}^{(2)} \tag{30.17}$$

The prime on B will be dropped hereafter.

Using the new notation, we write the full Hamiltonian H as

$$H = \sum_{k,s} \varepsilon_k^{(1)} n_{k,s}^{(1)} + \sum_{k,s} \varepsilon_k^{(2)} n_{k,s}^{(2)}$$

$$- \sum_k {}' \sum_q \sum_{k'} {}' v_0 \left[B_{kq}^{(1)\dagger} B_{k'q}^{(1)} + B_{kq}^{(1)\dagger} B_{k'q}^{(2)\dagger} + B_{kq}^{(2)} B_{k'q}^{(1)} + B_{kq}^{(2)} B_{k'q}^{(2)\dagger} \right] \tag{30.18}$$

The zero momentum pairons are often written in terms of b:

$$B^{(j)}_{k0} \equiv b^{(j)}_k \tag{30.19}$$

If we retain zero momentum pairons only, we obtain the BCS reduced Hamiltonian

$$H_0 = \sum_{k,s} \varepsilon^{(1)}_k n^{(1)}_{k,s} + \sum_{k,s} \varepsilon^{(2)}_k n^{(2)}_{k,s}$$

$$- {\sum_k}' {\sum_{k'}}' v_0 \left[b^{(1)\dagger}_k b^{(1)}_{k'} + b^{(1)\dagger}_k b^{(2)\dagger}_{k'} + b^{(2)}_k b^{(1)}_{k'} + b^{(2)}_k b^{(2)\dagger}_{k'} \right] \tag{30.20}$$

30.4.2
Summary of the Results

The cause of the superconductivity is the phonon-exchange attraction. Under favorable conditions (see below), electrons near the Fermi surface form Cooper pairs (pairons) by exchanging phonons. Let us take a typical elemental superconductor such as lead Pb, which forms an fcc lattice. The virtual phonon exchange between a pair of electrons can generate an attraction if kinetic energies of the electrons involved are all close to each other. This exchange can generate an attractive transition (correlation) between "electron" (or "hole") pair states whose energies are separated by twice the limit phonon energy $\hbar\omega_D$, where ω_D is the Debye frequency. Exchanging a phonon can also pair-create $\pm$ pairons from the physical vacuum. Phonons are electrically neutral, and hence, the states of two electrons between which a phonon is exchanged, must have the same net charge before and after the exchange. Because of this, if the Fermi surface is favorable, then equal numbers of $\pm$ pairons are formed in the conductor. The phonon-exchange attraction is a quantum field theoretical effect, and hence it cannot be explained by considering the potential energy alone. In fact, the attraction depends on the kinetic energies of electrons. Pairons move independently with a linear dispersion relation,

$$w_q = w_0 + cq < 0 \tag{30.21}$$

where c is $(2/\pi, 1/2)v_F$ for (2,3)D, v_F is the Fermi velocity (magnitude), and

$$w_0 \equiv \frac{-2\hbar\omega_D}{\exp(2/v_0\mathcal{N}(0)) - 1} \tag{30.22}$$

is the ground-state energy of a pairon. A pairon's motion is very similar to a photon's. Unlike photons, pairons have charges $\pm 2e$, and the total number of $\pm$ pairons in a superconductor is limited. At 0 K the superconductor may contain great and equal numbers of stationary $\pm$ pairons all condensed at zero momentum.

The most striking superconducting phenomenon is a never-decaying supercurrent ring. In the flux quantization experiment, a weak supercurrent goes around the ring, enclosing a number of fluxons (magnetic flux quanta). Here, macroscopic numbers of $\pm$ pairons are condensed at a momentum

$$q_n \equiv \frac{2\pi\hbar n}{L} \tag{30.23}$$

where L is the ring circumference and n a quantum number ($0, \pm 1, \pm 2, \ldots$) such that the flux Φ enclosed by the ring is $|n|$ times the flux quantum $\Phi_0 \equiv (h/2e)$:

$$\Phi = n\Phi_0 \equiv n\frac{h}{2e} \qquad (30.24)$$

The factor $2e$ means that the charge (magnitude) of the current-carrying particle is twice the electron charge e, supporting the BCS picture of a supercondensate composed of pairons of charge (magnitude) $2e$.

The macroscopic supercurrent generated by the supercondensate in motion is not stopped by impurities. This condition is somewhat similar to the situation in which a flowing river (big object) is perturbed but cannot be stopped by a stick (small object). The fact that small perturbations cause no energy loss arises from the quantum nature of the superconducting state. The change in the condensed system state requires redistribution of a great number of pairons. Furthermore, the supercurrent state can refocus by itself if the perturbation is not too strong. This is a bosonic effect peculiar to the condensed bosons moving with a linear dispersion relation. This self-focusing power is most revealing in the Josephson interference, where two supercurrents macroscopically separated up to 1 mm apart can interfere with each other just as two laser beams from the same source. Thus, there is a close similarity between a supercurrent and a laser.

In the steady state realized in a circuit containing superconductor, resistor, and battery, all flowing currents are supercurrents. These supercurrents run in the thin surface layer characterized by the penetration depth λ (~ 500 Å), and they keep the magnetic field off the interior (Meissner effect). The current density j in the superconductor can be represented by

$$j = \begin{cases} \frac{1}{2}en_0\left(v_F^{(2)} - v_F^{(1)}\right) & \text{(3-D)} \\ \frac{2}{\pi}en_0\left(v_F^{(2)} - v_F^{(1)}\right) & \text{(2-D)} \end{cases} \qquad (30.25)$$

where n_0 is the total density of condensed pairons. The $\pm$ pairons move in the same direction, but their speeds are different, and hence the net current does not vanish. If a "hole" were an antiparticle of an electron, then the hole and the electron would have had the same mass (magnitude). Hence the Fermi velocity $v_F^{(j)}$ must be the same. If so, the supercurrent vanishes according to (30.25). If a magnetic field $\mathbf{B}$ is applied, the Lorentz magnetic force tends to separate $\pm$ pairons. Hence, there is a critical magnetic field B_c. The supercurrent by itself generates a magnetic field, so there is a critical current. The picture of a neutral moving supercondensate explains why the superconducting state is not destroyed by the applied voltage. Since there is no net charge: $Q = 0$, no Lorentz electric force $\mathbf{F}_E$ can act on the supercondensate: $\mathbf{F}_E \equiv Q\mathbf{E} = 0$. Thus, the supercurrent is not accelerated, so it can gain no energy from the voltage.

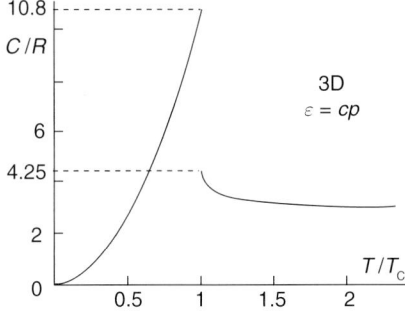

Fig. 30.14 The molar heat capacity C for 3D massless bosons rises like T^3, and reaches $10.8R$ at the transition temperature $T_c = 2.02\hbar c n^{1/3}$. It then drops abruptly by $6.57R$ and approaches the high-temperature limit $3R$.

The system of free pairons (bosons) moving with the linear dispersion relation undergoes a B–E condensation at the critical temperature T_c given by

$$k_B T_c = \begin{cases} 1.01\hbar v_F n_0^{1/3} & \text{(3-D)} \\ 1.24\hbar v_F n_0^{1/2} & \text{(2-D)} \end{cases} \tag{30.26}$$

The heat capacity in 3-D, shown in Figure 30.14, has a discontinuity at T_c, indicating a change of phase of second order. Below T_c there is a supercondensate made up of ± pairons condensed at zero momentum. The density of condensed pairons increases to the maximum n_0 as temperature is lowered toward 0 K. The quasielectron in the presence of the supercondensate has the energy $(\varepsilon_k^2 + \Delta^2)^{1/2}$. The gap Δ in this expression is temperature dependent, and it reaches its maximum Δ_0 at 0 K. The maximum gap Δ_0 can be connected to the critical temperature T_c by the famous BCS formula [18]

$$2\Delta_0 = 3.53 k_B T_c \tag{30.27}$$

in the weak-coupling limit. Phonon exchange is in action at all times and at all temperatures. Thus, two quasielectrons may be bound to form moving pairons. Since quasielectrons have an energy gap Δ, excited pairons also have an energy gap $\varepsilon_g(T)$, which is temperature dependent. The pairon energy gap $\varepsilon_g(T)$ grows to its maximum equal to the binding energy of a Cooper pair, $|w_0|$, as the temperature approaches 0 K.

The molar heat for 2D massless boson system is shown in Figure 30.15. The electronic heat capacity for $YBa_2Cu_3O_{7-\delta}$ [26] is shown in Figure 30.16. Clearly, the experiments support the BEC model of superconductivity.

Moving pairons have negative energies, while quasielectrons have positive energies. By the Boltzmann principle, moving pairons are therefore more numerous than quasielectrons, and these pairons are the predominant elementary excitations below T_c. Pairon energy gaps ε_g strongly influence the heat capacity $C(T)$ below T_c. The $C(T)$ far from T_c shows an exponential-decay-type T dependence due to the

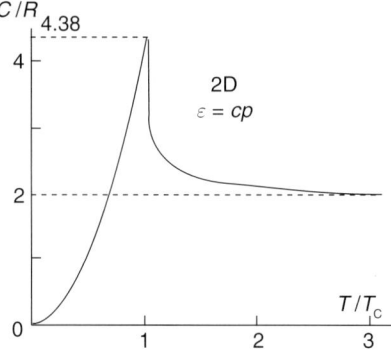

Fig. 30.15 The molar heat capacity C for 2D massless bosons rises like T^2, and reaches $4.38R$ at T_c. Then decreases to $2R$ in the high-temperature limit.

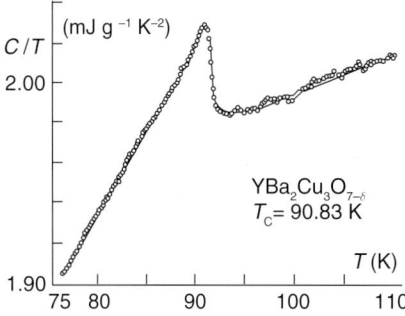

Fig. 30.16 The electronic heat capacity near the critical temperature in polycrystalline YBCO, after Fisher et al. [26].

energy gap ε_g; the maximum heat capacity $C_{\max}$ at T_c is modified by a small but nonnegligible amount. The energy gap ε_g between excited and condensed pairons can be probed by quantum-tunneling experiments. The threshold voltage V_a in the I–V curve for an S-I-S system can be connected simply with ε_g:

$$V_a = \varepsilon_g(T)/e \tag{30.28}$$

This allows a direct observation of the energy gap $\varepsilon_g(T)$ as a function of T.

Compound superconductors have optical and acoustic phonons. The Cooper pairs formed, mediated by optical phonons bridging between "electron"-like and "hole"-like Fermi surfaces, have smaller linear sizes ξ_0 (~ 50 Å). The critical temperature T_c tends to be higher for compound superconductors than for elemental ones, since smaller-size pairons can be packed more densely.

Cuprate superconductors have layered lattice structures. Conduction electrons move only in the CuO_2 planes. Since they are compounds, $\pm$ pairons can be generated with the aid of optical phonons bridging between "electron"-like and "hole"-like 2D Fermi surfaces. The pairon size is small (~ 14 Å for YBCO), and pairons may, therefore, be packed even more densely. The critical temperature T_c, based on the model of free massless bosons moving in 2D, is given by (30.26). The interpair-iron distance $r_0 \equiv n_0^{-1/2}$ is much smaller in cuprates than in elements, and the Fermi velocity v_F is smaller, making the critical temperature T_c higher. Mercury-based cuprates have T_c = 164 K [27]. In 2008 Zhao and his collaboraters found evidence of superconductivity in multiwalled carbon nanotubes with T_c = 1275 K [28]. This T_c is 8 times higher than the record T_c = 164 K for the cuprate superconductors. The nanotubes are extremely strong and technologically important materials. This ultrahigh T_c can also be described by the 2D formula in (30.26) [29].

References

1 Kamerlingh Onnes, H. (1911) *Akad. V. Wetenschappen*, Amsterdam, **14**, 113.
2 Meissner, W. and Ochsenfeld, R. (1933) *Naturwiss*, **21**, 787.
3 File, J. and Mills, R.G. (1963) *Phys. Rev. Lett.*, **10**, 93.
4 Deaver, B.S. and Fairbank, W.M. (1961) *Phys. Rev. Lett.*, **7**, 43.
5 Doll, R. and Näbauer, M. (1961) *Phys. Rev. Lett.*, **7**, 51.
6 Josephson, B.D. (1962) *Phys. Lett.*, **1**, 251.
7 Anderson, P.W. and Rowell, J.M. (1963) *Phys. Rev. Lett.*, **10**, 486.
8 Jaklevic, R.C. et al. (1965) *Phys. Rev.*, **140**, A1628.
9 Glover, III R.E. and Tinkham, M. (1957) *Phys. Rev.* **108**, 243.
10 Biondi, M.A. and Garfunkel, M. (1959) *Phys. Rev.*, **116**, 853.
11 Giaever, I. (1960) *Phys. Rev. Lett.*, **5**, 147, 464.
12 Phillips, N.E. (1959) *Phys. Rev.*, **114**, 676.
13 Giaever, I. and Megerle, K. (1961) *Phys. Rev.*, **122**, 1101.
14 Loram, J.W., Mirza, K.A., Cooper, J.R., and Liang, W.I. (1994) *J. Supercond.*, **7**, 347.
15 Bednorz, J.G. and Müller, K.A. (1986) *Z. Phys. B*, **64**, 189.
16 Wu, M.K. et al. (1987) *Phys. Rev. Lett.*, **58**, 908.
17 Cooper, L.N. (1956) *Phys. Rev.*, **104**, 1189.
18 Bardeen, J., Cooper, J.L., and Schriefer, J.R. (1957) *Phys. Rev.*, **108**, 1175.
19 Landau, L.D. (1956) *J. Exptl. Theoret. Phys.* (USSR), **30**, 1058.
20 Landau, L.D. (1957) *J. Exptl. Theoret. Phys.* (USSR), **32**, 59.
21 Fujita, S. (1991) *J. Supercond.*, **4**, 297.
22 Fujita, S. (1992) *J. Supercond.*, **5**, 83.
23 Fujita, S. and Watanabe, S. (1992) *J. Supercond.*, **5**, 219.
24 Fujita, S. and Watanabe, S. (1993) *J. Supercond.*, **6**, 75.
25 Fujita, S. and Godoy, S. (1993) *J. Supercond.*, **6**, 373.
26 Fisher, R.A., Gordon, J.E., and Phillips, N.E. (1988) *J. Supercond.*, **1**, 231.
27 Maeno, Y. et al. (1994) *Nature*, **372**, 532.
28 Zhao, G.M. and Beeli, P. (2008) *Phys. Rev. B*, **77**, 245433.
29 Fujita, Godoy, S., and Ito, K. (2009) *Quantum Theory of Conducting Matter. Superconductivity*, Springer, New York, p. 86.

31
Complex Numbers and Taylor Series

Complex numbers, complex functions, and Taylor series are discussed in this chapter.

31.1
Complex Numbers

A complex number

$$z = x + iy \tag{31.1}$$

can be represented as a point (x, y) on an xy-plane called a *Gaussian plane* as shown in Figure 31.1. A complex number can also be interpreted as a vector from the origin O to the point $P(x, y)$ with the coordinates (x, y).

The point P can alternatively be represented by the polar coordinates (r, θ) as shown in Figure 31.2. The coordinate set (x, y) and the set (r, θ) are related by

$$x = r\cos\theta, \quad y = r\sin\theta, \quad \tan\theta = \frac{y}{x} \tag{31.2}$$

The length

$$r \equiv |z| = \sqrt{x^2 + y^2} \tag{31.3}$$

is called the *absolute value* or the *magnitude* of z. The angle θ is called the *polar angle*.

The *real* and *imaginary* parts of a complex number $z = x + iy$ are denoted by

$$\text{Re}(z) = x \tag{31.4}$$

$$\text{Im}(z) = y \tag{31.5}$$

If $x = y = 0$, z is said to vanish, and is denoted by zero:

$$z = 0 \quad \text{if} \quad \text{Re}(z) = \text{Im}(z) = 0 \tag{31.6}$$

Mathematical Physics. Shigeji Fujita and Salvador V. Godoy
Copyright © 2010 WILEY-VCH Verlag GmbH & Co. KGaA, Weinheim
ISBN: 978-3-527-40808-5

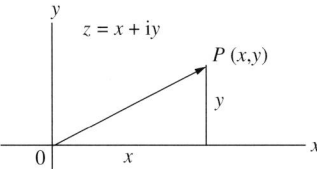

Fig. 31.1 Cartesian coordinates (x, y) of a complex number z.

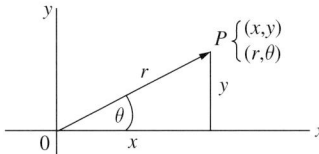

Fig. 31.2 The point $P(x, y)$ in the Gaussian plane can be represented by polar coordinates (r, θ) with $x + iy = r(\cos\theta + i\sin\theta)$, $x = r\sin\theta$ and $y = r\sin\theta$.

The product of two complex numbers z_1 and z_2 is a complex number. Using the polar representation, we obtain

$$z_1 z_2 = r_1 e^{i\theta_1} \cdot r_2 e^{i\theta_2} = r_1 r_2 e^{i(\theta_1 + \theta_2)}$$

$$= r_1 r_2 \left[\cos(\theta_1 + \theta_2) + i \sin(\theta_1 + \theta_2)\right] \tag{31.7}$$

If z_2 does not vanish: $z_2 \neq 0$, then the ratio of z_1 over z_2, z_1/z_2, calculated as can be defined, and is calculated as

$$\frac{z_1}{z_2} = \frac{r_1 e^{i\theta_1}}{r_2 e^{i\theta_2}} = \frac{r_1}{r_2} e^{i(\theta_1 - \theta_2)}$$

$$= \frac{r_1}{r_2} \left[\cos(\theta_1 - \theta_2) + i \sin(\theta_1 - \theta_2)\right] \tag{31.8}$$

For a real number a, de Moivre's theorem states that

$$z^a = [re^{i\theta}]^a = r^a e^{ia\theta} = r^a(\cos a\theta + i \sin a\theta) \tag{31.9}$$

If $a = 1/n$, where n is any positive integer, (31.9) can be written

$$z^{1/n} = r^{1/n} \left[\cos\left(\frac{\theta + 2k\pi}{n}\right) + i \sin\left(\frac{\theta + 2k\pi}{n}\right)\right] \tag{31.10}$$

where k is any integer. From this equation the nth root (there are n roots) of a complex number z can be obtained from (31.10) by putting $k = 0, 1, 2, \ldots, n - 1$.

31.2
Exponential and Logarithmic Functions

31.2.1
Laws of Exponents

In the following, p, q are real numbers, a, b are positive numbers and m, n are positive integers.

$$a^p a^q = a^{p+q} \tag{31.11}$$

$$a^p / a^q = a^{p-q} \tag{31.12}$$

$$(a^p)^q = a^{pq} \tag{31.13}$$

$$a^0 = 1 \quad a \neq 0 \tag{31.14}$$

$$a^{-p} = 1/a^p \tag{31.15}$$

$$(ab)^p = a^p b^p \tag{31.16}$$

$$\sqrt[n]{a} = a^{1/n} \tag{31.17}$$

$$\sqrt[n]{a^m} = a^{m/n} \tag{31.18}$$

$$\sqrt[n]{(a/b)} = \sqrt[n]{a}/\sqrt[n]{b} \tag{31.19}$$

31.2.2
Natural Logarithm

The *natural logarithm* of N is denoted by $\ln N$. Some important properties of logarithms are

$$\ln(mn) = \ln m + \ln n \tag{31.20}$$

$$\ln(m/n) = \ln m - \ln n \tag{31.21}$$

$$\ln(m^p) = p \ln m \tag{31.22}$$

31.2.3
Relationship between Exponential and Trigonometric Functions

Two fundamental relation are *Euler's identities*:

$$e^{i\theta} = \cos\theta + i\sin\theta \tag{31.23}$$

$$e^{-i\theta} = \cos\theta - i\sin\theta \tag{31.24}$$

We obtain

$$\sin\theta = \frac{e^{i\theta} - e^{-i\theta}}{2i} \qquad (31.25)$$

$$\cos\theta = \frac{e^{i\theta} + e^{-i\theta}}{2} \qquad (31.26)$$

$$\tan\theta = \frac{\sin\theta}{\cos\theta} = \frac{e^{i\theta} - e^{-i\theta}}{i(e^{i\theta} + e^{-i\theta})} \qquad (31.27)$$

The exponential function $e^{i\theta}$ has a period $2\pi i$:

$$e^{i(\theta + 2n\pi)} = e^{i\theta}, \quad n = \text{integer} \qquad (31.28)$$

The logarithm of a complex number $re^{i\theta}$ is multivalued:

$$\ln(re^{i\theta}) = \ln r + i\theta + 2\pi n i, \quad n = \text{integer} \qquad (31.29)$$

31.3
Hyperbolic Functions

31.3.1
Definition of Hyperbolic Functions

$$\text{Hyperbolic sine of } x = \sinh x \equiv \frac{e^x - e^{-x}}{2} \qquad (31.30)$$

$$\text{Hyperbolic cosine of } x = \cosh x \equiv \frac{e^x + e^{-x}}{2} \qquad (31.31)$$

$$\text{Hyperbolic tangent of } x = \tanh x \equiv \frac{\sinh x}{\cosh x} = \frac{e^x - e^{-x}}{e^x + e^{-x}} \qquad (31.32)$$

$$\text{Hyperbolic cotangent of } x = \coth x \equiv \frac{1}{\tanh x} = \frac{e^x + e^{-x}}{e^x - e^{-x}} \qquad (31.33)$$

$$\text{Hyperbolic secant of } x = \text{sech } x \equiv \frac{1}{\cosh x} = \frac{2}{e^x + e^{-x}} \qquad (31.34)$$

$$\text{Hyperbolic cosecant of } x = \text{csch } x \equiv \frac{2}{e^x - e^{-x}} \qquad (31.35)$$

31.3.2
Addition Formulas

Important addition formulas are

$$\sinh(x \pm y) = \sinh x \cosh y \pm \cosh x \sinh y \qquad (31.36)$$

$$\cosh(x \pm y) = \cosh x \cosh y \pm \sinh x \sinh y \qquad (31.37)$$

$$\tanh(x \pm y) = \frac{\tanh x \pm \tanh y}{1 \pm \tanh x \tanh y} \qquad (31.38)$$

31.3.3
Double-Angle Formulas

Having a double argument can be related to a single argument using

$$\sinh 2x = 2 \sinh x \cosh x \tag{31.39}$$

$$\cosh 2x = \cosh^2 x + \sinh^2 x = 2\cosh^2 x - 1 \tag{31.40}$$

$$\tanh 2x = \frac{2 \tanh x}{1 + \tanh^2 x} \tag{31.41}$$

31.3.4
Sum, Difference and Product of Hyperbolic Functions

$$\sinh x + \sinh y = 2 \sinh \tfrac{1}{2}(x+y) \cosh \tfrac{1}{2}(x-y) \tag{31.42}$$

$$\sinh x - \sinh y = 2 \cosh \tfrac{1}{2}(x+y) \sinh \tfrac{1}{2}(x-y) \tag{31.43}$$

$$\cosh x + \cosh y = 2 \cosh \tfrac{1}{2}(x+y) \cosh \tfrac{1}{2}(x-y) \tag{31.44}$$

$$\cosh x - \cosh y = 2 \sinh \tfrac{1}{2}(x+y) \sinh \tfrac{1}{2}(x-y) \tag{31.45}$$

$$\sinh x \sinh y = \tfrac{1}{2}[\cosh(x+y) - \cosh(x-y)] \tag{31.46}$$

$$\cosh x \cosh y = \tfrac{1}{2}[\cosh(x+y) + \cosh(x-y)] \tag{31.47}$$

$$\sinh x \cosh y = \tfrac{1}{2}[\sinh(x+y) + \sinh(x-y)] \tag{31.48}$$

31.3.5
Relationship between Hyperbolic and Trigonometric Functions

$$\sin(ix) = i \sinh x \tag{31.49}$$

$$\cos(ix) = \cosh x \tag{31.50}$$

$$\tan(ix) = i \tanh x \tag{31.51}$$

$$\sinh(ix) = i \sin x \tag{31.52}$$

$$\cosh(ix) = \cos x \tag{31.53}$$

$$\tanh(ix) = i \tan x \tag{31.54}$$

Problem 31.3.1

Define

$$\cos z \equiv \frac{e^{iz} + e^{-iz}}{2}, \quad \sin z \equiv \frac{e^{iz} - e^{-iz}}{2i}$$

Show that

1. $\cos^2 z + \sin^2 z = 1$
2. $\cos(z_1 + z_2) = \cos z_1 \cos z_2 - \sin z_1 \sin z_2$
3. $\sin(z_1 + z_2) = \sin z_1 \cos z_2 + \cos z_1 \sin z_2$
4. $\dfrac{d}{dz} \cos z = -\sin z$
5. $\dfrac{d}{dz} \sin z = \cos z$

Problem 31.3.2

Define

$$\cosh z \equiv \frac{e^{z} + e^{-z}}{2}, \quad \sinh z \equiv \frac{e^{z} - e^{-z}}{2}$$

Show that

1. $\cos z \equiv \cos(x + iy) = \cos x \cosh y - i \sin x \sin y$
2. $\sin z \equiv \sin(x + iy) = \sin x \cosh y + i \cos x \sinh y$
3. $\cosh z \equiv \cosh(x + iy) = \cosh x \cos y + i \sinh x \sin y$
4. $\sinh z \equiv \sinh(x + iy) = \sinh x \cos y + i \cosh x \sin y$

31.4
Taylor Series

31.4.1
Derivatives

Let us consider a function of a real variable x, $-\infty < x < \infty$. The first derivative $f'(x)$ is defined by

$$f'(x) \equiv \frac{df}{dx} = \lim_{\Delta x \to 0} \frac{f(x + \Delta x) - f(x)}{\Delta x} \tag{31.55}$$

where the increment Δx is positive or negative. If $f'(x)$ exists, the function f is said to be *differentiable*. The operation of taking a derivative is called *differentiation*.

The second derivative $f''(x)$ is defined by

$$f''(x) \equiv \frac{d^2 f}{dx^2} \equiv \frac{d}{dx} \frac{df}{dx} \tag{31.56}$$

The nth derivative $f^{(n)}(x)$ is defined by

$$f^{(n)}(x) \equiv \frac{d^n f}{dx^n} \equiv \frac{d}{dx}\left(\frac{d^{n-1} f}{dx^{n-1}}\right) \tag{31.57}$$

31.4.2 Taylor Series

If the function f is smooth, the infinite series

$$f(x) = f(a) + (x-a)f'(a) + \cdots + \frac{(x-a)^n}{n!} f^{(n)}(a) + \cdots \tag{31.58}$$

may exist. This is called the *Taylor series* for $f(x)$ about the *expansion point* $x = a$. If $a = 0$, the series is often called the *MacLaurin series*. These series, often called the *power series*, generally converge for all values of x in some interval, called the *interval of convergence*, and diverges for all x outside this interval.

31.4.3 Binomial Series

$$(a+x)^n = a^n + na^{n-1}x + \frac{n(n-1)}{2!} a^{n-2}x^2 + \cdots \tag{31.59}$$

is called a *binomial series*. Special cases are

$$(a+x)^2 = a^2 + 2ax + x^2 \tag{31.60}$$

$$(a+x)^3 = a^3 + 3a^2 x + 3ax^2 + x^3 \tag{31.61}$$

Since

$$(a+x)^p = a^p (1+X)^p, \quad X \equiv x/a \tag{31.62}$$

the binomial series in (31.59) can be reduced with $a = 1$ to

$$(1+x)^n = 1 + nx + \frac{n(n-1)}{2!} x^2 + \cdots \tag{31.63}$$

Special cases are

$$(1+x)^{-1} = 1 - x + x^2 - x^3 + \cdots \quad -1 < x < 1 \tag{31.64}$$

where the interval of convergence is $-1 < x < 1$.
Similarly

$$(1+x)^{-2} = 1 - 2x + 3x^2 - 4x^3 + \cdots \quad -1 < x < 1 \tag{31.65}$$

$$(1+x)^{-1/2} = 1 - \frac{1}{2}x + \frac{1 \cdot 3}{2 \cdot 4} x^2 - \cdots \quad -1 < x \leq 1 \tag{31.66}$$

$$(1+x)^{1/2} = 1 + \frac{1}{2}x - \frac{1}{2 \cdot 4} x^2 + \cdots \quad -1 < x \leq 1 \tag{31.67}$$

31.4.4
Series for Exponential and Logarithmic Functions

$$e^x = 1 + x + \frac{x^2}{2!} + \frac{x^3}{3!} + \cdots \quad -\infty < x < \infty \tag{31.68}$$

This series converges for all x-values.

$$\ln(1 + x) = x - \frac{x^2}{2} + \frac{x^3}{3} - \cdots \quad -1 < x \leq 1 \tag{31.69}$$

$$\frac{1}{2} \ln\left(\frac{1+x}{1-x}\right) = x + \frac{x^3}{3} + \frac{x^5}{5} + \cdots \quad -1 < x < 1 \tag{31.70}$$

31.5
Convergence of a Series

Let us consider some important series. The inverse integer series:

$$1 + \frac{1}{2} + \frac{1}{3} - \cdots \tag{31.71}$$

The alternating inverse integer series:

$$1 - \frac{1}{2} + \frac{1}{3} - \cdots = \sum_{n=0}^{\infty} (-1)^n \frac{1}{n} = \ln 2 \tag{31.72}$$

The oscillating series with $+1$ and -1

$$1 - 1 + 1 - 1 + \cdots \tag{31.73}$$

We consider an infinite sum (series)

$$s_1 + s_2 + s_3 + \cdots = S \tag{31.74}$$

The finite sum of the first n terms

$$S_n \equiv s_1 + s_2 + s_3 + \cdots s_n \tag{31.75}$$

is defined. If, for a given positive number ε, there exists the series S and an integer N such that

$$|S - S_n| < \varepsilon, \quad \text{for all } n > N \tag{31.76}$$

then the infinite sum S is said to *converge*. Otherwise, the series *diverges*. According to this definition, the series

$$\sum_{n=0}^{\infty} (-1)^n \, n^{-1}$$

converges to ln 2. On the other hand, the series

$$\sum_{n=0}^{\infty} n^{-1}$$

and the oscillating one

$$\sum_{n=0}^{\infty} (-1)^n$$

diverge.

The series

$$\sum_{n=0}^{\infty} (-1)^n \frac{1}{n^2} = 1 - \frac{1}{2^2} + \frac{1}{3^2} - \frac{1}{4^2} + \cdots \tag{31.77}$$

and the associated absolute (value) series with the omission of alternative signs $(-1)^n$:

$$\sum_{n=0}^{\infty} \frac{1}{n^2} = 1 + \frac{1}{2^2} + \frac{1}{3^2} + \cdots \tag{31.78}$$

are both convergent. We say that

$$\sum_{n=0}^{\infty} (-1)^n n^{-2}$$

is *absolutely convergent*. Since the series

$$\sum_{n=0}^{\infty} n^{-1}$$

diverges, the convergent alternating series

$$\sum_{n=0}^{\infty} (-1)^{n+1} n^{-1}$$

is not absolutely convergent.

The following two convergent tests are very useful.
1. Comparison Test
   Let us consider a series

$$S = \sum_{n=0}^{\infty} s_n \tag{31.79}$$

and a positive-number convergent series

$$C = \sum_{n=0}^{\infty} c_n < \infty, \quad c_n > 0 \tag{31.80}$$

If

$$|s_n| < c_n$$

then the series S is absolutely convergent.

2. Ratio Test

Consider the limit

$$\lim_{n \to \infty} \left| \frac{s_{n+1}}{s_n} \right| \equiv \lambda \tag{31.81}$$

If $\lambda < 1$, then the series $\sum_n S_n \equiv S$ is absolutely convergent. The series diverges if $\lambda > 1$. If $\lambda = 1$, the series may or may not converge. That is, the ratio test does not give definite answers.

32
Analyticity and Cauchy–Riemann Equations

Analytic functions, poles, branch points, continuous singularities and Cauchy–Riemann relations are discussed in this chapter.

32.1
The Analytic Function

Let us consider a real function $f(x)$ of a real variable x in the domain

$$a \leq x \leq b \qquad (32.1)$$

where (a, b) are real numbers. The numbers a and b are often called the lower and upper limits. The function $f(x)$ may be plotted as the ordinate in the xy-plane. If the curve $y = f(x)$ is smooth, as shown in Figure 32.1, the slope is given by the first derivative df/dx,

$$\frac{df}{dx} = \lim_{\Delta x \to 0} \frac{f(x + \Delta x) - f(x)}{\Delta x} \qquad (32.2)$$

where the increment Δx is positive or negative. The function f having a finite first derivative is called *differentiable*.

The *absolute value* of x is denoted by $|x|$, which is shown in Figure 32.2. The slope of $|x|$ is continuous everywhere except at the origin where the right slope is different from the left one. The function $|x|$ is *nondifferentiable* or *singular* at the origin: $x = 0$.

The first *derivative* of a complex function $f(z)$ of a complex variable z is defined similarly by

$$\frac{df(z)}{dz} \equiv \frac{df}{dz} \equiv f'(z) = \lim_{\Delta z \to 0} \frac{f(z + \Delta z) - f(z)}{\Delta z} \qquad (32.3)$$

where the increment Δz is taken in all directions (all angles) around the point z. If the function f has the first derivative at z, then f is called *analytic, differentiable* or *regular*. The term analytic is most often used, but other terms are also used. It is good to have all three adjectives.

Mathematical Physics. Shigeji Fujita and Salvador V. Godoy
Copyright © 2010 WILEY-VCH Verlag GmbH & Co. KGaA, Weinheim
ISBN: 978-3-527-40808-5

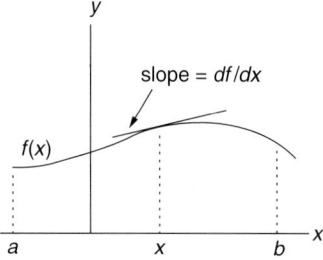

Fig. 32.1 The slope of a smooth function $y = f(x)$ is given by the first derivative df/dx.

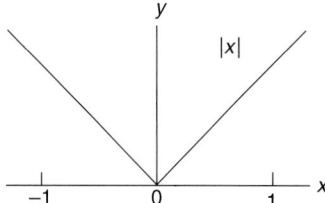

Fig. 32.2 The absolute value of x, $|x|$, is singular at $x = 0$, where the right slope is different from the left one.

Let us first choose $f = z$. Using (32.3) we obtain

$$\frac{d}{dz}z = \lim_{\Delta z \to 0} \frac{z + \Delta z - z}{\Delta z} = \lim_{\Delta z \to 0} 1 = 1 \tag{32.4}$$

Next, we choose $f = z^2$. Its derivative is:

$$\frac{d}{dz}z^2 = \lim_{\Delta z \to 0} \frac{(z + \Delta z)^2 - z^2}{\Delta z} = \lim_{\Delta z \to 0}(2z + \Delta z) = 2z \tag{32.5}$$

More generally, we obtain

$$\frac{d}{dz}z^n = nz^{n-1}, \quad n = 1, 2, 3, \ldots \tag{32.6}$$

This is a straightforward generalization of the results in a real function theory:

$$\frac{d}{dx}x^n = nx^{n-1}, \quad n = 1, 2, 3, \ldots \tag{32.7}$$

Theorem

Any polynomial $a_0 + a_1 z + a_2 z^2 + \ldots + a_n z^n$, where a_i are complex numbers, are analytic in the entire Gaussian plane.

This theorem follows directly from (32.6). The proof is omitted.

32.2
Poles

Let us consider the function

$$z^{-1} \equiv 1/z \tag{32.8}$$

This function z^{-1} is analytic along the real x-axis except at the origin 0, where it diverges to $+\infty$ or $-\infty$ depending on whether the point z approaches 0, from the right or from the left. Hence, this function is singular at the origin. In fact, the function z^{-1} diverges to $-i$ (i) ∞ if z approaches the origin from the positive (negative) imaginary y-axis. The z^{-1} has a first derivative

$$\frac{d}{dz}\frac{1}{z} = -\frac{1}{z^2} \tag{32.9}$$

and is therefore analytic, excluding the origin. We shall say that the function z^{-1} has a *polar singularity* at the origin $z = 0$.

Next, let us consider

$$f = \frac{1}{z - z_0}, \quad z_0 = \text{a complex number} \tag{32.10}$$

This function $(z - z_0)^{-1}$ has the same singular behavior as z^{-1} except that the singular point is located at z_0.

We consider

$$z^{-2} \equiv 1/z^2 \tag{32.11}$$

This function is singular at the origin and analytic everywhere else. Similar behaviors are observed for

$$z^{-n} \equiv 1/z^n, \quad n = 1, 2, 3, \ldots \tag{32.12}$$

The nature of the singularities is different for different n. We shall say that

$$\frac{1}{(z - z_0)^n} \equiv (z - z_0)^{-n} \tag{32.13}$$

has a pole of order n at z_0. If $n = 1$, the singular point is called a *simple pole*. Thus, the function $(z - z_0)^{-1}$ has a simple pole at $z = z_0$.

32.3
Exponential Functions

The real exponential function e^x can be expanded in power series:

$$e^x = 1 + x + \cdots + \frac{x^n}{n!} + \cdots \tag{32.14}$$

which converges for $-\infty < x < \infty$. The convergence may be checked by applying the ratio test (theorem). In fact,

$$\left| \frac{x^{n+1}}{(n+1)!} \middle/ \frac{x^n}{n!} \right| = \frac{|x|}{n+1} \to 0 \quad \text{as } n \to \infty. \tag{32.15}$$

Since the ratio is less than unity, the series converges.

The exponential function e^z of a complex z is defined by the series:

$$e^z \equiv 1 + z + \cdots + \frac{z^n}{n!} + \cdots \tag{32.16}$$

From this we obtain

$$\frac{d}{dz} e^z = e^z \tag{32.17}$$

Thus, we found that the function e^z is analytic in the whole Gaussian plane.

A function of z, which is analytic everywhere in the Gaussian plane is called an *entire function*. Hence, the exponential functions e^z and e^{iz} are entire functions. So are the polynomials $a_0 + a_1 z + \cdots + a_n z^n$. The following trigonometric and hyperbolic functions:

$$\sin z, \quad \cos z, \quad \sinh z \quad \text{and} \quad \cosh z$$

are also entire functions.

32.4
Branch Points

The *square-root function*

$$w(z) = \sqrt{z} \equiv z^{1/2} \tag{32.18}$$

of a complex variable z is defined as the roots of the quadratic equation:

$$w^2 = z \tag{32.19}$$

For $z = 1$, w is equal to ± 1. For $z = -1$, we obtain $w(-1) = \pm i$. The function $\sqrt{z}$ is *double-valued* except at the origin $z = 0$. The origin is a special point called a *branch point*. The function

$$\sqrt{z - z_0} \tag{32.20}$$

has a branch point at $z = z_0$.

The *cube-root function*

$$w(z) = \sqrt[3]{z} \equiv z^{1/3} \tag{32.21}$$

is defined as the roots of the cubic equation:

$$w^3 = z \tag{32.22}$$

The function $\sqrt[3]{z}$ is *triple-valued* except at the origin $z = 0$, a branch point.

Similarly, an *n*th-root function

$$\sqrt[n]{z} \equiv z^{1/n} \tag{32.23}$$

can be defined. It is a multivalued function with degree n, which has a branch point at $z = 0$.

We shall study the multivalue functions in detail in Chapter 35.

32.5
Function with Continuous Singularities

The *complex conjugate* of the variable z is defined by

$$z^* = (x + iy)^* \equiv x - iy \tag{32.24}$$

This is a well-defined function for the whole Gaussian plane. Let us calculate the first derivative of z^* at the origin. Along the real axis we obtain

$$\left(\frac{dz^*}{dz}\right)_{y=0} = \lim_{x \to 0} \left.\frac{(x+iy)^* - 0}{x+iy}\right|_{y=0}$$

$$= \lim_{x \to 0} \frac{x}{x} = 1 \tag{32.25}$$

Along the imaginary axis, we obtain

$$\left(\frac{dz^*}{dz}\right)_{x=0} = \lim_{y \to 0} \left.\frac{(x+iy)^* - 0}{x+iy}\right|_{x=0}$$

$$= \lim_{y \to 0} \frac{-iy}{iy} = -1 \tag{32.26}$$

Hence, we find that the first derivative dz^*/dz does not exist at the origin. (The derivative df/dz, if it exists, must have a unique value.)

☐ **Theorem**

The complex conjugate z^* is singular at all points in the Gaussian plane.

The proof of this theorem is left for the reader's exercises. The singularities are continuous, which are very different from the cases of the poles and branch points.

There are several "very bad" functions with continuous singularities. Examples are:

$$\text{Re}\{z\}, \quad \text{Im}\{z\}, \quad |z|, \quad zz^*, \quad \theta = \tan^{-1}(y/x) \tag{32.27}$$

Note that these functions are all well defined.

The real part function Re$\{z\}$ can be expressed as

$$\text{Re}\{z\} = \frac{1}{2}(z + z^*) \tag{32.28}$$

The singular (bad) behavior arises from z^*. Similarly, the imaginary part function Im$\{z\}$ can be written as

$$\text{Im}\{z\} = \frac{1}{2i}(z - z^*) \tag{32.29}$$

The *absolute square function*

$$zz^* = z^*z = x^2 + y^2 \ (\geqq 0) \tag{32.30}$$

is nonnegative. The magnitude function $|z|$ is the square root of nonnegative zz^*.

$$|z| = \sqrt{zz^*} \tag{32.31}$$

Both zz^* and $\sqrt{zz^*}$ are very bad functions.

Problem 32.5.1

Calculate the first derivative of z^* along the x-direction and y-direction
1. at $z = 1$
2. at $z = x + iy$

32.6 Cauchy–Riemann Relations

A complex function $f(z) = F(x, y)$ in general can be expressed as a sum of the real part u and the imaginary part v. In the Gaussian representation F, u and v are functions of (x, y). We may write this as

$$f(z) = u(x, y) + iv(x, y) \tag{32.32}$$

where u and v are real functions.

The differentiability of a complex function $f(z)$ requires that (i) the first derivative df/dz exists and (ii) its value is the same for all angles in approaching the point of differentiation. This differentiability requires that the components (u, v) must satisfy the following *Cauchy–Riemann* (CR) *relations (equations)*:

$$\boxed{\frac{\partial u(x, y)}{\partial x} = \frac{\partial v}{\partial y}, \quad \frac{\partial u}{\partial y} = -\frac{\partial v}{\partial x}} \tag{32.33}$$

We shall discuss these CR relations in this section.

The CR relations can be derived as follows. Let us consider a small increment $dz = dx$ along the x-axis. Using (32.32), we obtain

$$\left(\frac{df}{dz}dz\right)_y = \left(\frac{df}{dz}\right)_y dx = \left(\frac{\partial u}{\partial x} + i\frac{\partial v}{\partial x}\right)dx \qquad (32.34)$$

where the subscript y means that y = constant. By considering a small increment $dz = idy$ along the y-axis, we obtain

$$\left(\frac{df}{dz}dz\right)_x = \left(\frac{df}{idy}\right)_x (idy) = \left(\frac{\partial u}{\partial y} + i\frac{\partial v}{\partial y}\right)dy \qquad (32.35)$$

The analyticity requires that

$$\left(\frac{df}{dz}\right)_y = \left(\frac{df}{dz}\right)_x \qquad (32.36)$$

The $(df/dz)_y$ can be obtained from (32.34). Since $dx \neq 0$, we obtain

$$\left(\frac{df}{dz}\right)_y = \frac{\partial u}{\partial x} + i\frac{\partial v}{\partial x}$$

The $(df/dz)_x$ is obtained from (32.35)

$$\left(\frac{df}{dz}\right)_x = \frac{1}{i}\left(\frac{\partial u}{\partial y} + i\frac{\partial v}{\partial y}\right)$$

After introducing these two in (32.36), we compare the real and imaginary parts on both sides, and obtain the CR relation in (32.33).

Taking the x-derivative of the first of (32.33) and the y-derivative of the second, and adding together, we obtain

$$\frac{\partial^2 u}{\partial x^2} + \frac{\partial^2 u}{\partial y^2} \equiv \nabla^2 u = 0 \qquad (32.37)$$

Similarly we obtain

$$\nabla^2 v = 0 \qquad (32.38)$$

We, thus, see that the two functions u and v satisfy the 2D *Laplace equation*. Functions that satisfy the Laplace equation are called *harmonic functions*.

32.7
Cauchy–Riemann Relations Applications

The CR relations are derived from the analyticity of function $f(z)$. Hence, these relations are the necessary conditions for the analyticity. But, they are also sufficient conditions. If the real part u is given and it is harmonic, we can use the CR relations to determine the imaginary part v and obtain an analytic function $f(z)$ up to a complex constant. Similarly, if the imaginary part v is given, we can obtain the real part u and the analytic function f by using the CR relations. These important properties will be demonstrated in this section.

We take a few examples.

Let us consider the case of $u = x$. Then obviously $v = y$ and $z = x + iy$. If $u = y$, then we may obtain, by inspection, $v = -ix$ and $z = u + iv = y + (-ix) = -i(x + iy) = -iz$. Consider now

$$u = x^2 - y^2 \tag{32.39}$$

which is harmonic since

$$\frac{\partial^2 u}{\partial x^2} + \frac{\partial^2 u}{\partial y^2} = 1 - 1 = 0$$

We use the CR relations, (32.33) and obtain

$$\frac{\partial v}{\partial y} = \frac{\partial}{\partial x}(x^2 - y^2) = 2x$$

$$\frac{\partial v}{\partial x} = -\frac{\partial}{\partial y}(x^2 - y^2) = 2y$$

whose solution is

$$v = 2xy + c, \quad c = \text{constant} \tag{32.40}$$

Hence,

$$f = u + iv = x^2 - y^2 + i2xy + c = z^2 + c \tag{32.41}$$

Thus, we obtain an analytic function $z^2 + c$. If $u = x^2 + y^2$, then this u is not harmonic, and we cannot obtain an analytic function (Problem 32.7.1). Similarly, if the imaginary part v is given as a harmonic function, we can obtain the real part u and the analytic function f up to a constant.

Theorem

If the real (imaginary) part u (v) is given as a harmonic function, then an analytic function $f(z)$ is obtained by using the CR relations.

Problem 32.7.1

Show that $u = x^2 + y^2$ is not harmonic, that is, it does not satisfy the Laplace equation.

Problem 32.7.2

Find an analytic function of $z = x + iy$ whose real part is:
1. x
2. y
3. $x + y$

Problem 32.7.3

Find an analytic function of z whose imaginary part is:
1. x
2. $x + y$

First, check if $v = x$ (and $v = x + y$) satisfies the Laplace equation. Then, proceed to find the desired analytic functions.

Problem 32.7.4

If the derivative of a complex function $f(z)$ exists at z, then the function f at the neighboring points is given by

$$f(z + dz) = f(z) + dz\, f'(z)$$

Verify this equation.

33
Cauchy's Fundamental Theorem

Cauchy's fundamental theorem, line integrals, circular integrals, and Cauchy's integral formulas are discussed in this chapter.

33.1
Cauchy's Fundamental Theorem

The following theorem is regarded as one of the most fundamental theorems in complex-function theory.

Theorem C

The line integral of an analytic function $f(z)$ along a closed contour C vanishes if the contour C including its inside is entirely within a *regular* domain in which $f(z)$ is differentiable:

$$\oint_C dz\, f(z) = 0 \tag{33.1}$$

where the circle on the integral sign means a closed contour described *counterclockwise*. The precondition is shown in Figure 33.1.

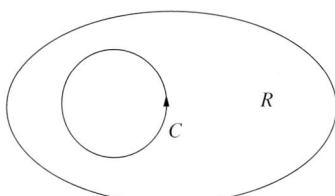

Fig. 33.1 The directed curve C and its interior is entirely within the regular domain R in which $f(z)$ is differentiable.

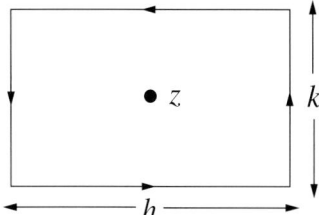

Fig. 33.2 A rectangle with small sides (h, k) and the center z. The contour runs counterclockwise.

The theorem may be proved in two steps as follows:

1. Let us take a small rectangle contour around the center z, as shown in Figure 33.2. We show that the closed contour integral I vanishes. For small sides (h, k), the integral becomes the sum of the four terms:

$$I \equiv \oint_\Box dz\, f(z)$$

$$= hf\left(z - \frac{1}{2}ik\right) + ikf\left(z + \frac{1}{2}h\right) - hf\left(z + \frac{1}{2}ik\right) - ikf\left(z - \frac{1}{2}h\right) \tag{33.2}$$

Since f is differentiable, we have

$$f'(z) \equiv \frac{df}{dz} = \frac{1}{dz}[f(z + dz) - f(z)]$$

or

$$f(z + dz) = f(z) + dz\, f'(z) \tag{33.3}$$

Using this relation we see from (33.2) that the terms in f vanish since $h + ik - h - ik = 0$. The terms in f' also vanish since

$$hk\left(-\frac{1}{2}i + \frac{1}{2}i - \frac{1}{2}i + \frac{1}{2}i\right) = 0$$

2. We decompose the area bounded by the contour C into small rectangles as shown in Figure 33.3, and apply the theorem to each rectangle. In the small-rectangle limits, the sum of the integrals over the decomposed rectangles, after the inner-line integral cancelation, becomes the original contour integral on the lhs of (33.1). Q. E. D.

Theorem C can be recast in an alternative form.

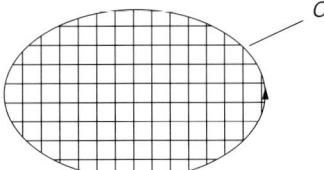

Fig. 33.3 The area enclosed by the curve C is decomposed in small rectangles. In the small-rectangle limit, the sum of the integrals over the decomposed rectangles becomes the original contour integral after inner-line integral cancelation.

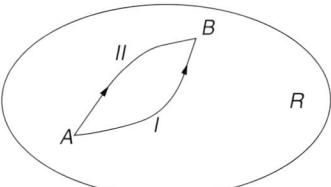

Fig. 33.4 Two path integrals over AIB and AIIB in the regular domain are equal to each other.

Theorem C*

We take two points (A, B) in the regular domain R. If A and B are connected by any two lines AIB and AIIB also within R, as in Figure 33.4 over paths, AIB and AIIB, then the two integrals are equal to each other

$$\int_{AIB} dz\, f(z) = \int_{AIIB} dz\, f(z) \equiv \int_A^B dz\, f(z) \tag{33.4}$$

The integral over the directed path BIIA is equal to the negative of the integral over the path AIIB. The integral over the combined paths AIB and BIIA vanishes because of Theorem C. Hence, (33.4) follows after the rearrangement of the integrals.

The two integrals in (33.4) depend only on the end-points (A, B). Such an integral will be denoted by

$$\int_A^B dz\, f(z) \tag{33.5}$$

as indicated on the third member of (33.4).

33.2
Line Integrals

We illustrate the meaning of Theorem C* by taking a few examples.
First, we take

$$f = z \tag{33.6}$$

which is analytic. We calculate the line integrals along the three paths $(0,0)$ (j) $(1, i)$, $j = 1, 2, 3$, as shown in Figure 33.5.

Along over $(0,0)$ (1) $(1, i)$, we have the sum of two line integrals:

$$I' = \int_{(0,0) \to (1,0)} dz\, z + \int_{(1,0) \to (1,i)} dz\, z \tag{33.7}$$

The first integral can be written as an x-integral ($dz = dx$) and the second as a y-integral ($dz = i\, dy$). We obtain

$$I' = \int_0^1 dx\, x + i \int_0^1 dy(1 + iy)$$

$$= \left.\frac{x^2}{2}\right|_0^1 + i\left(y + i\frac{y^2}{2}\right)\bigg|_0^1 = \frac{1}{2} + i - \frac{1}{2} = i \tag{33.8}$$

Along over the path $(0,0)$ (2) $(1, i)$, we have the sum of two integrals over $(0, 0) \to (0, i)$ and $(0, i) \to (1, i)$. After similar calculations we find, Problem 33.2.1, that

$$I'' \equiv \int_{(0,0)(2)(1,i)} dz\, z = i \tag{33.9}$$

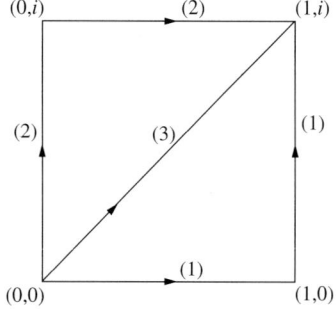

Fig. 33.5 The three directed paths from the origin $(0, 0)$ to $(1, i)$.

Along the path (0,0) (3) (1, i) we have

$$I^{(3)} \equiv \int_{0(3)(1,i)} dz\, z = \int_{0(3)(1,i)} (dx + i\, dy)(x + iy)$$

$$= \int_0^1 dx(x + iy) + i \int_0^1 dy(x + iy) \tag{33.10}$$

From Figure 33.4, we see that

$$y = x \tag{33.11}$$

Using this relation we obtain

$$I^{(3)} = \int_0^1 dx(x + iy) + i \int_0^1 dy(x + iy)$$

$$= \frac{1}{2}(x^2 + ix^2)\Big|_0^1 + i\frac{1}{2}(y^2 + iy^2)\Big|_0^1 = i \tag{33.12}$$

Thus, we obtain

$$I' = I'' = I^{(3)} = i \tag{33.13}$$

in agreement with Theorem C^*.

Next, we take

$$f = z^* \tag{33.14}$$

which is not analytic.

Along over (0,0) (1) (1, i), we obtain

$$J' = \int_{(0,0) \to (1,0)} dz\, z^* + \int_{(1,0) \to (1,i)} dz\, z^*$$

$$= \int_0^1 dx\, x + i \int_0^1 dy(1 - iy) = 1 + i \tag{33.15}$$

Along over (0,0) (2) (1, i), we obtain

$$J'' = \int_{0 \to (0,i)} dz\, z^* + \int_{(0,i) \to (1,i)} dz\, z^*$$

$$= i \int_0^1 dy(-iy) + \int_0^1 dx(x - i) = \frac{1}{2} + \frac{1}{2} - i = 1 - i \tag{33.16}$$

Along over (0,0) (3) (1, i), we obtain

$$J^{(3)} = \int_0^1 dx(x - iy) + i \int_0^1 dy(x - iy)$$

$$= \int_0^1 dx(x - ix) + i \int_0^1 dy(y - iy) = 1 \tag{33.17}$$

We, thus, found that J', J'' and $J^{(3)}$ are all different from each other.

Problem 33.2.1

Calculate the line integral

$$I'' = \int_{(0,0)(2)(1,i)} dz\, z$$

Decompose it into two line integrals, a y-integral and an x-integral.

Problem 33.2.2

Take $f = |z|$, which is not analytic. Calculate the line integrals of $|z|$ along the three paths (0,0) (1) (1, i), (0,0) (2) (1, i) and (0,0) (3) (1, i) shown in Figure 33.5.

33.3
Circular Integrals

Let us take an analytic function f within the regular domain R. The integral of f over the small *circle* surrounding the center z_0 in R, shown in Figure 33.6, is zero according to Cauchy's theorem. It is instructive to see this by explicitly calculating the integral

$$I = \oint_{\text{circle} \subset R} dz\, f(z) = 0 \tag{33.18}$$

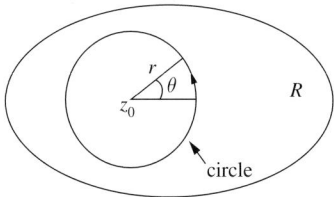

Fig. 33.6 The circle C of radius r surrounding z_0 is within the analytic domain R. The integral of $f(z)$ over the circle is zero according to Cauchy's theorem.

33.3 Circular Integrals

A point on the circle can be represented by

$$z_0 + re^{i\theta} \tag{33.19}$$

where r is the radius and θ is the polar angle shown in Figure 33.6. The circular integral I is given by

$$I = r \int_0^{2\pi} d\theta \, (ie^{i\theta}) f(z_0 + re^{i\theta})$$

For a very small r, we have from the differentiability

$$f(z_0 + re^{i\theta}) = f(z_0) + re^{i\theta} f'(z_0) \tag{33.20}$$

Using this, we obtain

$$I = ri f(z_0) \int_0^{2\pi} d\theta \, e^{i\theta} + ir^2 f'(z_0) \int_0^{2\pi} d\theta \, e^{i2\theta} = 0 \tag{33.21}$$

The two θ-integrals vanish since

$$\int_0^{2\pi} d\theta \, e^{in\theta} = 0, \quad n = 1, 2, \ldots \tag{33.22}$$

The function z^n, $n =$ nonnegative integer, is analytic, and therefore, its circular integral vanishes:

$$J_n = \oint_{\text{circle}} dz \, z^n = 0 \tag{33.23}$$

The function $z^{-1} \equiv 1/z$ has a pole at the origin. We calculate a circular integral of z^{-1} around the origin, and obtain

$$J_{-1} = \oint_{\text{circle}} dz \frac{1}{z} = r \int_0^{2\pi} d\theta \, (ie^{i\theta}) \frac{1}{re^{i\theta}} = 2\pi i \tag{33.24}$$

The function $z^{-n} \equiv 1/z^n$, $n \geq 2$, has a pole of the order n at the origin. Its circular integral vanishes.

$$J_{-n} \equiv \oint_{\text{circle}} dz \frac{1}{z^n} = r \int_0^{2\pi} d\theta \, (ie^{i\theta}) \frac{1}{r^n e^{in\theta}} = 0, \quad (n \geq 2) \tag{33.25}$$

where we used

$$\int_0^{2\pi} d\theta \, e^{-in\theta} = 0, \quad n = 1, 2, \ldots \tag{33.26}$$

We combine the results in (33.23)–(33.26), and obtain

$$\oint_{\text{circle}} dz\, z^p = \begin{cases} 2\pi i & \text{if } p = -1 \\ 0 & \text{otherwise} \end{cases} \qquad (33.27)$$

This is an important result, which will repeatedly be used later.

Problem 33.3.1

Evaluate directly the θ-integrals in (33.22) and (33.26).

33.4
Cauchy's Integral Formula

Let us consider a function $f(z)$ within a regular domain R. We take a closed contour C surrounding a point z, all of which are inside R. The function $f(z)$ can be expressed in terms of the following contour integral:

$$f(z) = \frac{1}{2\pi i} \oint_C \frac{f(\zeta)}{\zeta - z} d\zeta \qquad (33.28)$$

The point z and the contour C are shown in Figure 33.7. The theorem may be proved as follows. We first take a small circular contour of radius r, shown in the central part of Figure 33.8. From the figure

$$\zeta = z + r e^{i\theta}, \quad 0 \le \theta \le 2\pi \qquad (33.29)$$

Using the analyticity of f, we can write

$$f(z + re^{i\theta}) = f(z) + re^{i\theta} f'(z) \qquad (33.30)$$

where we retained the first-order term in r only. Using this expansion, we obtain

$$\oint_{\substack{\text{circle}\\|\zeta-z|=r}} d\zeta \frac{f(\zeta)}{\zeta - z} = f(z) i \int_0^{2\pi} d\theta + i r^2 f'(z) \int_0^{2\pi} d\theta\, e^{i\theta} = 2\pi i\, f(z) \qquad (33.31)$$

Next, we consider a contour C^*, which traces the contour C and the circular contour directed clockwise and runs the double-directed corridors joining the two,

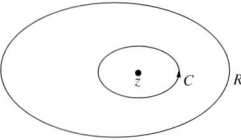

Fig. 33.7 The function $f(z)$ can be expressed in terms of Cauchy's integral formula (33.28), where z and C are all within the regular domain R.

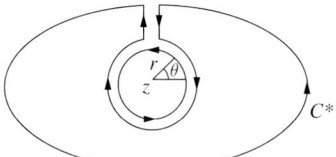

Fig. 33.8 The sum of the circular integral around z and the vanishing integral over the contour C* becomes the integral over the contour C shown in Figure 33.7, after cancelation of the contributions over the double-arrowed paths.

see Figure 33.8. There are no singular points inside C^*. Hence, the integral over C^* vanishes due to Cauchy's theorem. We add the integral over C^* to the small circular integral. After canceling the integral over the double-directed paths, we obtain (33.28). Q. E. D.

The theorem is quite remarkable. If the function f is analytic, then its value at z is determined entirely by its values on the contour C. If we take a small circular contour, then we can interpret it as follows: The function $f(z)$ is the average of its values on the circle: $|\zeta - z| = r$. We have seen earlier that the real and imaginary parts of the analytic z function $f(z)$ are both harmonic functions. Such harmonic functions have the same properties.

Using (33.28) and

$$\frac{d}{dz}\frac{1}{\zeta - z} = \lim_{\Delta z \to 0} \frac{1}{\Delta z}\left[\frac{1}{\zeta - z - \Delta z} - \frac{1}{\zeta - z}\right]$$

$$= \frac{1}{(\zeta - z)^2} \tag{33.32}$$

we obtain

$$f'(z) = \frac{1}{2\pi i}\oint_C d\zeta\, f(\zeta)\frac{d}{dz}\frac{1}{\zeta - z} = \frac{1}{2\pi i}\oint_C d\zeta\, \frac{f(\zeta)}{(\zeta - z)^2} \tag{33.33}$$

Repeating differentiation, we obtain

$$f^{(n)}(z) = \frac{n!}{2\pi i}\oint_C d\zeta\, \frac{f(\zeta)}{(\zeta - z)^{n+1}} \tag{33.34}$$

Thus, the first derivative exists, then derivatives of all orders exist, which is very remarkable. All arise from the analyticity of $f(z)$.

34
Laurent Series

Taylor series, uniform convergence, and Laurent series are discussed in this chapter.

34.1
Taylor Series and Convergence Radius

Let us consider the real power series

$$1 + x + x^2 + \cdots = (1-x)^{-1}, \quad -1 < x < 1 \tag{34.1}$$

Extending the real x to a complex variable z, we have

$$1 + z + z^2 + \cdots = \frac{1}{1-z} \equiv f(z) \tag{34.2}$$

This series converges absolutely for

$$|z| < 1 \tag{34.3}$$

We consider a general *power series*

$$a_0 + a_1 z + a_2 z^2 + \cdots = S \tag{34.4}$$

Extending the ratio test, (see Section 31.5) we may deduce that if

$$\left| \frac{a_{n+1} z^{n+1}}{a_n z^n} \right| = \left| \frac{a_{n+1}}{a_n} z \right| < 1 \tag{34.5}$$

then the series S converges. We introduce the *radius of convergence* or *convergence radius* r as

$$r \equiv \lim_{n \to \infty} \left| \frac{a_n}{a_{n+1}} \right| \tag{34.6}$$

We can then restate that the series S converges within the convergence radius r:

$$S \text{ converges for } |z| < r \tag{34.7}$$

The series $(1-z)^{-1}$ has the convergence radius one (1).

Mathematical Physics. Shigeji Fujita and Salvador V. Godoy
Copyright © 2010 WILEY-VCH Verlag GmbH & Co. KGaA, Weinheim
ISBN: 978-3-527-40808-5

Let us now consider an exponential function

$$e^z = 1 + z + \frac{1}{2}z^2 + \cdots + \frac{1}{n!}z^n + \cdots \tag{34.8}$$

The radius of convergence is

$$r = \lim_{n \to \infty} \left| \frac{(n+1)!}{n!} \right| = \lim_{n \to \infty} (n+1) = \infty \tag{34.9}$$

Thus, we see that the exponential function e^z converges in an entire Gaussian plane.

34.2
Uniform Convergence

A convergent series of functions,

$$f(z) = \sum_{n=0}^{\infty} f_n(z) \tag{34.10}$$

is said to *converge uniformly* in a region R if given a positive ε there exists an N_0 independent of z in R such that

$$\left| \sum_{n=0}^{N} f_n(z) - f(z) \right| < \varepsilon \tag{34.11}$$

for $N \geq N_0$. The function $(1-z)^{-1}$ is uniformly convergent at $z = 0$.

A uniformly convergent series has the useful property that it can be integrated term by term. This may be proved as follows:

We integrate (34.10) along the contour C and using (34.11), we obtain

$$\left| \int_C dz\, f(z) - \sum_{n=0}^{N} \int_C dz\, f_n(z) \right| = \left| \int_C dz \left\{ f(z) - \sum_{n=0}^{N} f_n(z) \right\} \right| < \varepsilon L \tag{34.12}$$

where L is the length of the contour. Since L can be made arbitrarily small, we obtain

$$\int_C dz\, f(z) = \sum_{n=0}^{\infty} \int_C dz\, f_n(z) \tag{34.13}$$

We use this property and Cauchy's integral formula (33.28) for a function $f(z)$ about a point z_0, see Figure 34.1. We take the closed contour C to be a circle cen-

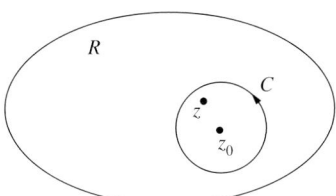

Fig. 34.1 Cauchy's integral formula (33.28) about a point z_0.

tered at z_0, and let z be an arbitrary point inside C. We then expand the denominator

$$\frac{1}{\zeta - z} = \frac{1}{\zeta - z_0 - (z - z_0)} = \frac{1}{\zeta - z_0} \left(1 - \frac{z - z_0}{\zeta - z_0}\right)^{-1}$$
$$= \frac{1}{\zeta - z_0} \sum_{n=0}^{\infty} \left(\frac{z - z_0}{\zeta - z_0}\right)^n \tag{34.14}$$

The geometric series here converges since $|(z - z_0)/(\zeta - z_0)| < 1$. Furthermore, it is uniformly convergent through the comparison test since for z inside C and ζ anywhere on C, there is a number ρ such that

$$\left|\frac{z - z_0}{\zeta - z_0}\right| < \rho < 1$$

and the series

$$1 + \rho + \rho^2 + \cdots = \sum_{n=0}^{\infty} \rho^n$$

converges. We can therefore introduce (34.14) in (33.28) and integrate term by term and obtain

$$f(z) = \sum_{n=0}^{\infty} a_n (z - z_0)^n \tag{34.15}$$

with

$$a_n = \frac{1}{2\pi i} \oint_C d\zeta \, \frac{f(z)}{(\zeta - z_0)^{n+1}} = \frac{1}{n!} f^{(n)}(z_0) \tag{34.16}$$

This is just the Taylor series for $f(z)$ about the point z_0.

34.3 Laurent Series

Assume that the function $f(z)$ is analytic in an annular region. The function $f(z)$ may or may not be analytic in the shaded region, see Figure 34.2. The contour

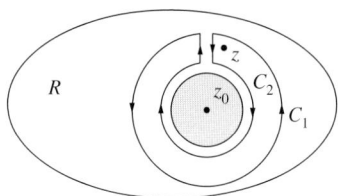

Fig. 34.2 Proof of the Laurent series.

integration is taken inside the regular region R. Take a point z in this region. We take a closed contour composed of: (a) a circle C_1 with the greater radius directed counterclockwise and centered on z_0 that is located in the shaded region, (b) a second circle C_2 with the smaller radius directed clockwise, and (c) the connecting double lines oppositely directed.

Assume that the point z is between the two circles C_1 and C_2 and use Cauchy's integral formula (33.28). After the cancelation of the double-directed integrals, we obtain

$$f(z) = \frac{1}{2\pi i} \oint_{C_1} d\zeta \, \frac{f(\zeta)}{\zeta - z} - \frac{1}{2\pi i} \oint_{C_2} d\zeta \, \frac{f(\zeta)}{\zeta - z} \tag{34.17}$$

where the negative sign in the second integral arises from the convention that the circular contour C_2 is directed counterclockwise.

The first integral of the rhs of (34.17) can be expanded as before and the result is given by (34.15). The second integral is calculated by expanding the denominator $(\zeta - z)^{-1}$ in a different series:

$$\frac{1}{\zeta - z} = \frac{1}{\zeta - z_0 - (z - z_0)} = -\frac{1}{(z - z_0) - (\zeta - z_0)}$$

$$= -\frac{1}{(z - z_0)} \left(1 - \frac{\zeta - z_0}{z - z_0}\right)^{-1}$$

$$= -\frac{1}{(z - z_0)} \sum_{n=0}^{\infty} \left(\frac{\zeta - z_0}{z - z_0}\right)^n \tag{34.18}$$

With ζ on the circle C_2, we observe that

$$\left|\frac{\zeta - z_0}{z - z_0}\right| < 1$$

This series is uniformly convergent. After a term-by-term integration, we obtain

$$-\frac{1}{2\pi i} \oint_{C_2} d\zeta \, \frac{f(\zeta)}{\zeta - z} = \sum_{n=1}^{\infty} b_n \frac{1}{(z - z_0)^n} = \sum_{n=-1}^{\infty} a_n (z - z_0)^n \tag{34.19}$$

where b_n is given by

$$b_n = a_{-n} = \frac{1}{2\pi i} \oint_{C_2} d\zeta (\zeta - z_0)^{n-1} f(\zeta) \tag{34.20}$$

34.3 Laurent Series

Combining the two expansions associated with the circles C_1 and C_2, we obtain

$$f(z) = \sum_{n=-\infty}^{\infty} a_n (z - z_0)^n$$
$$= a_0 + a_1(z - z_0) + a_2(z - z_0)^2 + \cdots$$
$$+ a_{-1} \frac{1}{z - z_0} + a_{-2} \frac{1}{(z - z_0)^2} + \cdots \qquad (34.21)$$

with

$$a_n = \frac{1}{2\pi i} \oint_C d\zeta \, \frac{f(\zeta)}{(\zeta - z_0)^{n-1}} \qquad (34.22)$$

where the closed contour C surrounds the point z in the annular regular region.

If $f(z)$ is analytic in the shaded region, then $a_n = 0$ for $n = -1, -2, \cdots$. If the series with negative n stops at a finite N, we say that we have a pole of order N at $z = z_0$.

Let us consider the function

$$e^{1/z} = 1 + \frac{1}{z} + \frac{1}{2!} \frac{1}{z^2} + \cdots \qquad (34.23)$$

This function contains all negative n. We say that the function $e^{-1/z}$ has an *essential singularity* at the origin ($z_0 = 0$).

35
Multivalued Functions

35.1
Square-Root Functions. Riemann Sheets and Cut

The *square-root function* w of a complex number z,

$$w = \sqrt{z} \equiv z^{1/2} \tag{35.1}$$

is defined as the solutions of

$$w^2 = z \tag{35.2}$$

If we write

$$z = re^{i\theta} \tag{35.3}$$

then the solutions of (35.2) are

$$z^{1/2} = r^{1/2} e^{i\theta/2} \quad \text{or} \quad -r^{1/2} e^{i\theta/2} \tag{35.4}$$

Thus, the function $\sqrt{z}$ is double-valued except at the origin $z = 0$. This special point is called a *branch point*.

Let us consider the case in which

$$r = 1 \tag{35.5}$$

The function $z = e^{i\theta}$ can be represented by the unit circle shown in Figure 35.1(a). As θ changes from 0 to 2π, the two functions

$$w_1 = e^{i\theta/2}, \quad w_2 = -e^{i\theta/2} = e^{i\pi} e^{i\theta/2} = e^{i(2\pi+\theta)/2} \tag{35.6}$$

change along the semicircles as shown in Figure 35.1(b). If we extend the range of angle θ to 0 to 4π, then we can regard the two functions $w_1 = e^{i\theta/2}$, and $w_2 = e^{i(2\pi+\theta)/2}$ as two sections of a single-valued function $e^{i\theta/2}$, $0 \leq \theta \leq 4\pi$.

Riemann devised an ingenuous scheme of constructing a single-valued function for the square-root function as follows.

Mathematical Physics. Shigeji Fujita and Salvador V. Godoy
Copyright © 2010 WILEY-VCH Verlag GmbH & Co. KGaA, Weinheim
ISBN: 978-3-527-40808-5

35 Multivalued Functions

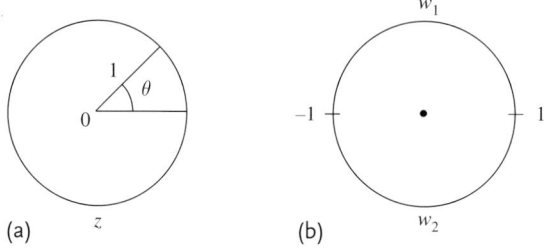

Fig. 35.1 (a) The unit circle for $z = e^{i\theta}$. (b) The semicircles for $w_1 = e^{i\theta/2}$ and $w_2 = -e^{i\theta/2} = e^{i(2\pi+\theta)/2}$.

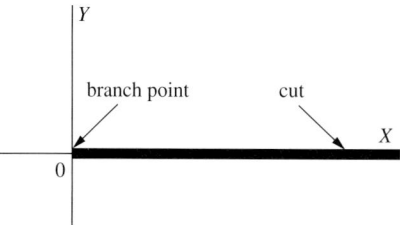

Fig. 35.2 A Riemann cut extends from the branch point 0 to ∞. The positive real axis is chosen in this figure.

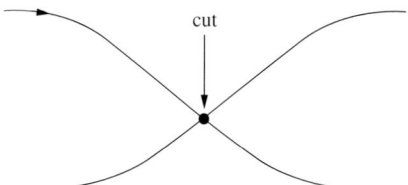

Fig. 35.3 Riemann cut.

We take two Gaussian planes, called *Riemann sheets*, joined at the origin (*branch point*). The two solutions (w_1, w_2) are represented by points in the upper and lower sheets. We make a Riemann cut extending along the real axis as shown in Figure 35.2. The continuous change of points z is represented by a line. If the line crosses the cut, then the line switches from the upper sheet to the lower sheet or vice versa, as indicated in Figure 35.3. Clearly this device works well for $\sqrt{z}$. We obtain a single analytic function $w = e^{i\theta/2}$, $0 \leq \theta \leq 4\pi$. The two branches of $\sqrt{z}$, w_1 and w_2, in (35.6) are represented by points in the upper and lower Riemann sheets. The device works in general. For example, the cut may be chosen along the negative axis (Problem 35.1.1).

Let us take the square-root function

$$w = \sqrt{z-a}, \quad a = \text{complex number} \tag{35.7}$$

The branch point is $z = a$. The Riemann sheets are joined at this point. A Riemann cut may be taken along *any* line extending from a to infinity (Problem 35.1.2).

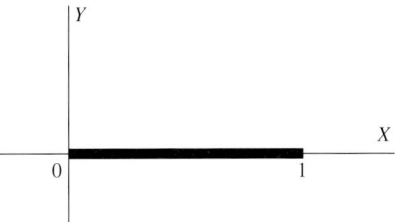

Fig. 35.4 A Riemann cut joining the two branch points (0, 1).

We now consider a square-root function

$$w = \sqrt{z(z-1)} \tag{35.8}$$

The branch points are $z = 0$ and $z = 1$. The two Riemann sheets are joined at $z = 0$ and $z = 1$. The Riemann cuts may be taken along two lines from the branch points to infinity. The cut may also be taken along the line joining the two branch points as shown in Figure 35.4.

Problem 35.1.1

Obtain a single-valued analytic function $w = e^{i\theta/2}$, $-\pi \le \theta \le 3\pi$, using a Riemann cut along the negative axis.

Problem 35.1.2

Obtain a single-valued analytic function by taking two Riemann sheets and a cut extending to $+\infty$.

35.2 Multivalued Functions

The *cube root function* $w = z^{1/3}$ of a complex number z, is defined as the solutions of

$$w^3 = z \tag{35.9}$$

If we write

$$z = re^{i\theta}$$

then the three solutions are

$$w_1 = r^{1/3}e^{i\theta/3}, \quad w_2 = r^{1/3}e^{i(2\pi+\theta)/3}, \quad w_3 = r^{1/3}e^{i(4\pi+\theta)/3} \tag{35.10}$$

By considering three Riemann sheets joined at the branch point $z = 0$ and a Riemann cut extending from the origin to ∞, we may construct a single-valued function:

$$w = r^{1/3} e^{i\theta/3}, \quad 0 \leqq \theta \leqq 6\pi \tag{35.11}$$

whose three branches are w_1, w_2, and w_3.

The *logarithmic* function of z, denoted by $\ln z$, is defined as the solutions of

$$z = e^w \tag{35.12}$$

We can write

$$z = e^{\ln r + i\theta + i2\pi n} \tag{35.13}$$

where n are integers: $0, \pm 1, \pm 2, \cdots$. Hence,

$$w = \ln r + i(\theta + 2\pi n) \tag{35.14}$$

is a multivalued function of an infinite degree. We may construct a single-valued function by using an infinite number of sheets and a cut.

The fractional power function

$$w = z^{p/q}, \quad p, q = \text{integer} \tag{35.15}$$

is a multivalued function of degree q. Introducing q Riemann sheets and a cut we can construct a single-valued function.

Let us consider a function

$$w = z^\alpha, \quad \alpha = \text{complex number} \tag{35.16}$$

Using

$$z = e^{\ln z} \tag{35.17}$$

we can define z^α by

$$z^\alpha = e^{\alpha \ln z} \tag{35.18}$$

The point $z = 0$ is a branch point. Except for this point, $\ln z$ is analytic and the exponential is analytic. Hence, z^α is analytic. If α is not a fraction, we may use the same infinite-sheeted Riemann surface used for $\ln z$. We can construct a single-valued function

$$z^\alpha = e^{\alpha(\ln r + i\theta)} \tag{35.19}$$

with θ varying continuously from $-\infty$ to $+\infty$.

The usefulness of the Riemann sheets and cuts will be demonstrated in the following chapter.

36
Residue Theorem and Its Applications

The residue theorem and its applications are discussed in this chapter.

36.1
Residue Theorem

Let us consider a function $f(z)$ of a complex variable z. We assume that the function f is analytic except for isolated poles at points z_j. The poles can be either of finite or infinite order. The integral of $f(z)$ over a closed counterclockwise contour C can be evaluated as

$$\oint dz\, f(z) = 2\pi i \sum_j \text{Res}(f, z_j) \qquad (36.1)$$

The *residue* of $f(z)$ at z_j, denoted by $\text{Res}(f, z_j)$, is the coefficient a_{-1} of the Laurent expansion of $f(z)$ about the point z_j:

$$f(z) = a_0 + a_1(z - z_j) + a_2(z - z_j)^2 + \cdots$$
$$+ a_{-1}\frac{1}{z - z_j} + a_{-2}\frac{1}{(z - z_j)^2} + \cdots \qquad (36.2)$$

Equation (36.1) is called the *residue theorem*.

The theorem may be proved as follows. First, we consider the case in which there is one singular point z_1 assumed to be inside the original closed contour C. We may add a vanishing integral of f over the closed contour C^* to the original integral over the contour C, see Figure 36.1.

The integral over C^* vanishes because of Cauchy's theorem. The sum of the two integrals becomes an integral around the circle with the center z_1 after the cancelation of the double-directed line integrals.

$$I = \oint_C dz\, f(z) = \oint_C dz\, f(z) + \oint_{C^*} dz\, f(z)$$
$$= \oint_{C_1} dz\, f(z) \qquad (36.3)$$

Mathematical Physics. Shigeji Fujita and Salvador V. Godoy
Copyright © 2010 WILEY-VCH Verlag GmbH & Co. KGaA, Weinheim
ISBN: 978-3-527-40808-5

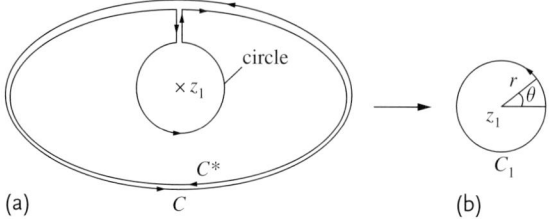

(a) C* C (b)

Fig. 36.1 Contour used to prove (36.3).

The circular integral over the circle $|z - z_1| = r$, where r is the radius and z_1 the center, can be calculated as demonstrated in Section 33.3.

$$\oint_{C_1} dz\, f(z) = \oint_{\text{circle:}\, |z-z_1|=r} dz\, f(z) = \oint_{\text{circle}} dz \left[\cdots + a_{-1}\frac{1}{z - z_1} + \cdots \right]$$
$$= 2\pi i a_{-1} = 2\pi i \operatorname{Res}(f, z_1) \tag{36.4}$$

If the singular points (poles) $z_1, z_2, \cdots$ are separated from each other, then, we can use the same techniques to reduce the original integral over the circles $C_1, C_2, \cdots$, each with the centers at $z_1, z_2, \cdots$, which establishes the residue theorem.

If the function f is very bad and contains a continuous singularity, the theorem does not hold. Such very bad functions that depend on z^* were discussed in Chapter 32.

Multivalued functions can be expressed in terms of a single-valued function by using Riemann sheets and cuts. The residue theorem can be used for the single-valued function, but the contour C must avoid the branch point.

Finding residues may be helped by using the following theorems:

Theorem

For a simple pole at z_0

$$\boxed{\operatorname{Res}(f, z_0) = \lim_{z \to z_0} (z - z_0)\, f(z)} \tag{36.5}$$

This may be proved as follows:

$$\lim_{z \to z_0} (z - z_0) f(z) = \lim_{z \to z_0} (z - z_0) \sum_{n=-1}^{\infty} a_n (z - z_0)^n$$
$$= \lim_{z \to z_0} \left(a_{-1} + \sum_{n=0}^{\infty} a_n (z - z_0)^{n+1} \right) = a_{-1}$$

Theorem

If $f(z)$ is a fractional form:

$$f(z) = \frac{P(z)}{Q(z)} \tag{36.6}$$

where $Q(z)$ has a simple zero at $z = z_0$, $Q'(z_0) \neq 0$, and $P(z)$ is analytic at $z = z_0$, then the residue can be found from

$$\boxed{\operatorname{Res}(f, z_0) = \frac{P(z_0)}{Q'(z_0)}} \tag{36.7}$$

This formula is good for a simple pole only. It may be proved by using (36.5) as follows. We expand $P(z)$ and $Q(z)$ in a Taylor series at $z = z_0$, and keep only the lowest terms:

$$\operatorname{Res}(f, z_0) = \lim_{z \to z_0} (z - z_0) \frac{P(z_0) + (z - z_0) P'(z_0) + \cdots}{(z - z_0) Q'(z_0) + \frac{1}{2}(z - z_0)^2 Q''(z_0) + \cdots}$$

$$= \frac{P(z_0)}{Q'(z_0)} \quad \text{(Q. E. D.)}$$

36.2
Integrals of the Form $\int_{-\infty}^{\infty} dx\, f(x)$

Let us consider an integral

$$\int_{-\infty}^{\infty} dx\, \frac{1}{x^2 + 1} = I_1 \tag{36.8}$$

The indefinite integral $\int dx\, (x^2 + 1)^{-1}$ is known:

$$\int dx\, \frac{1}{x^2 + 1} = \tan^{-1} x \tag{36.9}$$

Using this and taking the limits, we obtain

$$I_1 = \pi \tag{36.10}$$

We may obtain this result alternatively by using the residue theorem as follows. We add a semicircle C_2 to the straight line path C_1 and obtain a closed contour in the Gaussian plane as shown in Figure 36.2. The straight line integral C_1 along the real axis matches the integral I_1 with $R \to \infty$. The contribution from the semicircle C_2 with $R \to \infty$ vanishes, since

$$\int_{C_2} dz\, \frac{1}{z^2 + 1} = \int_0^{\pi} \frac{R i e^{i\theta} d\theta}{R^2 e^{i2\theta} + 1} \xrightarrow[R \to \infty]{} 0 \tag{36.11}$$

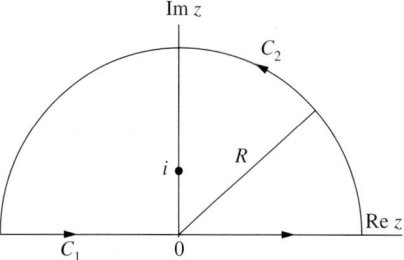

Fig. 36.2 The closed contour $C_1 + C_2$ for the integral in (36.8).

There is a simple pole at $z = i$. Hence, we obtain

$$I_1 = \oint_{C_1+C_2} dz \frac{1}{z^2+1} = 2\pi i \, \text{Res}[\frac{1}{z^2+1}, i] = 2\pi i \times \frac{1}{2i} = \pi \tag{36.12}$$

in agreement with (36.10).

Let us now consider

$$\int_{-\infty}^{\infty} dx \frac{1}{x^4+1} = I_2 \tag{36.13}$$

The indefinite integral is unknown. We can take the same contour to calculate I_2. The semicircle contribution vanishes with $R \to \infty$. There are two simple poles at $z_0 = e^{i\pi/4}$ and $e^{i3\pi/4}$ inside $C_1 + C_2$. We obtain, using (36.7)

$$I_2 = \oint_{C_1+C_2} dz \frac{1}{z^4+1} = 2\pi i \left\{ \text{Res}\left[\frac{1}{z^4+1}, e^{i\pi/4}\right] + \text{Res}\left[\frac{1}{z^4+1}, e^{i3\pi/4}\right] \right\} \tag{36.14}$$

Then, we obtain

$$I_2 = 2\pi i \left[\frac{1}{4e^{i3\pi/4}} + \frac{1}{4e^{i9\pi/4}} \right] = 2\sqrt{2}\pi \tag{36.15}$$

The two integrals (I_1, I_2) can also be obtained by using the residue theorem with the addition of the lower semicircular contour, see Problem 36.2.1.

Problem 36.2.1

Evaluate integrals I_1, and I_2, using the residue theorem and closing the contour with a lower-half semicircle.

36.3
Integrals of the Type $\int_{-\infty}^{\infty} dx\, e^{ix} f(x)$

Let us consider

$$I = \int_{-\infty}^{\infty} dx \frac{\cos ax}{x^2 + b^2}, \quad a, b > 0 \tag{36.16}$$

We note that

$$e^{iz} = e^{i(x+iy)} = e^{-y}(\cos x + i \sin x)$$

Let us take the integrand to be $e^{iaz}(z^2 + b^2)^{-1}$ and close the contour with the upper semicircle. The contour C_1 and C_2 are the same as shown in Figure 36.2. We obtain

$$\text{Re} \left\{ \oint_{C_1+C_2} dz \frac{e^{iaz}}{z^2 + b^2} \right\} = \text{Re} \left\{ \int_{-\infty}^{\infty} dx \frac{e^{iax}}{x^2 + b^2} + \int_{C_2} dz \frac{e^{iaz}}{z^2 + b^2} \right\} \tag{36.17}$$

The first integral on the rhs becomes I after taking the real part. The second integral vanishes with $R \to \infty$:

$$\int_{C_2} dz \frac{e^{iaz}}{z^2 + b^2} = \int_0^{\pi} d\theta\, iRe^{i\theta} \frac{e^{iaR\cos\theta - aR\sin\theta}}{R^2 e^{2i\theta} + b^2} \xrightarrow[R\to\infty]{} 0 \tag{36.18}$$

Note that if $a > 0$, $e^{-aR\sin\theta}$ approaches zero exponentially for $0 < \theta < \pi$. Using the residue theorem, we obtain

$$I = \text{Re}\left\{ \oint_{C_1+C_2} dz \frac{e^{iaz}}{z^2 + b^2} \right\} = 2\pi i \frac{e^{-ab}}{2ib} = \frac{\pi}{b} e^{-ab} \tag{36.19}$$

Sometimes we must be careful in replacing sines or cosines by exponentials. Let us take

$$\int_0^{\infty} dx \frac{\sin x}{x} = \frac{1}{2} \int_{-\infty}^{\infty} dx \frac{\sin x}{x}, \quad a > 0 \tag{36.20}$$

which is convergent, while

$$\int_0^{\infty} dx \frac{e^{ix}}{x} \tag{36.21}$$

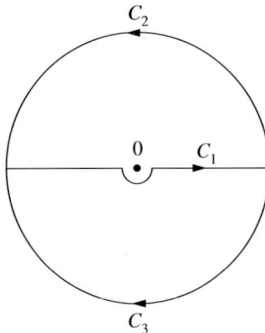

Fig. 36.3 Contour for (36.22).

diverges at $x = 0$. Since $\sin az/z$ is analytic near $z = 0$, we displace the contour away from $z = 0$ as shown in Figure 36.3. We obtain

$$I_2 = \frac{1}{2}\int_{C_1} dz \frac{\sin az}{z} = \frac{1}{4i}\int_{C_1} dz \left[\frac{e^{iaz}}{z} - \frac{e^{-iaz}}{z}\right]$$

$$-\frac{1}{4i}\oint_{C_1+C_2} dz \frac{e^{iaz}}{z} - \frac{1}{4i}\oint_{C_1+C_3} dz \frac{e^{-iaz}}{z}$$

$$= \frac{1}{4i} 2\pi i - 0 = \frac{\pi}{2} \tag{36.22}$$

36.4
Integrals of the Type $\int_0^{2\pi} d\theta\, f(\cos\theta, \sin\theta)$

If $f(\cos\theta, \sin\theta)$ is a rational function of $\cos\theta$ and $\sin\theta$, then the integral $\int_0^{2\pi} d\theta\, f$ can be done by the usual partial fractions decomposition method. If the limits on the θ-integral are $(0, 2\pi)$, the integral can be evaluated simply by using the residue theorem and the unit circular integration.

Let us consider

$$I = \int_0^{2\pi} d\theta\, \frac{1}{1 + a\cos\theta}, \quad a < 1 \tag{36.23}$$

We introduce

$$e^{i\theta} = z, \quad d\theta = \frac{dz}{iz}$$

$$\cos\theta = \frac{1}{2}(z + z^{-1}), \quad \sin\theta = \frac{1}{2i}(z - z^{-1}) \tag{36.24}$$

Then, we have

$$I = \oint_C \frac{dz}{iz} \frac{1}{1 + a(z + z^{-1})/2} = \frac{2}{ia}\oint_C \frac{dz}{z^2 + (2/a)z + 1} \tag{36.25}$$

where C is the unit circle. The roots (z_1, z_2) of the denominator are

$$z_1 = -\frac{1}{a} + \sqrt{\frac{1}{a^2} - 1}, \quad z_2 = -\frac{1}{a} - \sqrt{\frac{1}{a^2} - 1} \tag{36.26}$$

Note that z_1 is inside the unit circle and z_2 outside. Using the residue theorem, we obtain

$$I = 2\pi i \frac{2}{ia} \frac{1}{2z_1 + 2/a} = \frac{2\pi}{\sqrt{1 - a^2}} \tag{36.27}$$

Problem 36.4.1

Evaluate the following integrals using the residue theorem

1. $\displaystyle\int_0^{2\pi} d\theta \, \frac{1}{1 + a\cos\theta}, \quad 0 < a < 1$

2. $\displaystyle\int_0^{2\pi} d\theta \, \frac{1}{(1 + a\sin\theta)(1 + b\cos\theta)}, \quad 0 < a, b < 1$

36.5
Miscellaneous Integrals

Let us consider the integral

$$I = \int_0^\infty dx \, \frac{\sqrt{x}}{x^2 + 1} \tag{36.28}$$

The obvious candidate for the z-integrand is $\sqrt{z}(z^2 + 1)^{-1}$, which contains a double-valued function $\sqrt{z}$. We take a Riemann cut extending from the origin O to ∞ along the positive real axis, see Figure 36.4. We choose a closed contour $C_1 + C_2 + C_3 + C_4$. The integral over $C_1 \to I$ as the radius of the large circle, $R \to \infty$. The integral over C_3 becomes also equal to I with $R \to \infty$ because $e^{i\pi} = -1$ and the integration runs in the decreasing x-direction. The small semicircle C_4 around the origin gives a vanishing contribution. The large circle C_2 contributes nothing with $|z| = R \to \infty$. There are simple poles at $z_1 = i$ and $-i$. Thus, we obtain

$$I = \frac{1}{2} \oint_{C_1+C_2+C_3+C_4} dz \, \frac{\sqrt{z}}{z^2 + 1} = \frac{2\pi i}{2} \left[\frac{e^{i\pi/4}}{2e^{i\pi/2}} + \frac{e^{i3\pi/4}}{2e^{i3\pi/2}} \right] = \frac{\pi}{\sqrt{2}} \tag{36.29}$$

Using the same contour, we can obtain (Problem 36.5.2)

$$\int_0^\infty dx \, \frac{x^{t-1}}{x^2 + 1} = \frac{\pi}{\sin \pi t}, \quad 0 < \text{Re}(t) < 1 \tag{36.30}$$

Here, we used the Riemann cut for the nonintegral power z^t.

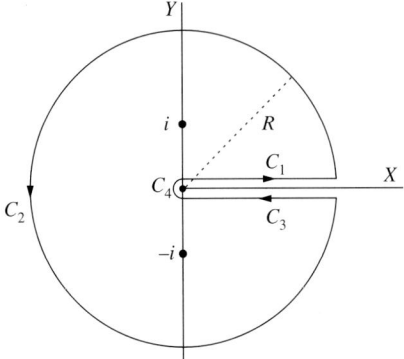

Fig. 36.4 A closed contour $C_1 + C_2 + C_3 + C_4$ for the integral in (36.29).

Problem 36.5.1

Use the residue theorem to evaluate

$$\int_0^\infty dx \, \frac{1}{x^{2/3}(x^2+1)} = \frac{2\pi}{\sqrt{3}}$$

Problem 36.5.2

Derive (36.30).

Appendix A
Representation-Independence of Poisson Brackets

By using the canonical variables (q, p), we obtain

$$\{q_r^n, p_r\}_{(q,p)} \equiv \sum_s \left[\frac{\partial q_r^n}{\partial q_s}\frac{\partial p_r}{\partial p_s} - \frac{\partial q_r^n}{\partial p_s}\frac{\partial p_r}{\partial q_s}\right] = nq_r^{n-1}$$

$$\{p_r^n, q_r\}_{(q,p)} \equiv \sum_s \left[\frac{\partial p_r^n}{\partial q_s}\frac{\partial q_r}{\partial p_s} - \frac{\partial p_r^n}{\partial p_s}\frac{\partial q_r}{\partial q_s}\right] = -np_r^{n-1}$$

In summary,

$$\{q_r^n, p_r\} = nq_r^{n-1}, \qquad \{p_r^n, q_r\} = -np_r^{n-1} \tag{A1}$$

These equations are valid in *any* representation. We will demonstrate this by taking the case of a single particle in linear motion. Dropping subscripts r from the first of (A1), we have

$$\{q^n, p\} = n\, q^{n-1} \tag{A2}$$

Let us prove this by mathematical induction. The proof proceeds in two steps:

(step 1) For $n = 1$, (A2) is true as it reduces to one of the fundamental Poisson bracket relations, (3.96).
(step 2) We take $\{q^{n+1}, p\}$, and obtain

$$\{q^{n+1}, p\} = \{q^n q, p\} = q^n\{q, p\} + \{q^n, p\}q$$
$$= q^n + nq^{n-1}q = (n+1)q^n \tag{A3}$$

where $\{q^n, p\} = nq^{n-1}$ and $\{q, p\} = 1$, valid in any representation, were used in the middle stages. Equation (A3) has the same form as (A2) except that $n + 1$ enters in place of n. Combination of the two steps extends the validity of (A2) from $n = 1$ to any positive integer. Clearly, a similar method can be used to prove the second of (A1).

The following equations also hold independent of the representation:

$$\{u, P_r\} = \frac{\partial u}{\partial Q_r}, \qquad \{u, Q_r\} = -\frac{\partial u}{\partial P_r} \tag{A4}$$

Mathematical Physics. Shigeji Fujita and Salvador V. Godoy
Copyright © 2010 WILEY-VCH Verlag GmbH & Co. KGaA, Weinheim
ISBN: 978-3-527-40808-5

To see this, we may expand u in powers of Q_r:

$$u = u_0 + u_1 Q_r + u_2 Q_r^2 + \cdots \tag{A5}$$

where the coefficients $u_0, u_1, u_2, \cdots$, are independent of Q_r but can depend on any other canonical variables. Dropping the subindex r for the moment, we get

$$\begin{aligned} \{u, P\} &\equiv \{u_0 + u_1 Q + u_2 Q^2 + \cdots, P\} \\ &= \{u_0, P\} + u_1\{Q, P\} + u_2\{Q^2, P\} + \cdots \\ &= 0 + u_1 + 2Q u_2 + \cdots = \frac{\partial u}{\partial Q} \end{aligned} \tag{A6}$$

which establishes the proof of the first equation. In a similar manner we can establish the second equation [with the aid of the expansion

$$u = u_0' + u_1' P_r + u_2' P_r^2 + \cdots \tag{A7}$$

where the coefficients $u_0', u_1', u_2', \cdots$ are functions of canonical variables excluding P_r].

We can further show (Problem A.1) that

$$\begin{aligned} \{u, P_r^n\} &= n P_r^{n-1}\{u, P_r\} = n P_r^{n-1} \frac{\partial u}{\partial Q_r} \\ \{u, Q_r^n\} &= n Q_r^{n-1}\{u, Q_r\} = -n Q_r^{n-1} \frac{\partial u}{\partial P_r} \end{aligned} \tag{A8}$$

For the moment, let us take the case of one degree of freedom. The dynamic function v depends on Q and P, and can be expanded in a double power series:

$$v(Q, P) = \sum_{m=0}^{\infty} \sum_{n=0}^{\infty} v_{m,n} Q^m P^n \tag{A9}$$

We now take the Poisson brackets of u and $v = v(Q, P)$, and calculate it as follows:

$$\begin{aligned} \{u, v\} &= \sum_{m=0}^{\infty} \sum_{n=0}^{\infty} v_{m,n}\{u, Q^m P^n\} \\ &= \sum_m \sum_n v_{m,n} \left[Q^m\{u, P^n\} + \{u, Q^m\} P^n \right] \\ &= \sum_m \sum_n v_{m,n} \left[Q^m n P^{n-1}\{u, P\} + m Q^{m-1} P^n\{u, Q\} \right] \\ &= \frac{\partial v}{\partial P}\{u, P\} + \frac{\partial v}{\partial Q}\{u, Q\} = \frac{\partial u}{\partial Q}\frac{\partial v}{\partial P} - \frac{\partial v}{\partial Q}\frac{\partial u}{\partial P} = \{u, v\}_{(Q, P)} \end{aligned} \tag{A10}$$

establishing that the Poisson bracket, $\{u, v\}$, has a value independent of representation. The proof for the general case is left as exercises for the reader (Problem A.2).

Problem A.1

Prove (A8). Be careful to proceed to each step in a representation-independent manner.

Problem A.2

Extend (A10) for the general system of arbitrary number of degrees of freedom.

Appendix B
Proof of the Convolution Theorem

The convolution theorem is

$$\mathcal{L}\left\{\int_0^t d\tau\, f(t-\tau)g(\tau)\right\} = \mathcal{L}\{f\}\mathcal{L}\{g\} = F(s)G(s) \tag{B1}$$

We introduce

$$f(t-\tau) = \frac{1}{2\pi i}\int_{c_1-i\infty}^{c_1+i\infty} ds_1 e^{s_1(t-\tau)} F(s_1) \tag{B2}$$

$$g(\tau) = \frac{1}{2\pi i}\int_{c_2-i\infty}^{c_2+i\infty} ds_2 e^{s_2\tau} G(s_2) \tag{B3}$$

in the τ-integral, and obtain

$$\int_0^t d\tau\, f(t-\tau)g(\tau)$$
$$= \frac{1}{(2\pi i)^2}\int ds_1 \int ds_2\, F(s_1)G(s_2) e^{s_1 t}\, \frac{e^{(s_2-s_1)t}-1}{s_2-s_1} \tag{B4}$$

We calculate the Laplace transform of this double integral and obtain

$$I \equiv \mathcal{L}\left\{\int_0^t d\tau\, e^{-st}(B4)\right\}$$

$$= \frac{1}{(2\pi i)^2}\int_{c_1-i\infty}^{c_1+i\infty} ds_1 F(s_1) \int_{c_2-i\infty}^{c_2+i\infty} ds_2 G(s_2)$$

$$\times \frac{1}{s_2-s_1}\left(\frac{1}{s-s_2}-\frac{1}{s-s_1}\right) \tag{B5}$$

where we assumed that

$$\text{Re}(s-s_2) > 0, \quad \text{and} \quad \text{Re}(s-s_1) > 0 \tag{B6}$$

Mathematical Physics. Shigeji Fujita and Salvador V. Godoy
Copyright © 2010 WILEY-VCH Verlag GmbH & Co. KGaA, Weinheim
ISBN: 978-3-527-40808-5

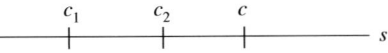

Fig. B.1 Numbers c_1, c_2, and $\mathrm{Re}(z) = c$.

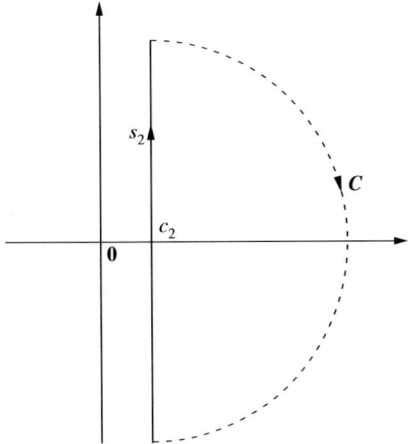

Fig. B.2 We calculate the s_2-integral in a semicircular path C.

We have two cases: (a) $c_2 > c_1$ and (b) $c_1 > c_2$, and calculate separately:

(a) The numbers c_1, c_2, and $\mathrm{Re}(z) = c$ are ordered: $c_1 < c_2 < c$ as in Figure B.1. We calculate the s_2-integral by adding a semicircular path, closing the contour, as shown in Figure B.2, and using the residue theorem. We obtain

$$\frac{1}{2\pi i}\int_{c_2-i\infty}^{c_2+i\infty} ds_2\, G(s_2)\frac{-1}{s_2-s_1}\left(\frac{1}{s_2-s}-\frac{1}{s_1-s}\right) = \frac{1}{s-s_1}G(s) \tag{B7}$$

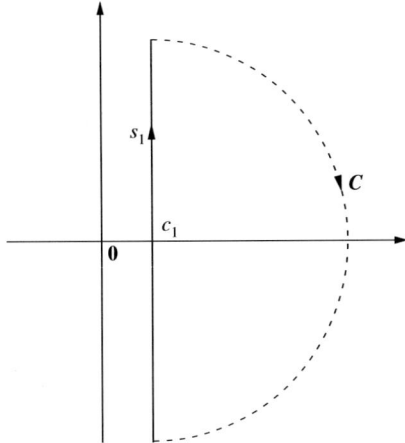

Fig. B.3 We calculate the s_1-integral in a semicircular path C.

We further integrate with respect to s_1 and obtain

$$I = \frac{1}{2\pi i} \int_{c_1-i\infty}^{c_1+i\infty} ds_1\, F(s_1) G(s) \frac{1}{s-s_1} = F(s)G(s) \tag{B8}$$

where we added the semicircle integration path shown in Figure B.3, and closed the contour.

(b) We may reverse the order of the integrations and calculate similarly. We obtain the same result.

Appendix C
Statistical Weight for the Landau States

The statistical weight W for the Landau states in 3-D and 2-D will be calculated in this appendix.

The Poisson's sum formula is

$$\sum_{n=-\infty}^{\infty} f(2\pi n) = \frac{1}{2\pi} \sum_{m=-\infty}^{\infty} F(m) \equiv \frac{1}{2\pi} \sum_{m=-\infty}^{\infty} \int_{-\infty}^{\infty} d\tau \, f(\tau) e^{-im\tau} \quad \text{(C1)}$$

where F is the Fourier transform of f, and the sum

$$\sum_{n=-\infty}^{\infty} f(2\pi n + t) \quad 0 \leq t < 2\pi$$

is periodic with the period 1. The sum is by assumption uniformly convergent.

We write the sum in (26.85) as

$$2 \sum_{n=0}^{\infty} \sqrt{\varepsilon - (2n+1)\pi} = (\varepsilon - \pi)^{1/2} + \phi(\varepsilon; 0) \quad \text{(C2)}$$

$$\phi(\varepsilon; x) \equiv \sum_{n=-\infty}^{\infty} (\varepsilon - \pi - 2\pi |n + x|)^{1/2} \quad \text{(C3)}$$

Note that $\phi(\varepsilon; x)$ is periodic in x with the period 1, and it can, therefore, be expanded in a Fourier series. After the Fourier series expansion, we set $x = 0$ and obtain (C2). By taking the real part (Re) of (C2) and using (C1) and (26.85), we obtain

$$\left(A \frac{(\hbar \omega_c)^{3/2}}{\sqrt{2\pi}} \right)^{-1} W(E) = \frac{1}{\pi} \int_0^\epsilon d\tau (\varepsilon - \tau)^{1/2}$$

$$+ \frac{2}{\pi} \sum_{m=1}^{\infty} (-1)^m \int_0^\epsilon d\tau (\varepsilon - \tau)^{1/2} \cos m\tau \quad \text{(C4)}$$

where we assumed

$$\varepsilon \equiv \frac{2\pi E}{\hbar \omega_c} \gg 1 \quad \text{(C5)}$$

Mathematical Physics. Shigeji Fujita and Salvador V. Godoy
Copyright © 2010 WILEY-VCH Verlag GmbH & Co. KGaA, Weinheim
ISBN: 978-3-527-40808-5

and neglected π against ε. The integral in the first term in (C4) yields $(2/3)\varepsilon^{3/2}$, leading to W_0 in (26.88). The integral in the second term can be written after integrating by parts, changing the variable $(m\varepsilon - m\tau = t)$, and using $\sin(A - B) = \sin A \cos B - \cos A \sin B$ as

$$\frac{1}{2m^{3/2}} \left(\sin m\varepsilon \int_0^{m\varepsilon} dt \frac{\cos t}{\sqrt{t}} - \cos m\varepsilon \int_0^{m\varepsilon} dt \frac{\sin t}{\sqrt{t}} \right) \tag{C6}$$

We use asymptotic expansions for $m\varepsilon = x \gg 1$:

$$\int_0^{m\varepsilon} dt \frac{\sin t}{\sqrt{t}} \sim \sqrt{\frac{\pi}{2}} - \frac{\cos x}{\sqrt{x}} - \cdots$$

$$\int_0^{m\varepsilon} dt \frac{\cos t}{\sqrt{t}} \sim \sqrt{\frac{\pi}{2}} + \frac{\sin x}{\sqrt{x}} + \cdots \tag{C7}$$

The second terms in the expansion lead to W_L in (26.89), where we use $\sin^2 A + \cos^2 B = 1$ and

$$\sum_{m=1}^{\infty} \frac{(-1)^{m-1}}{m^2} = \frac{\pi^2}{12} \tag{C8}$$

The first term leads to the oscillatory term W_{osc} in (26.90).

Appendix D
Useful Formulas

$$\mathbf{A} \cdot (\mathbf{B} \times \mathbf{C}) = \begin{vmatrix} A_1 & A_2 & A_3 \\ B_1 & B_2 & B_3 \\ C_1 & C_2 & C_3 \end{vmatrix}$$

$$= A_1(B_2C_3 - B_3C_2) + A_2(B_3C_1 - B_1C_3) + A_3(B_1C_2 - B_2C_1)$$

$$= \mathbf{C} \cdot (\mathbf{A} \times \mathbf{B}) = \mathbf{B} \cdot (\mathbf{C} \times \mathbf{A}) = (\mathbf{A} \times \mathbf{B}) \cdot \mathbf{C}$$

$|\mathbf{A} \cdot (\mathbf{B} \times \mathbf{C})|$ = volume of parallelepiped with sides $\mathbf{A}$, $\mathbf{B}$, $\mathbf{C}$

$\mathbf{A} \times (\mathbf{B} \times \mathbf{C}) = \mathbf{B}(\mathbf{A} \cdot \mathbf{C}) - \mathbf{C}(\mathbf{A} \cdot \mathbf{B})$

$(\mathbf{A} \times \mathbf{B}) \times \mathbf{C} = \mathbf{B}(\mathbf{A} \cdot \mathbf{C}) - \mathbf{A}(\mathbf{B} \cdot \mathbf{C})$

$(\mathbf{A} \times \mathbf{B}) \cdot (\mathbf{C} \times \mathbf{D}) = (\mathbf{A} \cdot \mathbf{C})(\mathbf{B} \cdot \mathbf{D}) - (\mathbf{A} \cdot \mathbf{D})(\mathbf{B} \cdot \mathbf{C})$

$(\mathbf{A} \times \mathbf{B}) \times (\mathbf{C} \times \mathbf{D}) = \mathbf{C}\{\mathbf{A} \cdot (\mathbf{B} \times \mathbf{D})\} - \mathbf{D}\{\mathbf{A} \cdot (\mathbf{B} \times \mathbf{C})\}$
$\qquad = \mathbf{B}\{\mathbf{A} \cdot (\mathbf{C} \times \mathbf{D})\} - \mathbf{A}\{\mathbf{B} \cdot (\mathbf{C} \times \mathbf{D})\}$

$\nabla(U + V) = \nabla U + \nabla V$

$\nabla \cdot (\mathbf{A} + \mathbf{B}) = \nabla \cdot \mathbf{A} + \nabla \cdot \mathbf{B}$

$\nabla \times (\mathbf{A} + \mathbf{B}) = \nabla \times \mathbf{A} + \nabla \times \mathbf{B}$

$\nabla \cdot (U\mathbf{A}) = \nabla U \cdot \mathbf{A} + U(\nabla \cdot \mathbf{A})$

$\nabla \times (U\mathbf{A}) = \nabla U \times \mathbf{A} + U(\nabla \times \mathbf{A})$

$\nabla \cdot (\mathbf{A} \times \mathbf{B}) = \mathbf{B} \cdot (\nabla \times \mathbf{A}) - \mathbf{A} \cdot (\nabla \times \mathbf{B})$

$\nabla \times (\mathbf{A} \times \mathbf{B}) = (\mathbf{B} \cdot \nabla)\mathbf{A} - \mathbf{B}(\nabla \cdot \mathbf{A}) - (\mathbf{A} \cdot \nabla)\mathbf{B} + \mathbf{A}(\nabla \cdot \mathbf{B})$

$\nabla(\mathbf{A} \cdot \mathbf{B}) = (\mathbf{B} \cdot \nabla)\mathbf{A} + (\mathbf{A} \cdot \nabla)\mathbf{B} + \mathbf{B} \times (\nabla \times \mathbf{A}) + \mathbf{A} \times (\nabla \times \mathbf{B})$

$\nabla \times (\nabla U) = 0$

$\nabla \cdot (\nabla \times \mathbf{A}) = 0$

$\nabla \times (\nabla \times \mathbf{A}) = \nabla(\nabla \cdot \mathbf{A}) - \nabla^2 \mathbf{A}$

Mathematical Physics. Shigeji Fujita and Salvador V. Godoy
Copyright © 2010 WILEY-VCH Verlag GmbH & Co. KGaA, Weinheim
ISBN: 978-3-527-40808-5

References

Mathematical Physics

1. Arfken, G.B. and Weber, H.J. (1995) *Mathematical Methods for Physicists*, 4th edn, Academic, San Diego.
2. Bradbury, T.C. (1984) *Mathematical Methods with Applications to Problems in the Physical Sciences*, John Wiley & Sons, Ltd, New York.
3. Butkov, E. (1968) *Mathematical Physics*, Addison-Wesley, Reading.
4. Chattopadhyay P.K. (1990) *Mathematical Physics*, John Wiley & Sons, Ltd, New York.
5. Chow, T.L. (2000) *Mathematical Methods for Physicist: A Concise Introduction*, Cambridge University Press, Cambridge.
6. Cohen, H. (1992) *Mathematics for Scientists and Engineers*, Prentice Hall, Englewood Cliffs.
7. Courant, R. and Hilbert, D. (1953) *Methods of Mathematical Physics*, John Wiley & Sons, Ltd, New York.
8. Dennery, P. and Krzywicki, A. (1967) *Mathematics for Physicists*, Harper and Row, New York.
9. Hassani S. (1999) *Mathematical Physics*, Springer, New York.
10. Jeffreys, Sir H. and Jeffreys, B.S. (1962) *Mathematical Physics*, Cambridge University Press, Cambridge.
11. Margenau, H. and Murphy, G.M. (1956) *Methods of Mathematical Physics*, Van Nostrand, Princeton.
12. Mathews, J. and Walker, R.L. (1964) *Mathematical Methods of Physics*, W.A. Benjamin, New York.
13. McQuarrie D.A. (2003) *Mathematical Methods For Scientists and Engineers*, University Science Books, Sausalito, CA.
14. Morse, P.M. and Feshbach, H. (1953) *Methods of Theoretical Physics*, McGraw-Hill, New York.
15. Kasse, B. and Westwig, E. (1998) *Mathematical Physics*, John Wiley & Sons, Ltd, New York.
16. Lea, S.M. (2004) *Mathematics for Physicists*, Thomson, Belmont, CA.
17. Riley, K.F., Hobson M.P., and Bence S.J. (2006) *Mathematical Methods for Physics and Engineering*, 3rd edn, Cambridge University Press, New York.
18. Wyld, H.W. (1999) *Mathematical Methods for Physics*, Perseus, Reading, MA.

Physics

Quantum Mechanics

19. Alonso, M. and Finn, E.J. (1989) *Fundamental University Physics, III Quantum and Statistical Physics*, Addison-Wesley, Reading, MA.
20. Dirac, P.A.M. (1958) *Principles of Quantum Mechanics*, 4th edn, Oxford University Press, London.
21. Gasiorowitz, S. (1974) *Quantum Physics*, Wiley, New York.
22. Liboff, R.L. (1992) *Introduction to Quantum Mechanics*, Addison-Wesley, Reading, MA.
23. McGervey, J.D. (1971) *Modern Physics*, Academic, New York.
24. Pauling, L. and Wilson, E.B. (1935) *Introduction to Quantum Mechanics*, McGraw-Hill, New York.
25. Powell, J.L. and Crasemann, B. (1961) *Quantum Mechanics*, Addison-Wesley, Reading, MA.

Mechanics

26. Goldstein, H. (1950) *Classical Mechanics*, Addison-Wesley, Reading, MA.
27. Kibble, T.W.B. (1966) *Classical Mechanics*, McGraw-Hill, London.
28. Marion, J.B. (1965) *Classical Dynamics*, Academic, New York.
29. Symon, K.R. (1971) *Mechanics*, 3rd edn, Addison-Wesley, Reading, MA.

Electricity and Magnetism

30. Griffiths, D.J. (1989) *Introduction to Electrodynamics*, 2nd edn, Prentice-Hall, Englewood Cliffs, NJ.
31. Lorrain, P. and Corson, D.R. (1978) *Electromagnetism*, Freeman, San Francisco.
32. Wangsness, R.K. (1979) *Electromagnetic Fields*, John Wiley & Sons, Ltd, New York.

Thermodynamics

33. Andrews, F.C. (1971) *Thermodynamics: Principles and Applications*, John Wiley & Sons, Ltd, New York.
34. Bauman, R.P. (1992) *Modern Thermodynamics with Statistical Mecanics*, Macmillan, New York.
35. Callen, H.B. (1960) *Thermodynamics*, John Wiley & Sons, Ltd, New York.
36. Fermi, E. (1957) *Thermodynamics*, Dover, New York.
37. Pippard, A.B. (1957) *Thermodynamics: Applications*, Cambridge University Press, Cambridge, England.

Statistical Physics (undergraduate)

38. Baierlein, R. (1999) *Thermal Physics*, Cambridge University Press, Cambridge, UK.
39. Carter, A.H. (2001) *Classical and Statistical Thermodynamics*, Prentice-Hall, Upper Saddle River, NJ.
40. Fujita, S. (1986) *Statistical and Thermal Physics*, I and II, Krieger, Malabar, FL.
41. Kittel, C. and Kroemer, H. (1980) *Thermal Physics*, Freeman, San Francisco, CA.
42. Mandl, F. (1971) *Statistical Physics*, John Wiley & Sons, Ltd, London.
43. Morse, P.M. (1969) *Thermal Physics*, 2nd edn, Benjamin, New York.
44. Reif, F. (1965) *Fundamentals of Statistical and Thermal Physics*, McGraw-Hill, New York.
45. Rosser, W.G.V. (1982) *Introduction to Statistical Physics*, Horwood, Chichester, England.
46. Terletskii, Y.P. (1971) *Statistical Physics*, Froman, N. trans., North-Holland, Amsterdam.
47. Zemansky, M.W. (1957) *Heat and Thermodynamics*, 5th edn, McGraw-Hill, New York.

Statistical Physics (graduate)

48. Davidson, N. (1969) *Statistical Mechanics*, McGraw-Hill, New York.
49. Feynman, R.P. (1972) *Statistical Mechanics*, Benjamin, New York.
50. Finkelstein, R.J. (1969) *Thermodynamics and Statistical Physics*, Freeman, San Francisco, CA.
51. Goodstein, D.L. (1975) *States of Matter*, Prentice-Hall, Englewood Cliffs, NJ.
52. Heer, C.V. (1972) *Statistical Mechanics, Kinetic Theory, and Stochastic Processes*, Academic, New York.
53. Huang, K. (1972) *Statistical Mechanics*, 2nd edn, John Wiley & Sons, Ltd, New York.
54. Isihara, A. (1971) *Statistical Physics*, Academic, New York.
55. Kestin, J. and Dorfman, J.R. (1971) *Course in Statistical Thermodynamics*, Academic, New York.
56. Landau, L.D. and Lifshitz, E.M. (1980) *Statistical Physics*, 3rd edn, Part 1, Pergamon, Oxford, England.
57. Lifshitz, E.M. and Pitaevskii, L.P. (1980) *Statistical Physics*, Part 2, Pergamon, Oxford, England.
58. McQuarrie, D.A. (1976) *Statistical Mechanics*, Harper and Row, New York.
59. Pathria, R.K. (1972) *Statistical Mechanics*, Pergamon, Oxford, England.
60. Robertson, H.S. (1992) *Statistical Thermodynamics*, Prentice Hall, Englewood Cliffs, NJ.
61. Wannier, G.H. (1966) *Statistical Physics*, John Wiley & Sons, Ltd, New York.

Solid State Physics

62. Ashcroft, N.W. and Mermin, N.D. (1976) *Solid State Physics*, Saunders, Philadelphia.
63. Harrison, W.A. (1979) *Solid State Theory*, Dover, New York.

64 Haug, A. (1972) *Theoretical Solid State Physics*, I, Pergamon, Oxford, England.
65 Kittel, C. (1986) *Introduction to Solid State Physics*, 6th edn, John Wiley & Sons, Ltd, New York.

Superconductivity

Introductory and Elementary Books
66 Feynman, R.P., Leighton R.B., and Sands, M. (1965) *Feynman Lectures on Physics*, Vol. 3, Addison-Wesley, Reading, MA, pp. 1–19.
67 Feynman, R.P. (1972) *Statistical Mechanics*, Addison-Wesley, Reading, MA, pp. 265–311.
68 Lynton, E.A. (1962) *Superconductivity*, Methuen, London.
69 Rose-Innes, A.C. and Rhoderick, E.H. (1978) *Introduction to Superconductivity*, 2nd edn, Pergamon, Oxford, England.
70 Vidali, G. (1993) *Superconductivity*, Cambridge University Press, Cambridge, England.

More Advanced Texts and Monographs
71 Abrikosov, A.A. (1988) *Fundamentals of the Theory of Metals*, A. Beknazarov, trans., North Holland-Elsevier, Amsterdam.
72 Anderson, P.W. (1997) *Theory of Superconductivity in High-T_c Cuprates*, Princeton University Press, Princeton, NJ.
73 Burns, G. (1992) *High-Temperature Superconductivity, an Introduction*, Academic, New York.
74 Gennes, P. (1966) *Superconductivity of Metals and Alloys*, Benjamin, Menlo Park, CA.
75 Ginsberg, D.M. (ed.) (1989) *Physical Properties of High-Temperature Superconductors*, World Scientific, Singapore (series).
76 Halley, J.W. (ed.) (1988) *Theory of High-Temperature Superconductivity*, Addison-Wesley, Redwood City, CA.
77 Kresin, V.Z. and Wolf, S.A. (1990) *Fundamentals of Superconductivity*, Plenum, New York.
78 Kresin, W.Z. (1989) *Novel Superconductivity*, Plenum, New York.
79 Lindquist, S. et al. (eds) (1988) *Towards the Theoretical Understanding of High-T_c Superconductivity*, Vol. 14, World Scientific, Singapore.
80 Lynn, J.W. (ed.) (1990) *High Temperature Superconductivity*, Springer, New York.
81 Owens, F.J. and Poole, C.P. (1996) *New Superconductors*, Plenum, New York.
82 Phillips, J.C. (1989) *Physics of High-T_c Superconductors*, Academic, San Diego, CA.
83 Poole, C.P., Farach, H.A., and Creswick, R.J. (1995) *Superconductivity*, Academic, New York.
84 Rickayzen, G. (1965) *Theory of Superconductivity*, Interscience, New York.
85 Saint-James, D., Thomas, E.J., and Sarma, G. (1969) *Type II Superconductivity*, Pergamon, Oxford, England.
86 Schafroth, M.R. (1960) *Solid State Physics*, Vol. 10 (eds F. Seitz and D. Turnbull), Academic, New York, p. 488.
87 Schrieffer, J.R. (1964) *Theory of Superconductivity*, Benjamin, New York.
88 Sheahen, T.P. (1994) *Introduction to High-Temperature Superconductivity*, Plenum, New York.
89 Tilley, D.R. and Tilley, J. (1990) *Superfluidity and Superconductivity*, 3rd edn, Adam Hilger, Bristol, England.
90 Tinkham, M. (1975) *Introduction to Superconductivity*, McGraw-Hill, New York.
91 Waldram J.R. (1996) *Superconductivity of Metals and Cuprates*, Intitute of Physics Publishing, Bristol, UK.

Index

a

Absolute convergence   383
Absolute square function   390
Absolute value   375, 385
Absolute zero temperature   114
Absorption   365
Action integral   333
Addition law in probability theory   148
Additivity of entropy   127
adjoint   191
Amplitude relations   46
Analytic function   385
Angular frequency   259
Annihilation and creation operators   337
Antisymmetric function   269
Antisymmetric ket   278
Antisymmetric kets   340
Antisymmetrizing operator   278
Average collision duration   102
Average position   206

b

Bardeen Cooper and Schrieffer   366
Bernoulli distribution   151
Bernoulli walks   156
Binomial distribution   150
Binomial expansion   151
Binomial series   381
Bohr magneton   307
Boltzmann equation   99
Bose commutation rules   335
Bose distribution function   298
Bose-Einstein condensation   297
Bosons   231, 345
Bosons in condensed phase   301
Box-like boundary condition   173
Boyle's law   93
Bra and ket notations   215
Bra symbol   186

Branch point   388–389, 411
Bulk force   90
Bulk limit   285

c

Canonical equations of motion   30
Canonical momentum   20, 30, 50, 314, 319
Canonical transformation   32–33
Canonical variables   30
Carnot cycle   109–110
Carnot engine   110
Carnot's theorem   111
Cartesian components   4
Cartesian unit vectors   3
Cauchy–Riemann relations   390
Cauchy's fundamental theorem   395
Central force   29
Characteristic frequencies   46
Characteristic frequency equation   45
Chemical potential   284, 293, 322
Circular integrals   400
Classical harmonic oscillator   259
Classical Indistinguishable Particles   273
Clausius statement of the Second Law   108
Clausius's theorem   114–115, 125
Coefficient of thermal expansion   138
Collision rate   101, 104
Commutation relation   203
Commutators   221
Commutators of angular momentum   221
Comparison Test   383
Completeness relation   198, 209, 216
Complex conjugate   389
Complex numbers   375
Conservative force   28
Constant tensor   14
Continuity equation   89
Convergence interval   381
Convolution Theorem   252

Index

Cooper pair 365–366
Cube-root function 388, 413
Cuprate superconductors 365
Cyclotron frequency 311
Cyclotron radius 311
Cyclotron resonance 317
Cylindrical coordinates 7
Cylindrical unit vectors 8

d

dc Josephson effect 360
Degeneracy 317
Degrees of freedom 37
de Haas–van Alphen (dHVA) oscillation 325
Densities of states for up- and down-spins 308
Density of states 287, 290, 309
Density operator 239
Deviation from the average 154
Diagonal matrix 189
Different-fermion composite 348
Diffusion 97
Diffusion coefficient 97
Dirac delta function 202, 325
Dirac ket 241
Dirac observable 241
Dirac picture 240, 253
Dirac's Delta Function 158, 160
Direct product 11
Divergence of series 382
Dot product 3
Dot product of two tensors 13
Double-valued function 388
Drift velocity 313
Dyadic 12
Dynamical functions 36
Dynamical state 205
Dynamical variables 36

e

Effective mass 325
Efficiency of a heat engine 110
Ehrenfest–Oppenheimer–Bethe (EOB) rule 345
Eigenvalue 16, 192
Eigenvalue equation 15, 202
Eigenvalue problem 192
Eigenvector 16, 192
Electrical conductivity 104
Electromagnetic Potentials 313
Electronic heat capacity 292
Emission of a boson 365
Energy conservation law 107
Energy eigenvalue equation 211
Energy eigenvalue problem 211
Energy eigenvalues 255
Entire function 388
Entropy 107, 119, 126
Equation of continuity 93
Equation of motion for a fluid 92
Equation of state 93, 107
Essential singularity 409
Euler's identity 377
Euler-Lagrange Equation 329, 331
Evolution operator 240
Exact Differential 122
Exact differential 123
Expectation value 206
Exponential function 378, 387

f

Fermat's principle 331
Fermi anticommutation rules 340
Fermi–Dirac statistics 291
Fermi distribution function 284
Fermi energy 284, 308
Fermi momentum 285
Fermi sphere 286
Fermi temperature 286, 295
Fermion–boson composite 349
Fermions 231, 345
Fermi's golden rule 244
First Law of Thermodynamics 107, 126
Fixed-end boundary conditions 44
Fluid Dynamics 87, 92
Fluid Equation of Motion 89
Fluid velocity 87
Flux quantum 359
Fourier inverse transformation 216
Fourier Series 213
Fourier transformation 216
Fourier transforms 214
Fourier's law 100
Free-electron model 283
free energy 322

g

g-factor 235
Gaussian Plane 375
Gauss's divergence theorem 88
Generalized coordinates 28
Generalized Lagrangian function 314
Generalized vectors 183
Gibbs ensemble 173–175
Gibbs free energy 130
Grand partition function 326
Ground-state wavefunction 257

h

Hall current 313
Hamiltonian (function) 21
Hamiltonian description 20
Hamiltonian formulation 17
Hamilton's equations of motion
   22, 165, 169, 175
Hamilton's Principle 333
Harmonic force 19
Harmonic functions 391
Harmonic oscillator 255
Heat capacity 108, 293
Heat capacity at constant volume 136
Heat conductivity 100
Heat engine 110, 114
Heisenberg equation of motion 342
Heisenberg picture 245, 341
Heisenberg's uncertainty principle 219, 317
Helmholtz equation 213
Helmholtz free energy 128–129
Hermitean conjugate 191
Hermitian operator 192
High temperature superconductor 365
Hilbert space 196
Homogeneous stationary state 180

i

Identical fermion composite 346
Identity permutation 265
Incompressible fluid 93
Interchange 263
Invariance under rotation 180
Invariance under translation 180
Inverse permutation 266
Irreversible Processes 97, 125
Isothermal compressibility 93, 138
Isotropic state 181

j

Jacobian of transformation 33
Josephson interference 361
Josephson junction 360
Josephson tunneling 360

k

Kelvin's statement of the Second Law 108
Ket symbol 186
Kinetic momentum 319

l

Lagrange's equation 19, 29
Lagrange's field equation 333–334
Lagrange's formulation 19
Lagrangian density 334

Lagrangian description 19
Lagrangian function 19
Lambda-transition 304
Landau diamagnetism 307, 325
Landau states 316–317
Laplace Equation 391
Laplace Transformation 249
Laurent series 405
Law of conservation of energy 19
Line integrals 398
Linear Operator Algebras 253
Linear operators 188
Liouville Equation 165, 171–172, 174, 176
Liouville Theorem 165–166
Liquid He I 304
Liquid He II 304
Liquid Helium 297
Logarithmic function 414
Lorentz force 313
Lower critical field 364
Lowering operator 226

m

MacLaurin series 381
Magnetic flux 359
Magnetic moment 234
Magnetization 310
Magneto-mechanical ratio 235
Magnitude of the quantum angular
   momentum 227
Magnitude of the vector 4
Mapping 189
Matrices product 184
Matrices sum 184
Matrix 13
Matrix definition 184
matrix hermitean conjugate 185
Matrix transpose 185
Matthiessen's rule 104
Maxwell relations 132, 135
Mean free path 102
Mean free time 103
Meissner effect 357
Microscopic Density 156
Moivre's theorem 376
Molar heat capacity 292
Momentum eigenstates 283
Momentum eigenvalue problem 207
Motional diamagnetism 321
Multiplication law of probability 148
Multivalued functions 413
Mutually exclusive events 148

n

Natural Logarithm 377
Newtonian description 17
Newton's equation of motion 17, 91
Normal coordinates 49
Normal modes 46
Normalization condition 149, 170, 289
Nuclear magnetic resonance 237
Nuclear magneton 236
Null vector 3
Number density 157, 293
Number representation 281, 335

o

Observables 338
Occupation-Number Representation 280, 335
Odd and Even Permutations 267
Ohm's law 104
Orbital angular momentum 229
Order of the group 266
Orthogonal functions 213
Orthogonal vectors 4
Orthonormal vectors 195
Orthonormalities 216
Orthonormality 3
Orthonormality relation 202

p

Parallel translation 179
Paramagnetic Resonance 238
Parity of the permutation 278
Pauli Paramagnetism 307–308, 310, 325
Pauli spin operator 231
Pauli's Exclusion Principle 278, 335
Pauli's principle 351
Pauli's spin matrices 232
Penetration depth 359
Periodic boundary 213
Periodic boundary condition 208, 289
Permutation 263
Permutation Group 263–264
Perturbation Theory 239
Phase space 22, 165
Point transformations 36
Poisson bracket 36–37, 40, 178
Polar angle 375
Polar Coordinates 375
Polar singularity 387
Position representation 203
Position vector 2
Potential energy 28
Power series 381, 405

Principal-axis transformation 51
Probabilities 147
Probability distribution function 170, 205
Properties of Angular Momentum 224

q

Quantum Angular Momentum 221
Quantum harmonic oscillator 255, 259
Quantum Liouville equation 246–247
Quantum Liouville operator 246
Quantum-statistical postulate 277
Quantum Statistics 273
Quantum statistics of composites 345
Quasielectron energy gap 367
Quasielectrons 367

r

Radius of convergence 405
Raising operator 226
Random walks 155
Rate of collision 104
Ratio test 384
Real matrix 185
real part 375
Refrigerator 114
Relative probability 157
Representation 197
Residue 415
Residue theorem 415
Reversible Motion 22
Reversible motion 27
Reversible path 126
Riemann sheets 412
Root mean square deviation 154

s

Scalar product 3, 183
Scattering Problem 242
Schrödinger equation of motion 205, 239
Schrödinger picture 205, 241, 341
Second Intermediate Picture 245
Second Law 108
Second quantization 335, 341
Second quantization formulation 365
Self adjoint operator 192
Simple pole 387
Simultaneous eigenstates 224
Snell's law 331–332
Spherical coordinates 6
Spherical unit vectors 6
Spin angular momentum 229
Spin degeneracy factor 286
Spin paramagnetism 308
Square-root function 388, 411

Squared deviation   154
Standard deviation   154
State of maximum entropy   126
State variables   153
States for Bosons   276
States for Fermions   278
Statistical physics   92
Stress tensor   90
Stretched string Lagrangian   44
Superconducting Quantum Interference Device   361
Superconductivity   357
Superconductors   357
Superfluid phase   297
Surface force   90
Susceptibility   310
Symmetric function   269
Symmetric ket   278, 337
Symmetric matrix   192
Symmetric tensor   15
Symmetrizing operator   278
Symmetry-breaking state   182
System of Many Particles   28
System-points density   174

t

Taylor series   381
Tensor product   11
Tensors definition   11
The Bose–Einstein condensation   297
The Dirac Picture   239
The Entropy   107, 119
The Euler–Lagrange equation   329, 331
Theorem C   396
Theory of Variations   329
Thermodynamic Inequalities   125
Thermodynamic state   124
Three-Dimensional Delta Function   161
Time-Dependent Pertubation Theory   239
Total angular momentum   223, 230

Total reflection   332
Transformation functions   208
Transition probability   240, 242
Transition rate   244
Transposition   263
Two-particle composites   346

u

Uncertainty principle   217
Uncorrelated events   148
Unit dyads   13
Unit matrix   189
Unit tensor   14, 91
Unpaired electrons   367
Upper critical field   364

v

Vacuum ket   337
Van Leeuwen's theorem   321
Vector addition   1
Vector definition   1
Vector direction   1, 183
Vector magnitude   1, 183
Vector potential fields   313
Vector product   4
Viscosity coefficient   95
Viscous fluid   95
Viscous resistance   94
Viscous stress   97, 99
Vortices   364

w

Wave equation   213
Wave function   205

z

Zero-momentum bosons   301
Zero-point energy   258
Zero resistance   357